AF575326

Combustion Instabilities in Liquid Rocket Engines: Testing and Development Practices in Russia

Combustion Instabilities in Liquid Rocket Engines: Testing and Development Practices in Russia

Mark L. Dranovsky
Research Institute of Chemical Engineering, NIICHIMMASH
Moscow, Russia

Edited by
Vigor Yang
Pennsylvania State University
University Park, Pennsylvania

Fred E. C. Culick
California Institute of Technology
Pasadena, California

Douglas G. Talley
Air Force Research Laboratory
Edwards Air Force Base, California

Volume 221
PROGRESS IN ASTRONAUTICS AND AERONAUTICS

Frank K. Lu, Editor-in-Chief
University of Texas at Arlington
Arlington, Texas

Published by the
American Institute of Aeronautics and Astronautics, Inc.
1801 Alexander Bell Drive, Reston, Virginia 20191-4344

American Institute of Aeronautics and Astronautics, Inc., Reston, Virginia

1 2 3 4 5

ISBN-10 1-56347-921-4
ISBN-13 978-1-56347-921-2

Foreword

In 1993 the international meeting Liquid Rocket Engine Combustion Instability was held at the Pennsylvania State University. The proceedings were subsequently published as Vol. 169 of the AIAA Progress in Astronautics and Aeronautics Series. There were seven participants from the former Soviet Union, the first extensive contacts between the Soviet and U.S. rocket communities. Out of that meeting came two books translated from their Russian editions: the first by M. S. Natanzon, treating largely theoretical work on combustion instabilities in liquid rockets, and the present volume covering experimental matters and applications. The author, Mark L. Dranovsky, has long enjoyed a leading position in the field of liquid rockets in Russia. Although the author makes no attempt to compare or contrast practices in the East and West, the material covered is sufficient to allow experienced readers to do so.

Much, if not most, of the original Western work was available in the open literature. It was translated for Russian engineers and scientists, as attendees to the meeting just mentioned confirmed. Yet in certain respects, Soviet strategies to developing liquid rockets seem to have differed noticeably from those in the West. See, for example, the book *Korolev*, by James Harford; several postings on the Web; and to some extent the introductory chapter to this book. There exists no comprehensive assessment comparing the fields as they developed in the East and West. Whatever the underlying reasons might be, there are also differences at the deeper technical levels. There is much to be learned from the detailed material in this book.

Following the Introduction and Chapter 2 on Terms and Definitions, Chapter 3 covers the three general categories of excitation of instabilities in combustion chambers: "soft" excitation, "hard" excitation, and growth of instabilities out of noise. Some aspects of noise in combustion chambers are introduced, beginning a subject and possible connections between random (probabilistic) behavior and well-defined instabilities, which is an important recurring theme in the book.

Whereas Dranovsky expertly uses well-developed mathematical methods for correlating experimental data, and for working out bases of formal interpretation, he is sparing in his explanations of physical behavior. There are, for example, no analyses of flowfields or of the behavior of injected liquids. Any guidelines for design, or for modifications of a design directed to improve stability, must be based, so far as the materials in this book are concerned, entirely on experimental results. Thus, systematic testing is evidently even more important than in the West for developing new combustion chambers. Experimental results are first discussed in Chapter 4 (for the RD-217 engine), and thereafter the discussion is never far from the test stand. There are considerable materials for the engines RD-120, RD-170, RD-172, and RD-180, all using the combination of oxygen and kerosene.

As Dranovsky emphasizes in his Preface, the main theme of the book is concerned with methods of ensuring adequate margin for stability of high-frequency oscillations. A major concern in Russian work has long been to provide sufficient margin in the presence of random disturbances, always present and often with considerable amplitudes in chambers generating high thrust. The subject is covered thoroughly in this book. Comparable discussions do not exist in the Western literature.

The last four chapters particularly contain a great deal of materials relating to production engines and tests performed during their development. As true of most of the book, the text tends to be quite dense and detailed. Practicing engineers and researchers will find that Dranovsky's presentation will amply repay their efforts to absorb the materials.

It has been a genuine pleasure for us to become better acquainted with Dranovsky and his work during our period of editing his book. We expect that professionals in the field of liquid rockets will be pleased to join us in thanking him for making his experience and expertise accessible to the general community.

Vigor Yang
Fred E. C. Culick
Douglas G. Talley
June 2007

Table of Contents

Preface

The technology ensuring the required quantitative margin of operating process stability to high-frequency pressure oscillations in liquid rocket engine (LRE) combustion chambers and gas generators is described in this book. Operating process stability margin is provided in the engine design specifications. The technology involves the application of a number of solutions guaranteeing stability at all stages of engine development.

Although the book is not aimed at integrating data for LRE stability, it presents results that substantiate the technology for development of LRE combustion chambers and gas generators with required process stabilities. Studies carried out over a long period of time resulted in the procedure that offers the same quantitative characteristics of stability for engines operating with any type of propellant combinations, including liquid–liquid, gas–liquid, and gas–gas systems.

The procedure for estimating process stability for high-frequency pressure oscillations is based upon the quantitative characteristics determined from combustion chamber "noise" and on pressure oscillation decrements determined from artificial disturbances of strictly defined values introduced into a combustion chamber. Over 1000 tests of model combustors, full-scale combustors, and engines have been conducted with different configurations over a wide range of operating conditions and parameters in order to quantitatively determine the stability characteristics for required engine reliability.

The procedure ensures obtaining reliable quantitative stability data from tests of two or three combustion chambers of a given design. Determination of quantitative stability values involves recording oscillations, introducing disturbances into the combustion chamber, and processing the recorded parameters using appropriate programs. The data obtained make it possible to choose the best system to provide the desired margin of stability to high-frequency pressure oscillations. A vast amount of experimental data is given in this book, confirming the effectiveness of the procedure for quantitative estimation of stability in LRE combustion chambers and gas generators. The methods for estimating stability margins described in the book are required in Russia as mandatory in the development of new engines.

The book is organized in 17 chapters. Chapter 1 covers the development of the technologies to ensure the required stability margins in LRE combustion chambers and gas generators. The important terms in treating various cases of process instabilities observed in actual LREs are presented in Chapter 2. Terms such as "operating process stability" and "process stability margin" are discussed.

Mechanisms of transition from noise to organized pressure oscillations having high frequencies and amplitudes in a combustion chamber are discussed in Chapter 3. The mechanisms of soft and hard excitations of oscillations in combustion chambers are discussed. Experimental results are presented for the realization of probabilistic-mechanism-induced instabilities from the noise in a combustion chamber.

Examples of combustion instabilities are given in Chapter 4. Experimental data on the instability of the RD-217 engine mixture formation system are given. Cases of instabilities caused by the random behavior of flame stabilization in injectors for gas–liquid LRE are covered. Chapter 5 briefly describes the technology for development of combustion chambers and gas generators with required stability margins. The technology incorporates solutions at every stage of engine development, from the design to commercial product supplies, in order to maintain consistent stability characteristics in production.

Chapter 6 contains theoretical and experimental results for characterizing the stability behavior of LRE combustion chambers. Two parameters are taken as quantitative measures of stability: δT, the pressure oscillation decrement determined from natural noise in a combustion chamber, and the process relaxation time τ_r obtained from the data taken following the introduction of an artificial disturbance. The defined stability characteristics are then used in the following chapters of the book.

The acoustic properties of a combustion chamber for the stability problem are described in Chapter 7. The major contents include 1) identification of the modes of oscillations; 2) evaluation of the effects of various factors on the acoustic properties; and 3) evaluation of oscillation energy losses when using, for example, vibration baffles, resonance absorbers, adjusted injectors (for the gas–liquid configuration), and porous combustor walls.

Chapter 8 is dedicated to the development of optimal methods for determination of pressure oscillation decrements from natural disturbances (noises) in a combustion chamber under steady-state and transient operating conditions. Chapter 9 contains information concerning development of an optimal method for estimating stability with the use of artificial pressure pulses. The magnitude and shape of a pulse that ensure the most valid data for stability to finite disturbances are justified theoretically and experimentally. The data-analysis procedure is also discussed providing the maximum accuracy for evaluating the stability margin to hard excitation.

Information on the modeling of processes determining the stability characteristics for closed-circuit engines (gas–liquid configuration) is included in Chapter 10. The method using subscale firings is based on the following conditions: 1) a low pressure level practically equal to atmospheric pressure; 2) a small mass flow rate of propellant (two and more orders of magnitude lower than actual flow rates); and 3) use of gaseous species as surrogates of liquid propellants to obtain the actual volumetric flow rates at low weight flow rates. It is shown that model tests, especially on a unit with a single injector, allow effective selection of injectors with the highest damping of high-frequency oscillations. The method makes it possible to perform studies of selected physical processes leading to combustion instabilities in LREs.

Chapters 11 and 12 deal with experimental results for estimation of high-frequency stability in combustion chambers with liquid–liquid propellant combinations. Two methods based on quantitative stability characteristics are developed, both making use of pressure oscillation decrements evaluated from combustion chamber noise and artificial pressure pulses. The effectiveness of the stability characteristics used for selection of an optimal mixture formation is demonstrated.

Chapters 13 and 14 contain results for estimation of the stability of gas–liquid combustion processes. They are obtained from pressure-oscillation decrements subject to the introduction of artificial pressure pulses. Testing of full-scale and experimental combustion chambers allowed determination of the effects of structural parameters and operating conditions on stability.

The results of studies carried out during the development of the RD-170 engine combustion chamber and its modifications are given in Chapter 15, as an example of the estimation of LRE combustion stability with the help of measured stability characteristics. The chapter contains the test results for various injector heads. The selection of the optimal design based on quantitative stability characteristics is also discussed. In addition, the stability test results for combustion chambers used as part of the RD-170 engine are provided.

Quantitative stability characteristics obtained during tests of the RD-208 engine are given in Chapter 16. The propellants are nitrogen tetroxide (NTO) and unsymmetrical dimethylhydrazine (UDMH) components, with injectors having adjustable geometries. Vibration baffles are made of injectors protruding into the combustion chamber. The effectiveness of using the established procedures is confirmed by tests carried out with the combustion chambers of the RD-170, RD-172, RD-180, and RD-120 engines operating on oxygen-kerosene propellants and of engines operating on NTO-UDMH propellants. The quantitative values of stability characteristics obtained in experiments discussed in Chapters 11–16 made it possible to determine the quantities required for ensuring adequate stability in LRE combustion chambers and gas generators.

Finally, the data confirming the inconsistency of the process stability margin control in serial production of engines are provided and discussed in Chapter 17.

This book is of great interest for research workers, academicians, engineers, and post-graduate students, as well as for the fourth- and fifth-year students of higher technical education establishments dealing with the LRE development.

I would like to express my cordial thanks to Vigor Yang, Fred E. C. Culick, and Douglas G. Talley for their participation in editing this book and for their assistance in publication of the book.

I am especially grateful to my wife Faina Dranovskaya and to my granddaughters Alena Timusheva and Tatiana Lupenkova who made great contributions to the creation of this book.

Mark L. Dranovsky
June 2007

Acknowledgments

Publication of this volume was made possible through the substantial contributions of a number of individuals and organizations. We owe a large debt of gratitude to Danning You, Yanxing Wang, and Tao Liu of the Pennsylvania State University for providing the technical drawing services, organizing the book materials, and proofreading the manuscript. Mary Newby, also of the Pennsylvania State University, deserves special thanks for her administrative help. The invaluable assistance of Rodger Williams, Heather Brennan, and Janice Saylor of AIAA in preparation of the book for publication is gratefully acknowledged.

Acronyms and Abbreviations

AFC	= amplitude-frequency characteristic
BC	= blasting cap
CC	= combustion chamber
CR	= carrier rocket
DB	= design bureau
DD	= disturbance device
DW	= development work
EDB	= experimental design bureau
EX	= explosive
GOST	= State All-Union Standard
IBR	= intercontinental ballistic rocket
IDD	= in-chamber disturbance device
IH	= injector head
IST	= inspection-sampling test
ITT	= inspection-technological test
KTAF-10	= special coating material
LRE	= liquid rocket engine
LVD	= low-velocity detonation
MDD	= multipulse disturbance device
NA-27I	= oxidizer: nitric acid with 27% of nitrogen tetroxide and inhibitor admixture
ND	= normal detonation
NIICHIMMASH	= Research Institute of Chemical Machine Building
NTO	= oxidizer: nitrogen tetroxide
PC	= personal computer
PG-2	= starting fuel which self-ignites with oxygen
R	= rocket
RE	= rocket engine
RSC-N1	= rocket and space complex N1
RVP	= rapidly varying parameters
RW	= research work
ST	= shock tube
SVT	= special verification test
TAT-1-1	= beam blasting cap
TG-02	= fuel: mixture of 50% triethylamine and 50% xylidine
TPU	= turbopump unit
UDMH	= fuel: unsymmetrical dimethylhydrazine
1UKS, 2UK, 2UKS	= units of development combustion chambers for engine RD-170

Chapter 1

Introduction

EXAMINATION of the development of liquid-propellant rocket engines (LRE) all over the world indicates that high-frequency instabilities in LRE combustion chambers and gas generators are one of the constituent structural-and-functional elements limiting LRE reliability. (Statements about oscillations and their properties in combustion chambers apply equally to gas generators; hence, unless otherwise required, the term “gas generator” will not be included.) It is the cause of the greatest number of LRE failures at the development stage and even during manufacture.

It is known from the history of the A. M. Isayev Design Bureau (DB) [1] that in 1946–1947 significant advances were made in many DBs. In the A. M. Isayev DB rather simple and reliable expendable single-mode propulsion systems were developed. Conical combustion chambers with flat heads and with interconnected inner and outer shells made from sheet material were used in the engines. The engines were used for a flying model of a supersonic airplane, for a sea torpedo, for an air-to-sea rocket, and for an anti-aircraft rocket. The parameters of combustion chambers were small, and engines had not yet faced this disastrous danger, a “monstrous beast.” They did not doubt that making a combustion chamber with a thrust of 10–15 tons or greater was as easy as making a combustion chamber with a thrust of 2 tons; they believed that the main problem consisted in production capabilities. Thus, with a light heart they designed an 8-ton combustion chamber and fabricated the first specimens. During the very first test, because of high-frequency pressure oscillations in the combustion chamber, a strong breakdown of the engine occurred. The window panes were blown out in all of the neighboring buildings, and a roof of the assembly shop located close to the testing stand nearly went down [1].

From this point on, extensive studies of the problem of stabilities of LRE combustion chambers and gas generators were started. It should be recognized that the DB led by the chief designer A. M. Isayev was the first to introduce, in 1951, vibration baffles for high-frequency pressure oscillation control.

The astronautics encyclopedia, *Sovietskaya Encyclopedia*, [2] gives the following definition of operation process instability in LRE:

> Instability of a rocket engine operation process, instability of combustion in a rocket engine, manifests itself as spontaneous oscillations with a wide range of parameters governing the rocket engine operation process (pressure, speed, gas and liquid temperatures, and so on) with respect to average values. Unstable

operating conditions of a rocket engine are characterized by a range of oscillation process development and by a range of self-oscillations, that is, self-sustained non-linear periodic oscillations with constant amplitudes. In almost all cases, the oscillation process is maintained directly due to thermal energy released during propellant combustion. The energy entering the oscillating system is controlled by a feedback loop between the oscillating system and the energy source. There are many complicated mechanisms of this feedback. High-frequency instability of the operation process is determined by relations between pressure oscillations and heat and mass input processes during combustion.

A whole range of different sorts of pressure oscillations is realized in a combustion chamber and in a gas generator. There are many sources of oscillations: propellant conversion in a combustion chamber and in a gas generator to combustion products, pulsations induced by pumps and turbines, disturbances in a cooling circuit, etc., resulting in a complicated feedback system.

All pressure oscillations in a combustion chamber should be divided into two main groups: natural and forced pressure oscillations. Forced oscillations are generated by sources located outside the combustion chamber volume: They might have an influence upon the process of propellant conversion to combustion products. Taking into account the characteristics of physical phenomena inducing natural oscillations in a combustion chamber, they are classified into low-, high-, and intermediate-frequency oscillations. This division is important because it is not a formal division by the frequency of pressure oscillations, but results from physical phenomena governing these oscillations.

Low-frequency oscillations in a LRE combustion chamber are pressure oscillations with frequencies below the minimum natural acoustic frequency. Generally, a low-frequency instability is defined by the relation between the combustion and hydrodynamic processes in the system feeding the propellant to the combustion chamber. Low-frequency instability of the operation process might be responsible for dangerous longitudinal elastic oscillations of a rocket airframe (natural frequencies of longitudinal oscillations for heavy rockets are 5–20 Hz) as well as for fluctuations of thrust delivered by a rocket engine. Structural oscillations can be enhanced by cavitation oscillations of LRE centrifugal pumps. Instability of automatic control systems also is a type of low-frequency instability of the operation process.

As just stated, apart from low-frequency and high-frequency instabilities, the process instability at intermediate frequencies should be distinguished. These pressure oscillations are characterized by the existence of longitudinal oscillations in such circuits as gas-generator inlet lines (injectors), between gas duct and combustion chamber, and so on. The process instability at intermediate frequencies is mainly determined by the relation between the combustion and hydrodynamic processes in the propellant feed system as well as the mixing process in the combustion chamber (including injectors). So-called entropy waves also fall into the region of intermediate frequencies. The entropy waves are generated when the gas flow with longitudinal gradients of mixture composition crosses sonic surface within the nozzle. The resulting pressure waves propagate upstream from the nozzle. The waves reflected from the injector head influence the ratio of propellant components and move at the local gas velocity along the flow as entropy breaks. The probability of exciting instabilities at intermediate frequencies is especially high in LREs with large cavities in the injector head, in the feed system, or in the inlet lines. In addition, instabilities at intermediate frequencies are dangerous because they often result in high-frequency instabilities.

High-frequency oscillations are characterized as pressure oscillations in a LRE combustion chamber with frequencies equal to or greater than the lowest acoustic frequency. Depending on the direction of oscillatory motion, high-frequency pressure oscillations in a LRE combustion chamber are classified according to the mode shape into longitudinal oscillations along the combustion-chamber axis and transverse oscillations perpendicular to the combustion chamber axis. They comprise tangential, radial, and mixed oscillations. In addition, mixed longitudinal and transverse pressure modes of oscillations can arise in a combustion chamber.

The preceding definition provides a classification of pressure oscillations, which can occur in a LRE combustion chamber. However, in designing LRE combustion chambers and gas generators there arise a number of questions that should be solved to provide stability of the operation process in all required regimes and under all conditions of engine operation. The attained process stability margins in combustion chambers and gas generators should be controlled and retained over the period of engine delivery to the rocket.

When designing, engineers use all achievements available at the present time to provide process stability. When testing engines, they observe the parameters of a running chamber and gas generator (pulsation and oscillation frequencies and amplitudes), but their main interest is to get answers to the following questions:

1) Will the observed pulse and oscillation amplitudes change to a new, higher level over the required engine operation range according to the performance specification during testing of the engine manufactured to the same drawings but with different tolerances?

2) Within what limits of external and internal factor variations should an engine be tested to ensure its serviceability as regards the operation process stability?

3) Several versions of the combustion chamber (mixing head) are considered in the design; thus, how should one select the version that is optimal as regards the operation process stability on the basis of a minimum number of tests (e.g., two to three tests)?

4) What kind of inspection system used in serial production will provide control over quantitative stability margins attained during engine development?

The necessity of solving the problem of combustion-chamber process stability became especially urgent in the 1960s when the need for development of high-rated, high-efficiency engines was envisaged. Instability of high-frequency operation processes in LRE combustion chambers and gas generators became the cause of a considerable number of accidents both in engineering development of engines and during flight tests of space rockets. So, for the period of 1957–1969, one-third of the total number of LRE failures were caused by initiation of destructive high-frequency pressure oscillations. Material loss caused by those accidents was extremely high, especially in satellite launchs. Because of the lack of a procedure for quantitative evaluation of the operation process stability margin, LREs with inadequate stability margins were introduced in serial production. Since 1958, the development of rockets R-14 and R-16 was started in the Soviet Union [3]. The flight range of rocket R-14 is up to 4500 km, and that of R-16 is up to 13,000 km. The engines for those rockets were developed on the components UDMH (unsymmetric dimethylhydrazine) and NA-271 (nitric acid). Starting with the development of R-14, UDMH found a wide applicaton in rocketry, later with nitrogen teroxide (NT), which acts as an oxidizer. The engines for the just-mentioned rockets were developed in DB "Energomash" named after V. P. Glushko [3].

The engines were developed on a modular principle; each module was subsequently integrated into two- or three-module arrays, which form the first-stage engine of the rocket. The same engine module with a slightly modified chamber (application of a high-expansion nozzle) and engine guide frame was used for the second stage of the same rocket. The design features of engine modules are as follows: use of two combustion chambers; one turbopump unit located between the chambers; a gas generator producing the reducing gas; and automatic control units. On this principle were developed the engines for rocket R-14; engines RD-216 consisting of two modules RD-215 for rocket R-16; the first-stage engine RD-218, consisting of two modules RD-217; and the second-stage engine RD-219.

However, in engine serial production the failures of the combustion chambers in the first stage RD-217 occurred periodically as a result of high-frequency pressure oscillations. The occurrence of high-frequency pressure oscillations and the ensuing failure of engine RD-217 were observed during rocket tests as well. Before being mounted on the rocket, all engines were subjected to technological inspection tests. That is, the fact of operability in proof tests is not evidence of sufficiency of the operation process stability margin. Studies revealed the causes of the operation process instability (mixing system instability) and corresponding modifications were made in the design of the mixing heads (see Sec. 4.1).

In 1962, the development of engines for rocket R-36 was started: RD-251 consisting of three engines modules RD-250 for the first stage of the rocket and RD-252 for the second stage. Oxidizer for these engines was changed to NT instead of NA 27, and the pressure in combustion chambers was increased as compared to the engine for rockets R-14 and R-16.

For increasing the pace of R-36 development, design documents were submitted to the factory long before the development of engines. However, during static firing tests of the engines, the combustion chamber process instability occurred quite often. In the first-stage engines RD-250, instability occurred at launch only; in the second-stage engines RD-252, instability occurred in the main operation mode only, with the mixing heads of these engines being identical.

High-frequency pressure oscillations resulted in failure of the combustion chambers, and immediately the question of serial-production suspension, and taking a decision of using the ready-made engines, arose. In Energomash intensive studies were started on stands and in the laboratories for providing the process stability in combustion chambers and in gas generators. This was the starting point of developing the procedure for quantitative evaluation of the margin of process stability for high-frequency pressure oscillations.

A number of solutions for ensuring stability in RD-250 and RD-252 engine combustion chambers were suggested: complete replacement of the mixing system, installation of vibration baffles on the combustion chamber head (see Chapter 12), and others.

Ultimately, taking into account that the engines were in serial production, it was decided [3] to introduce an additional resistance in the hydraulic circuit of

RD-250 engine combustion chambers, which provided ignition in the combustion chamber extended in time, and consequently, more smooth pressure growth. The stability of engine RD-252 was improved through decreasing the fuel heating temperature by applying a heat-resistant coat on the inner wall of the combustion-chamber nozzle. However, according to the data of the studies, the operation process in the combustion chambers of engines RD-250 and RD-252 remained highly sensitive to disturbances.

The process instability in combustion chambers and gas generators was also observed in other engines being developed in the same period.

Research institutes and divisions of the Soviet Academy of Science were recruited to the studies with the intent to solve the problem of providing process stability. The specialists suggested the following items:

1) Theoretical methods of problem solution is the first item. It was assumed that the sets of equations would describe all physical processes taking place in the combustion chamber. The equations should comprise particular dimensions of the mixing system structural elements and combustion chamber design features.

2) The second item is the solution of the problem of operation process stability in the combustion chamber by simulating the process as a whole and its individual elements. Both electrical models and firing simulation methods were used for the process simulation [4].

3) Another way of solving the operation process stability problem was the development of a procedure for quantitative evaluation of the operation process stability margin. For this purpose, it was necessary to work out relevant criteria quantitatively characterizing the stability margins, which could ensure engine performance in all required modes and under the required external conditions according to the engine performance specification. A method similar to that used in mathematics for solving sets of equations, the disturbance method, was proposed to solve this problem: introduce artificial disturbances into a running combustion chamber or into a gas generator, or, use natural operation process disturbances, "combustion-chamber noise." As shown next, application of the combustion-chamber process noise as disturbance is of great importance for gaining information on the operation process stability margins. A scope of information on process stability obtained with the help of artificial disturbances is essentially limited as compared to the data obtained from analysis of combustion chamber noise. In addition, in serial production of engines it is impossible to control the production stability in respect to retaining the stability margin by introducing artificial disturbances into the chamber.

As will be shown next, for comprehensive assessment of the process stability margin, quantitative evaluations of stability margins on the basis of both the operation process noise and artificial disturbances introduced into the combustion chamber need to be carried out. Each of the procedures mentioned will be described in detail in relevant sections of the book.

It was necessary to carry out a thorough examination of physical processes in the combustion chambers of different engines for all three trends. Each engine type has essential differences in the system of mixing the propellant components and, hence, in the whole operation process. In a liquid–liquid engine both components are fed to the chamber as liquids; in a gas–liquid engine one of the components is fed to the chamber as gas; in a gas–gas engine both components are fed to the chamber as gases. It is evident that for each of these mixing diagrams there is a specific set of equations describing the operation process physical phenomena. The characteristics of propellant components used in the engine (high-boiling and low-boiling components, and so on) also introduce certain features into the sets of equations. Therefore, it was evident a priori that quantitative values of criteria defining the operation process stability margin can differ for different types of engines. Regretfully, studies on the first and second trends did not give adequate practical data required by engine designers.

In the development of the third trend, solutions have been gained that make it possible to develop engines with the desired operation process stability margins in combustion chambers and gas generators for all types of engines. A lot of problems had to be solved for the development and implementation of the procedure for quantitative evaluation of the margin of the operation process stability with respect to high-frequency pressure oscillations.

The book *50 Years Ahead of Its Age* published in 1998 [5] gives the following description of the success achieved by the Department NIO-512 of Energomash led by the author over 30 years:

> NIICHIMMASH is a leading institute in the space-rocket industry of Russia on evaluation of high-frequency stability of combustion in LRE (NIO-512). The developed procedure for evaluation of the operation process stability with respect to high-frequency pressure oscillations in LRE combustion chambers and gas generators provides determination of the stability characteristics of the operating process at natural acoustic frequencies from the damping characteristics of pressure oscillations at a natural disturbance level, and from the response of the operation process to pulse disturbances. Generators have been developed for producing pulse disturbances; they make it possible to generate in the course of testing combustion chambers and gas generators up to five pressure pulses of a specified value at preset moments of time....

The present book is not aimed at generalization of all materials on operation process stability in LRE; it contains mainly the materials required for justification of the procedure for quantitative evaluation of the operation process stability margins in LRE combustion chambers and gas generators. Results of the evaluation of the stability margin obtained in experimental setups and for actual engines operating on NT and UDMH, oxygen–kerosene, and oxygen–hydrogen are presented in the book.

Chapter 2

Terms and Definitions

IN PREPARING documents on evaluation of process stability in LRE, the application of clear terms and definitions has been a serious problem. This chapter reviews the most important terms and notions, as well as problems arising from the uncertainty of treating the operation process instability phenomena observed in an actual LRE from different standpoints. The basic terms used in the description of operation process instability in LRE combustion chambers are contained in the state standards "Liquid Propellant Rocket Engines. Terms and Definitions" [6] and "Oscillation. Terms and Definitions" [7]. (Throughout this book the term "combustion chamber" means gas generator as well.) It is difficult to present in a few words a clear definition of operation process stability and instability due to the wide variety and complexity of physical phenomena occurring in the combustion chamber of an actual LRE.

The definition of operating process instability taken from an encyclopedia is given in the Introduction [2]. This definition, however, does not include all necessary notions to describe various manifestations of operating process instability.

Let us take a close look at the notions of operation process stability, transition of oscillations to a new level, and stability margin. The operation process stability with respect to high-frequency pressure oscillations should mean a monotone change of pulsation amplitude in the combustion chamber, within the bounds limited by engine operating range. There should be no transitions to a new level of pressure oscillations, which is not specified by the design documentation, as a result of inadequate operation process stability margin or as a result of the unsteadiness of the conversion process from propellants to combustion products. The transition of pressure noise oscillations to self-oscillations, or to a new noise level of pressure oscillations, in an LRE combustion chamber can proceed both monotonically and in spurts. For high-frequency acoustic oscillations occurring in a combustion chamber, there are soft and hard excitation modes as well as transitions caused by unsteadiness of the conversion process from propellants to combustion products.

High-frequency instability is a complicated and insufficiently studied problem, and quite often it has been a formidable obstacle to developing rocket engines. In addition, working out a clear description of operation process stability presents a problem for the following reasons. To provide combustion chamber (and gas generator) reliability, a designer must know not only the nature of pressure oscillations and their modes (noise oscillations vs self-oscillations), but also the level of pressure oscillation amplitude defining its serviceability. However, the amplitude of pressure oscillations on transition to a new level, rather than a level of pressure oscillations actually observed and known, is of great importance for ensuring operating capability of the combustion chamber. Therefore, it is very important to make an evaluation of operation process uncertainties because determination of pressure oscillation amplitudes realized on transition to a new limit cycle presents serious difficulties.

At the present time the development of theoretical evaluation methods is practically impossible because of the complexity of the physical processes in combustion chambers. The following experimental methods can be used to determine possible amplitudes of limit-cycle pressure oscillations, depending on the combustion chamber operating mode:

1) The first method is inducing hard excitation of pressure oscillations when the combustion chamber is running in a hysteresis loop, by introducing a finite disturbance and subsequently determining the limit-cycle oscillation amplitude.

2) Determination of the amplitude dependence of the damping factor change by introducing artificial disturbances during operation of a combustion chamber is the second method. The amplitude of the self-oscillation limit cycle, which arises because of changes of the engine operating mode, corresponds to the minimum oscillation decrement (see Chapter 6).

Both of these methods can be used in practice during research studies only.

Parametrical or statistical methods of estimating operation process uncertainty and stability within a specified range of variations of the combustion-chamber operating parameters, and stability margin value, are used for evaluation of an actual LRE.

Figure 2.1 shows the scheme of possible transitions from high-frequency pressure oscillations (or noise) to self-oscillations (or noise oscillations) with a new level of amplitude. The transition from an operating mode with noise pressure oscillations in the combustion chamber to a self-oscillation mode can occur for different reasons: 1) changes in the LRE operating conditions and reaching the stability boundary and 2) hard excitation of the operation process under the action of a random pulse.

The unsteadiness of propellant conversion to combustion products can be manifested in different ways.

1) *Mixing system unsteadiness* will be discussed first. The term "mixing system uncertainty" means an unplanned change of mixing element or injector characteristics, unspecified by the design documentation. Transitions of the mixing system from some characteristics to others can be caused by a change of operating mode or by the action of natural or artificial disturbances. With the tendency of the mixing system to unsteadiness, such a transition can be provoked also by launching conditions within the specified variations of operating conditions as well as by the difference between engines made within tolerances specified by documentation. Depending on the physical phenomenon initiating the unauthorized change of mixing characteristics, and on the mixing head design, transition to a new mode can be rather smooth, if the transition occurs in each mixing element in sequence, or in spurts, if characteristics of the entire mixing system or a large group of mixing elements change simultaneously. Mixing system ambiguity was revealed in a number of engines, including LRE accepted

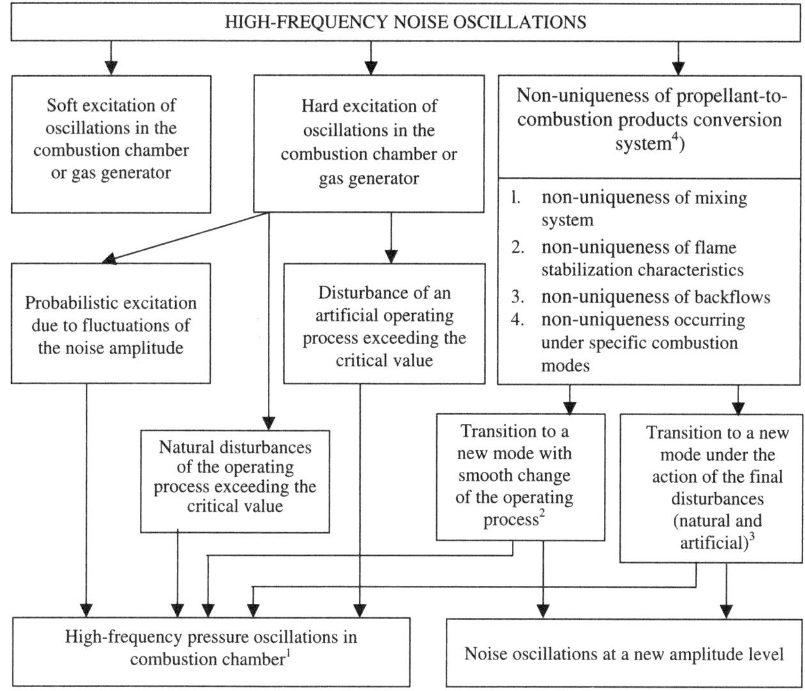

Notes:

1) Here and subsequently, high-frequency oscillations in a LRE combustion chamber mean acoustic modes of oscillations. However, in special cases, a rotating detonation wave is recorded on transition from noise to oscillations. Researchers encountered this specific oscillation mode in LRE combustion chambers of large diameter (for instance, in the D-480 mm T170.000 and T180.000) where both components were fed as liquids (liquid–liquid LRE).
2) The just-mentioned transition to a new mode can be both monotone and uneven, depending on peculiarities of the nonuniqueness of propellant-to-combustion products conversion system.
3) Probabilistic excitation caused by fluctuations of noise amplitude in the combustion chamber is considered in Fig 2.1 as one of the hard excitation mechanisms.
4) The term "nonuniqueness of propellant-to-combustion products conversion system" means an unplanned change of characteristics of propellant conversion to combustion products, which is not specified by the design documentation.

Fig. 2.1 Diagram of possible transitions from noise to oscillations or to noise at a new level.

for production. When testing a steering engine (see Sec. 11.6), a decrease in oxidizer consumption took place in the second-stage engine block during launching a mid-flight engine as a result of a sharp pressure reduction at the extraction point, which in a number of cases caused high-frequency pressure oscillations. During special tests of engine combustion chambers, there was found a mixing system unsteadiness, which during firing tests caused a change of the first tangential mode oscillation decrement of up to 25% and a change of pressure specific pulse of up to 10–25%.

An unsteadiness of mixing characteristics of a slightly different kind was revealed during study of the hydrodynamic characteristics of engine jet-centrifugal injectors. The unsteadiness was manifested in the fact that on jet-centrifugal injectors at some outflow velocities, with an increase in the medium density, the jet adhered to an atomizing cone. This adhesion was one sided and resulted in a substantial change of propellant distribution over the atomizing cone and, hence, over the combustion chamber volume (see Sec. 4.1).

2) Another type of nonuniqueness in propellant conversion to combustion products is the *nonuniqueness of combustion stabilization characteristics*—realization of two more flame stabilization modes at the combustion-chamber injector head. It is to be kept in mind that the just-mentioned changes are caused by physical phenomena not associated with mixing system unsteadiness. As in the case of a mixing system nonuniqueness, the transition from a specific flame stabilization mode to another one can be caused by the change of combustion-chamber operating mode or by an action of natural or artificial disturbances. The transition from one flame stabilization system to the other, depending on the physical phenomenon initiating the unauthorized change of flame stabilization mode, and on the design features of mixing elements, can proceed quite smoothly or in spurts (see Sec. 4.2).

3) One of the phenomena causing nonuniqueness in propellant conversion to combustion products is *unsteadiness of flow reversal in the combustion chamber*. As flow reversal occurs at the combustion chamber head end during a launch or change during engine transition from one mode to another, the nonuniqueness in flow organization, depending on the peculiarities of a specific change in engine parameters on reaching the main mode, can influence the operation process stability.

4) In the course of the development of combustion chambers with different mixing elements carried out in our country and in the United States, there were *mixing systems that caused specific forms of transition to self-oscillations* under certain operating conditions of the mixing element.

In the case of nonuniqueness in the mixing or flame-stabilization system, the operating process in the combustion chamber can change not only to self-oscillations, but also to noise oscillations with another level of amplitudes. In this case, the preceding statements on the pattern of transitions from noise oscillations to self-oscillations hold true for transitions of the operating process from some noise level to another.

However, in practice for most combustion chambers, mainly soft or hard excitations occurred on transition from noise oscillations to high-frequency self-oscillations of different acoustic modes. It follows from experience with improvement in the stability of the operating process in LRE combustion chambers that in virtually all engines the operating modes are located in the hysteresis zone. The detailed description of different manifestations of the nonuniqueness of propellant conversion to combustion products in this chapter is explained by the lack of relevant terms and definitions used in the operations documents. When describing soft and hard excitations of oscillations in the operating process in LRE combustion chambers, it is advisable to use the term "operating process instability." When describing a transition of the operating process from noise pressure oscillation level to self-oscillations, or to a new level of amplitudes as a result of an unauthorized

change in the process of propellant conversion to combustion products, it is advisable to use the term "operating process nonuniqueness."

It is indicated in note 1 to Fig. 2.1 that high-frequency self-oscillations in LRE combustion chambers are hereafter meant as acoustic modes of oscillation. However, in very rare cases the transition from noise pressure oscillations to a rotational detonation wave was observed [8].

In the combustion chamber T-170/180 of the RD-216 series of engines, acoustic oscillations mainly of the first tangential mode with a frequency of 1200 Hz were observed. Acoustic oscillations of large amplitude occurred both with self-excitation and with artificial excitation by disturbing devices. However, when an

Table 2.1 Terms used in methodical guidelines

No.	Term	Definition
1	Self-oscillations	System oscillations arising from self-excitation.
2	Self-oscillating system	The system wherein oscillations arise in the absence of external periodic force; their frequency and amplitude depend on the system parameters.
3	Amplitude spectrum	Frequency spectrum of oscillations when their amplitudes are the quantities characterizing harmonic components of oscillations.
4	Amplitude of harmonic oscillations on vibrations; Amplitude	Highest value of the quantity characterizing harmonic oscillations or vibrations.
5	Rapidly varying parameters (RVP)	——
6	Time of relaxation of damped oscillations after a pulse disturbance	The time of damping caused by pulse disturbance of oscillations in the system, on expiration of which the highest peak of the signal of pressure oscillations decreases e times. Note: e is the base of natural logarithm ($e = 2.72...$).
7	Forced oscillations (vibrations)	Oscillations (vibrations) induced and maintained by power and (or) kinematic excitation.
8	Harmonic oscillation (vibration)	Pressure oscillation (vibration) wherein the value of the variable quantity varies in time according to $A \sin(\omega t + \varphi)$, where t is time; A, ω, φ are constant parameters; A is the amplitude; $\omega t + \varphi$ is the phase; φ is the initial phase; ω is the angular phase.
9	Stability margin of the combustion-chamber (gas-generator) operating process Stability margin	Relation (modulus of the difference or the ratio) between the actual value of the stability factor and its specified limit value.
10	Identification of frequencies in the oscillation spectrum	Identification of frequencies corresponding to the maxima of the amplitude spectrum (or the energy spectrum) with natural frequencies of the system or the frequencies of forced oscillations.

(Continued)

Table 2.1 (Continued) Terms used in methodical guidelines

No.	Term	Definition
11	Oscillating system	——
12	System damping factor Damping factor Relaxation factor	Ratio of resistance coefficient to twice the mass or moment of inertia.
13	Logarithmic oscillation decrement	Natural logarithm of the ratio between two successive maximum values of the amplitude at times separated by T, the period of oscillation.
14	Unstable operating process in the combustion chamber	Operating process in the combustion chamber with self-oscillations caused by the combustion process.
15	Peak value of an oscillating quantity	Highest absolute value of an oscillating quantity within the time interval considered.
16	Operating process stability characteristic	Quantitative characteristic of stability of the operating process for a combustion chamber.
17	Operable condition of an LRE	LRE condition ensuring the execution of operations required for creating thrust or for changing its value and (or) direction; or for ensuring operating conditions for the component parts of a movable vehicle in accordance with the prescribed requirements.
18	Reaction volume of the combustion chamber (gas generator) Reaction volume	Volume of combustion chamber (gas generator) where mixing and combustion processes take place.
19	Regularity of the oscillating process Regularity	Relative frequency of realization of the maximum value of the oscillation amplitude in the given frequency band during the analyzed time interval.
20	Free oscillations (vibrations)	System oscillations occurring without external action and without energy supply from the environment.
21	Natural frequency of linear system oscillations (vibrations)	Any of the frequencies of free oscillations of a linear system.
22	Frequency spectrum	A set of frequencies of oscillating harmonic components arranged in increasing order.
23	Stable operation process in the combustion chamber	Operating process in the combustion chamber without oscillations caused by the combustion process.
24	Phase spectrum	Oscillation spectrum when the initial phases of the oscillation are the quantities characterizing the harmonic components.
25	Combustion-chamber noise	Noise created by the combustion chamber operation.
26	Energy spectrum	Oscillation spectrum in which squares of the velocity amplitudes characterize the specific energies of harmonic components.

artificial disturbance of sufficient intensity was introduced into the combustion chamber, tangentially to the chamber inner shell, a rotational detonation wave was generated in the combustion chamber. Four transducers fitted around the chamber recorded the growth of the detonation wave pressure. Attention must be paid to different fracture patterns of the combustion chambers because of acoustic oscillations of large amplitude and to a rotational detonation wave. Rupture of the combustion chamber inner shell near the head end and wear hardening at the points of attachment of the combustion chamber to the engine frame were observed in cases of failures caused by acoustic oscillations of the first tangential mode. The pressure growth near the combustion chamber shell up to a level resulting in local tear of a part of inner and outer shells was observed in cases of the rotational detonation wave. Studies of the stability of combustion chambers T-170/180 and their characteristics are described in Chapters 11 and 12.

Definitions of the terms most widely used in the documentation of process stability are given in Table 2.1 [6, 7, and 9].

The definition of the stability margin of the operating process for a LRE combustion-chamber (gas-generator) operation is missing in GOSTs, but there is more general definition: the difference between the LRE operability parameter margin in testing (or in use) and its critical value. Because LRE combustion chamber operating process is one of the critical systems responsible for engine reliability, the evaluation of the operating process stability margin is set off as a separate system in evaluating engine reliability.

The term "sufficient stability margin" means that no transition of oscillations to a new level will occur during engine operation throughout the range of its parameters specified by the operating documentation. That is, these are quantitative characteristics (in parametric estimation), which ensure adequate engine reliability as regards organization of the operating process.

Chapter 3

Mechanisms of Transition from Noise to High-Frequency Oscillations or to Noise at a New Level

THE mechanisms of transition from noise to well-organized high-frequency oscillations with large amplitudes, which result in the failure of LRE combustion chambers or gas generators, are listed in Fig. 2.1 of Chapter 2. Let us consider the main transition mechanisms.

I. Soft Excitation in Combustion Chambers and Gas Generators

In the 1950s and 1960s, the margin of combustion-chamber stability with respect to high-frequency pressure oscillations was estimated on the basis of the departure of the operating modes from the regimes in which high-frequency oscillations of large amplitudes occurred (i.e., from stability boundaries). During testing, the combustion chamber pressure (and the flow rates of propellants fed to the combustion chamber) was increased to a level at which high-frequency, self-excited pressure oscillations occurred. The effectiveness of various measures aimed at stability improvement was determined from the location of stability boundaries, that is, the departure of the operating mode from self-excited, high-frequency oscillation boundary. The effects of modifications made in the injector head and combustion-chamber geometry on the stability were evaluated from the increase in the operating mode departure from the stability boundary. With an increase in the combustion-chamber pressure (according to the chamber pressure pulsation records), the growth of pressure oscillation amplitude was observed at a frequency at which high-frequency oscillations can occur in the future (see Sec. III in Chapter 6). A rather smooth transition to self-oscillations with large amplitudes was then observed. Such a type of transition from noise in a combustion chamber to self-oscillations is commonly referred to as "soft excitation of self-oscillations" [4, 10, 11]. Naturally, because the evaluation was based on stability boundaries, the theoretical studies were aimed at development of methods for calculating stability boundaries. Comprehensive reviews of theoretical models are given in [10], and descriptions of various theoretical models are presented in [4 and 11].

The phase shift φ between fluctuating heat-release rate and pressure oscillation in a combustion chamber is used in all known theoretical models for calculating stability boundaries. This parameter depends on the relation between a specific process time delay in the combustion zone τ and the period of acoustic oscillation. The occurrence of self-oscillations, however, is determined to a greater extent by energy considerations, that is, the relation between the energy for generating pressure oscillations and energy losses. By increasing the energy loss in the oscillatory fields, self-oscillation can be terminated. At the present time, it is practically impossible to calculate the difference between energy generation and energy losses caused by the complexity of the processes involved in a combustion chamber. Therefore, the recommendations obtained on the basis of theoretical studies often do not find support in practice.

All theories developed so far, which take into account many of the processes occurring in combustion chambers and specific design elements, ultimately attempt to define the stability boundaries. But the questions "What is the stability margin under all the required engine operating regimes? and Is it sufficient for providing the required reliability?" are of concern to designers of combustion chambers and gas generators. Also, no theoretical study on defining the stability boundaries provides the data on how far the engine operating regimes should depart from the stability boundaries.

Practically, it makes no sense to find experimentally the stability boundaries for contemporary heavy-duty high-thrust combustion chambers, for example, for the engines RD-170, RD-172, and RD-180. It is even more senseless to determine the influence of various design and operating factors on the stability margin based on the stability boundaries of operating processes, considering the poor accuracy of stability boundary determination, because of the action of "probabilistic mechanism of transition from noise to oscillations" (see Sec. III).

II. Hard Excitation of Oscillations in Combustion Chambers

Hard excitation of oscillations in a combustion chamber is associated with nonlinearity of the stability characteristics of the operating processes. During the development of LREs, designers often encountered situations in which some of the combustion chambers in the engine experienced spontaneous pressure disturbances, in the form of local explosions. With the introduction of hard excitations, the initial disturbances can result in high-frequency oscillations, if the combustion chamber does not have a sufficient stability margin.

The following model describes quite well the nature of many phenomena observed in practice during the investigation of operation process instabilities in combustion chambers and gas generators. The model describes the operating process as an oscillating system with distributed parameters. Using this model, it is possible to explain a number of observed manifestations of combustion instabilities: soft and hard excitations of pressure oscillations; hysteresis of the pressure oscillation with variations of the combustion chamber operating parameters; and probabilistic occurrence of operation process instability, which is generally accompanied by poor reproducibility. In addition, such a model explains the changes of the probability density distribution of noise amplitude during the transition from stable to unstable operation mode of a combustor.

Possible mechanisms for hard excitation of triggered instabilities, methods of estimating the stability margin with respect to the final disturbance, selection of the characteristics artificial disturbances, and other related results are described in detail in Chapter 9.

III. Probabilistic Excitation of an Instability Caused by Fluctuations of Noise Amplitude

In practical LRE testing, so-called probabilistic occurrence of instabilities was often observed. It was manifested in the fact that under engine steady-state operation there occurred a spontaneous development of pressure oscillations of large amplitude. In other words, under steady-state engine conditions the combustion chamber process becomes unstable. A unique feature of probabilistic excitation of high-frequency pressure oscillations is poor reproducibility, even in testing of the

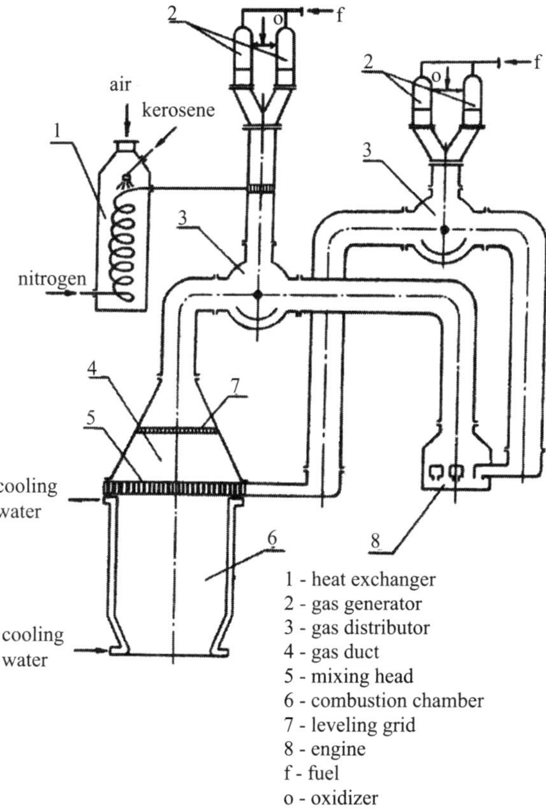

Fig. 3.1 Diagram of a model setup with full-scale combustion chamber running by the method outlined in Sec. III: $P \approx 1.0$ atm (abs) with variable $\dot{m}$.

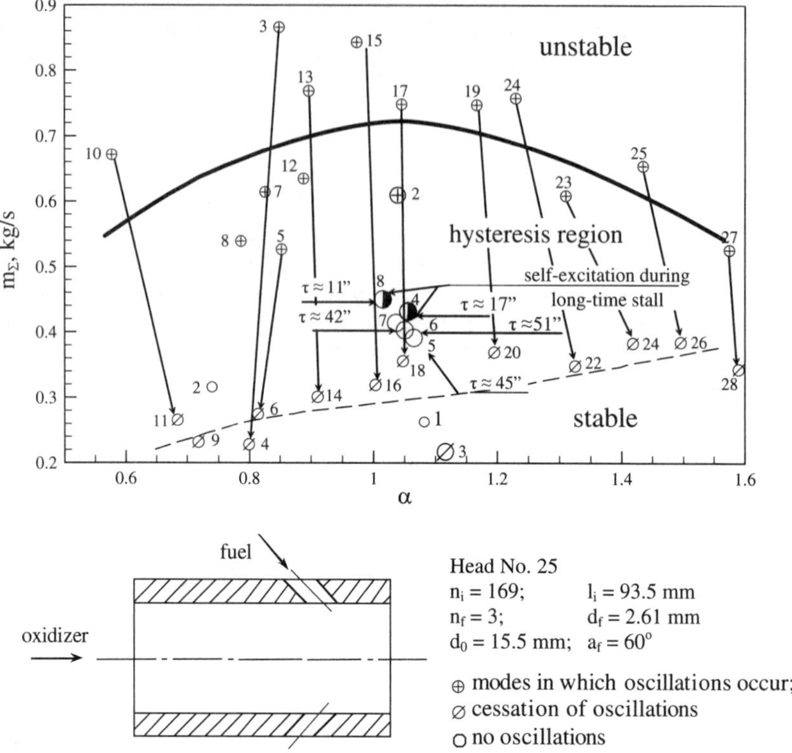

Fig. 3.2 Boundaries of unstable oscillation with frequency f_1T = 2.0 kHz (——) and hysteresis (------): τ = test duration in constant mode, s.

same combustion chamber under identical conditions. The cause of nonreproducibility is fluctuation of the "noise" amplitude.

When an engine is running, the maximum amplitude of natural pressure oscillations in the combustion chamber has a probabilistic nature. When the combustion-chamber (or gas-generator) operation approaches the oscillation boundary, the pressure oscillation amplitude for that particular mode which will occur increases (see Sec. III of Chapter 6). Accordingly, the value of the total maximum amplitude grows. Therefore, when approaching the excitation boundary, the number of cases for which the maximum amplitude crosses the lower boundary of the bifurcation curve (in the hysteresis region) increases. Because the probability of the realization of the maximum amplitude of noise oscillations in the combustion chamber is time related, the occurrence of probabilistic excitation of oscillations is determined by the time during which the combustion chamber operates at conditions near the stability boundary. A model combustor operating at a reduced pressure, $P_{ch} \sim 1$ atm (abs), was used for experimental verification of the mechanisms of probabilistic excitation of high-frequency pressure oscillations, as shown in Fig. 3.1. A total of

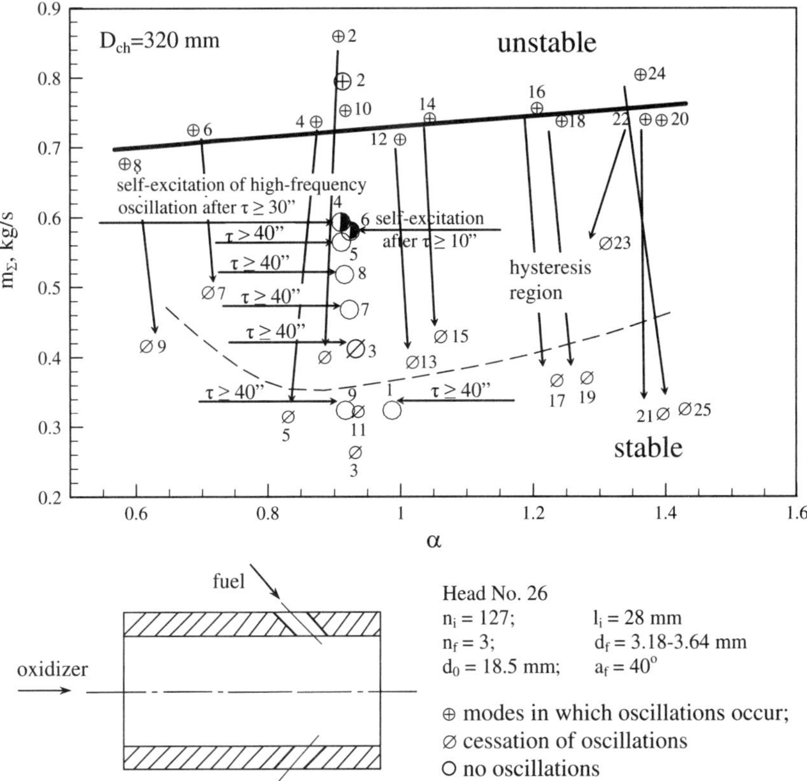

Fig. 3.3 Boundaries of unstable oscillations with frequency f_{1T} = 2.1 kHz (——) and hysteresis (------): τ = test duration in constant mode, s.

169 injectors were installed. The overall mass flow rate varied in the range of 0.2–2.5kg/s. The simulator was designed for investigation of the operating process stability in the combustion chamber of engine RD-263 (15D117) [12]. The engine was developed in 1969–1973 for the first stage of the intercontinental ballistic rocket R-36M/36 MU. Four LRE RD-263 form DU-RD-264. The LRE operates with afterburning of the oxidizing gases from the gas generators. The engine characteristics are as follows: thrust (ground) P_g = 100tf (1040kN), R_m = 2.67, specific impulse (ground) J_g = 293 s; and thrust (vacuum) P_v = 115tf (1130kN), D_{ch} = 320mm, specific impulse (vacuum) J_v = 318 s.

The same propellants employed in the engine—NTO oxidizer and UDMH fuel—were used in the model combustor. The model unit consisted of two gas generators operating with an excess of oxidizer and with an excess of fuel, respectively. The generator operating with an excess of fuel was used for feeding "cold" gas instead of liquid propellant to the injectors (T_g ~ 70°C). Gas distributors were used to vary the propellant mixture ratio (see Fig. 3.1). The ratio of injector velocity heads was varied by diluting the oxidizing gas with nitrogen preheated in

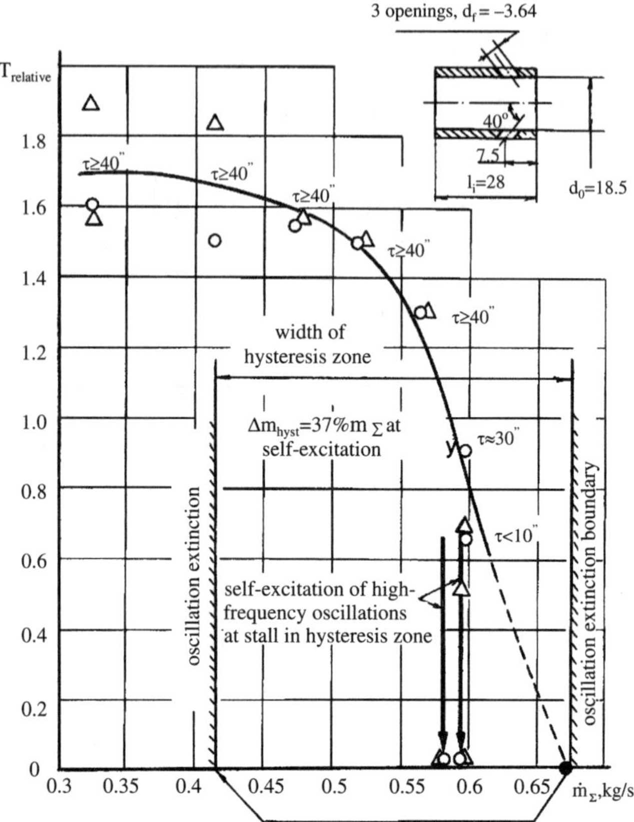

Fig. 3.4 Change of decrement δT when approaching the boundary of unstable oscillations: τ = time of stable operation in constant mode, s; $\alpha = 0.93$; O, Δ = data of transducers; 1,2 = head no. 26.

the heat exchanger. To meet the modeling conditions (see Chapter 10), the fuel injectors of the model unit had a slightly larger diameter of holes compared to those in an actual injector, as shown in Figs. 3.2 and 3.3, where α is the oxidizer excess coefficient, defined as the ratio of the actual mass ratio of the oxidizer to fuel $\mathcal{K}^{\cdot}$ to the ratio for a stoichiometric mixture. Figures 3.2 and 3.3 indicate the unit operating modes in which the boundaries of pressure oscillations and the boundaries of hysteresis were defined. The boundaries of unstable oscillations were obtained by rapidly changing the setting of the gas distributors. (These are spontaneous oscillations; hence, in the Russian text they are called “self-excited” oscillations.) Figures 3.2 and 3.3 also show the characteristics of the injectors tested.

Large circles used in the figures denote the model combustor operation in the hysteresis zone that remained for a prescribed period of time. It is seen from the experiment that in the hysteresis zone transition takes place from the stable operating process to oscillations at a distance of ~14–20% in terms of the propellant flow rate from the "provisional" boundary of self-excitation stability at respective periods of operation in a given mode. This boundary is called provisional, as it should be remembered that the unstable boundary is obtained at a certain rate of changing the setting of the gas distributors. In spite of the fact that pressure oscillation records have a characteristic tulip-like pattern typical for systems with soft excitation of oscillations, in this case there was hard excitation of a random nature. This is evidenced by rather large values of the pressure oscillation decrement at the moment when oscillations originated; thereafter, a sharp transition to the zero value occurred. Under soft excitation, smooth transition of the pressure oscillation decrement to zero level takes place. Development of tulip-pattern oscillations is accounted for by a slight excess of the initial pulse over the noise level in the combustion chamber.

The pattern of relative decrement variation for head no. 26 at different locations from the stability boundaries with $\alpha = 0.93$ is shown in Fig. 3.4, where the relative decrement is defined as the ratio of the decrement in an arbitrary mode to the decrement at a flow rate of 0.58 kg/s. Figure 3.4 also shows the modes in which the probabilistic excitation of high-frequency pressure oscillations occurred.

The results obtained are supported by numerous experiments carried out in the 1960s. A wide scatter of experimental data essentially exceeding the measurement accuracy was observed in estimating the operating process stability at locations on the excitation boundaries. In many cases, it precluded revealing the effectiveness of the modifications made in the combustion-chamber mixing system to improve stability. The mechanisms of random excitation are quite frequently encountered in LRE development. For example, the random excitation of high-frequency pressure oscillations in combustion chambers occurred in some cases during serial production of the Proton second- and third-stage engines (see Chapter 17).

Chapter 4

Uncertainty in Conversion of Propellant to Combustion Products

AS STATED in Chapter 2, the term "uncertainty in the mixing system" means an unauthorized change of the characteristics of propellant conversion to combustion products in a combustion chamber, which is not specified by the design documentation. It is impossible to manufacture two absolutely identical engines according to the same design documentation because of manufacturing tolerances specified by a designer. In addition, it is impossible to carry out identical tests, as it is impossible to adjust the engine to the same mode and exactly observe all external factors such as ambient temperature, temperature of the engine and of propellant components, gas saturation of components, and so on. These remarks point to the necessity of using the quantitative characteristics of stability margins for estimating combustion-chamber process stability. The selected values of stability characteristics must ensure reliable performance of the engines with due regard to tolerance differences and under all combinations of external and internal factors. The quantitative stability characteristics must define steady operation of the combustion-chamber operation process. Uncertainty in the conversion of propellant to combustion products was encountered in a number of engines during LRE development. Some of the observed cases are described in this chapter and in Sec. VI of Chapter 11.

I. Uncertainty in Mixing System

Events of high-frequency instability occurring in the combustion chambers of the first-stage engines RD-217 were repeatedly noted during flight-engineering tests of rocket R-16 [3]. Cases of instability were observed during stand tests in serial production of these engines at different factories. Engine RD-217 is a module for engine RD-218 consisting of three two-chamber modules [3]. The engine runs on the following propellant components: nitric acid (NA-271) and UDMH. Engine thrust in vacuum $P_v = 88.6$ tf (869 kN); thrust at ground $P_g = 75.5$ tf (740 kN); specific impulse in vacuum J_v = 289 s; specific impulse at ground J_g = 246 s; combustion chamber pressure P_{ch} = 7.335 MPa; and propellant mixture ratio Km = 2.6 [12].

The combustion chamber of engine RD-217 had a diameter of 480 mm. The throat diameter is 205.5 mm. The cylindrical chamber is 300 mm long.

The combustion chamber mixing head contains 793 single-component injectors, including 480 oxidizer injectors and 313 fuel injectors. The layout of the injectors is similar to a honeycomb pattern, with a fixed number of injectors arranged in concentric circles, the spacing between injectors increasing from 12 mm at the center of the combustion chamber to 18 mm at the periphery. Naturally, certain distortions were introduced in the relative position of the injectors as compared to a strictly honeycomb pattern; however, they did not affect the combustion efficiency.

A three-level mixing system is provided for operating process stability. Some oxidizer and fuel injectors have atomizing cones of 95–125 μg (the first level); the remaining injectors have atomizing cones of 30–50 μg (the second level). Injectors with large angles of the atomizing cone have central holes in a swirler, which provide jet-type injection of a part of the components (the third level). The flow rates through the injectors are the same, with jet-type injection through the injectors having large atomization angles making up about 50% of the total flow rate through those injectors. A diagram of the three-level mixing of the propellant components is shown in Fig. 4.1. Three oxidizer injectors with a large angle of the atomization cone (no. 1) and three injectors with a small angle of the atomization cone (no. 3) are located around fuel injectors (nos. 2 and 4) (Fig. 4.1).

A design of the fuel and oxidizer injectors is shown in Fig. 4.2; the layout of the injectors is shown in Fig. 4.3.

Such a mixing system ensures energy release throughout the entire volume of the combustion chamber. This reduces interaction with oscillations of the first

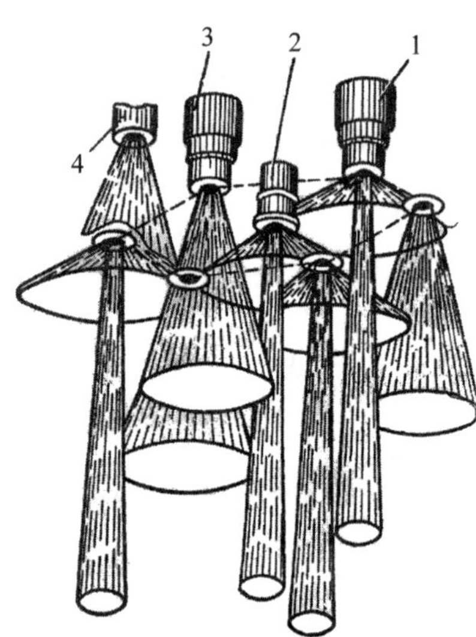

Fig. 4.1 Diagram of three-level mixing of fuel components: 1, oxidizer jet–centrifugal injector; 2, fuel jet–centrifugal injector; 3, oxidizer centrifugal injector with a small atomization angle; and 4, fuel centrifugal injector with a small atomization angle.

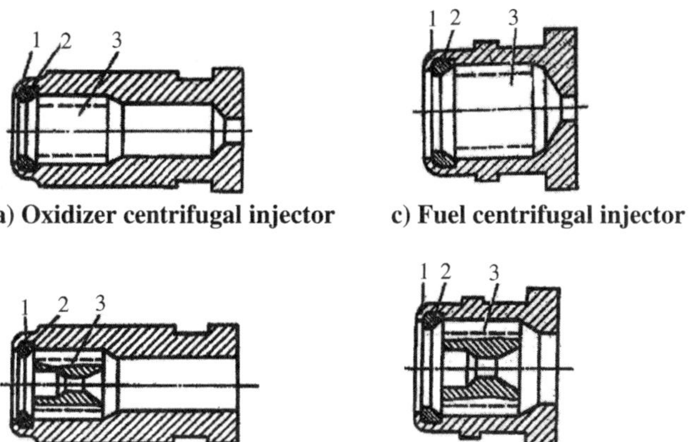

Fig. 4.2 Injectors for oxidizer and fuel before design modifications were made: 1, body; 2, insert; and 3, swirler.

tangential mode, which was observed in the combustion chamber. However, with this injector design mixing system unsteadiness was revealed, which resulted in instability of the operating process. High-frequency pressure oscillations of the first tangential mode occurred in the combustion chamber, resulting in its failure. Unsteadiness of mixing was revealed during study of the hydrodynamic characteristics of jet–centrifugal injectors. This unsteady phenomenon consists of the

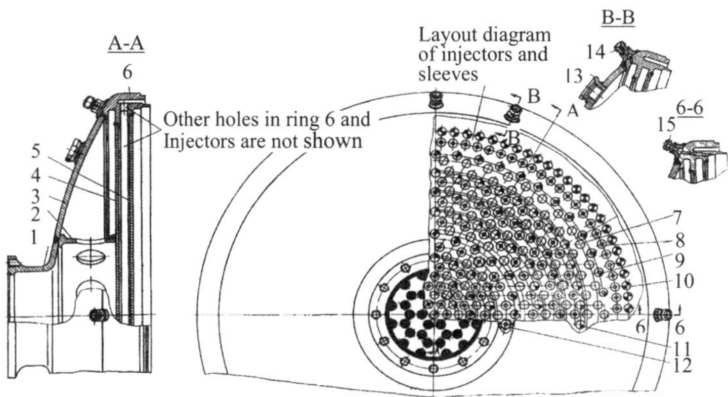

Fig. 4.3 Injector head: 1, flange; 2, shell; 3, outer bottom; 4, middle bottom; 5, inner bottom; 6, ring; 7, oxidizer injector with small beam angle of spraying; 8, jet–centrifugal oxidizer injector with large beam angle of spraying; 9, jet–centrifugal injector with large beam angle of spraying; 10, peripheral fuel injector with a reduced flow; 11, oxidizer injector with small beam angle of spraying; 12, spacing sleeve; 13, boss; and 14, 15, nipples.

fact that under some flow conditions in an injector the central jet adheres to the atomization cone. Because this is a unilateral adherence, it results in an essential change of component distribution along an atomization cone and consequently over the combustion-chamber volume. The number of injectors with adhered jets is unpredictable.

Before being installed in the mixing head, each injector is subjected to factory test of discharge into the atmosphere. Studies were conducted under conditions simulating the operation of injectors in the combustion chamber for finding more accurate hydrodynamic characteristics of injectors. Tests were carried out with discharge into the environment levels, the backpressure created by air and by helium, and with different turbulence intensity of the inlet flow. Disintegration of the central jet flowing out of a jet–centrifugal injector essentially depends on the flow turbulence at an injector inlet. Figure 4.4 shows the values of jet opening-angle tangent against the intensity of flow turbulence. The discharges were performed with water at a rate from 80 to 170 g/s. Flow turbulence at an injector inlet was measured by a hot-wire anemometer. A swirler in the oxidizer injector was located quite far from the outlet. Hence, when the level of turbulence in the initial inlet flow was $\varepsilon = 30\%$ ($\overline{u}_2 \sim 0.9$ m/s), the trajectories of liquid particles of the disintegrating jets, even during discharge into the atmosphere, touched the inner surface of the circular liquid flow in the injector nozzle. In testing the injectors with backpressure discharge into the environment with helium and without artificial turbulization at the injector inlet, a separate outflow of a jet and of a cone persisted up to a backpressure of 40 atm. In testing with increased backpressure of the turbulent flow at the injector inlet with a pressure drop of $\Delta P = 6$ atm, full adhesion of the central jet and atomization cone occurred under a pressure of about 10 atm. This result suggests the essential effect of flow turbulence at the inlet on uncertainty of the injector characteristics. Disintegration of the central jet was promoted by the diffuser part of the central opening.

Figure 4.5 shows the dependence on the chamber pressure of the mass flow rate through an injector, where G stands for the injector flow rate normalized by that

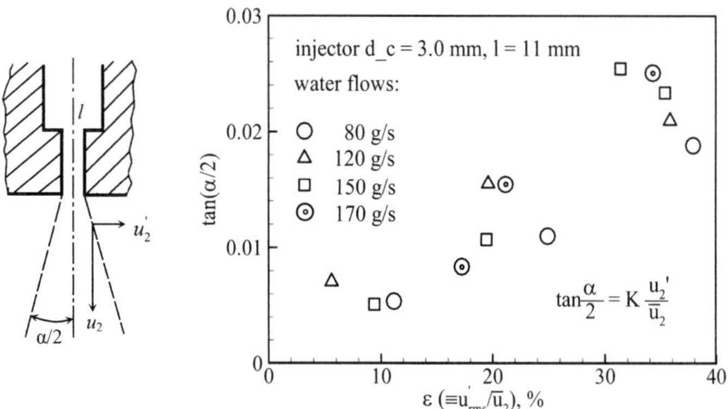

Fig. 4.4 Experimental dependence of jet spreading angle on inlet flow turbulence intensity: u_2-axial-flow velocity.

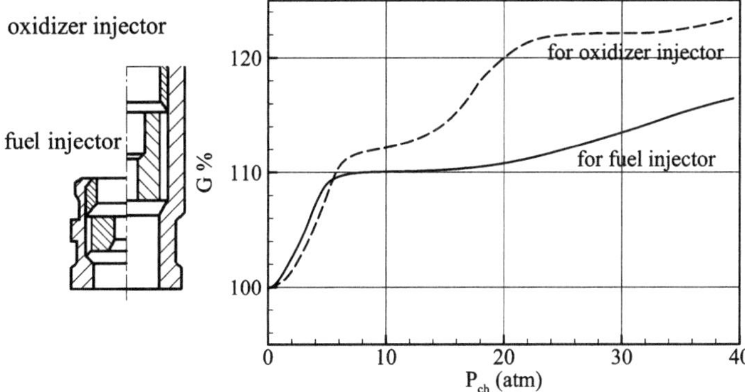

Fig. 4.5 Characteristic of a jet–centrifugal injector (initial version).

at 1 atm. The pressure drops in the fuel and oxidizer injectors are 0.8 and 0.7 MPa, respectively. It is seen from the figure that adhesion of the central jet and atomization cone increases the flow coefficient μ (defined as the ratio of the measured to the calculated injector flow rate), quite essential for oxidizer injectors. The following modifications were made in the design of injectors using the results obtained: the oxidizer injector swirler was positioned closer to the mixing head combustion bottom; a converging shape of the swirler inlet was applied; and the outlet diffuser shape was eliminated.

Use of the converging shape of the inlet essentially reduces flow turbulence. It is shown in Fig. 4.6 that dependence of the flow variation on the backpressure disappeared during discharges. By this means, the mixing system uncertainty responsible for operating process instability was revealed and eliminated.

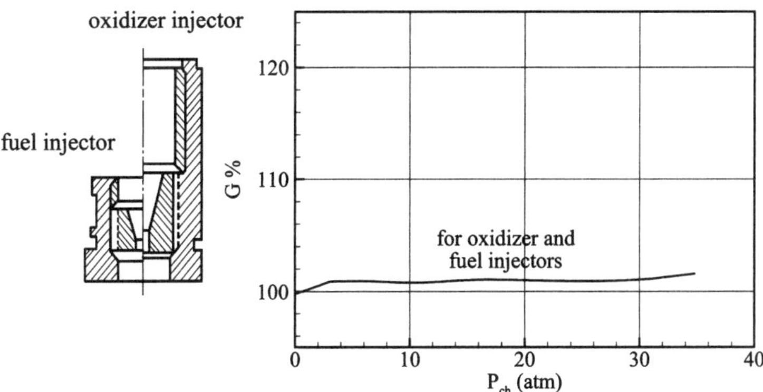

Fig. 4.6 Characteristic of a jet–centrifugal injector with a convergent central opening in the swirler.

The quantitative stability characteristics were not yet developed in the proceeding studies.

The results of a study of engine 8D69, combustion chamber U104-000, with the help of improved operating stability characteristics, are described in Sec. VI of Chapter 11. A number of researchers studied the problems of mixing system stabilization [13–15]. Their results made it possible to solve this problem.

II. Operating Process Uncertainty Caused by Loss of Flame Stabilization in Injectors of Gas–Liquid Combustion Chambers

During the development of combustion chambers and gas generators for closed-circuit gas–liquid engines, the uncertainty of the quantitative stability characteristics associated with the unsteadiness of flame stabilization in the injectors was encountered. In frequent cases such a problem arises for single-zone generators, where a zone with an elevated temperature of ~2000 K and a zone with a reduced temperature exist in the injectors. Unsteadiness of flame stabilization in gas generators results in elevated levels of high-frequency pressure pulsations and vibrations. In a single-zone generator, all of the propellants are injected through the mixing head. In a two-zone generator, part of the excessive propellant (either fuel or oxidizer) is injected downstream of the mixing head in order to reduce the generator gas temperature to a prespecified level.

Two experimental setups were used for studying this phenomenon applicable to gas–liquid engines.

A. First Setup

An experimental gas–liquid combustion chamber was used in this setup. Gaseous oxygen and kerosene were used as propellant components. The combustion chamber had a diameter D_{ch} = 104 mm; a diameter of the nozzle throat D_{cr} = 37 mm; and a length of cylindrical part ~125 mm. The combustion chamber was cooled with water, as shown schematically in Fig. 4.7. The injector head had four emulsion injectors with a gas passage diameter of 10 mm. Each injector had four openings for fuel injection with a diameter of 0.8 mm at a distance of 5.2 mm from injector exit section.

A temperature sensor was mounted in the gas passage of the injector, in the region where liquid fuel was introduced (see Fig. 4.7). The data taken in one of the experiments are in a report [13] and are shown in Fig. 4.8. The ratio of components was varied at practically constant pressure in the combustion chamber P_{ch} = 0.5 MPa. When the oxidizer excess coefficient α was reduced to 0.57, high-frequency pressure oscillations were excited in the combustion chamber. The temperature in the injection zone increased sharply, indicating that the flame was shifted to the injector. Further reduction of α to 0.518 interrupted high-frequency oscillations. The ensuing decrease of the temperature in the injector suggested that the flame was moved to the downstream region of the injectors in the combustion chamber. Similar changes of the flame stabilization pattern were observed when an artificial pressure disturbance was introduced into the combustion chamber to excite oscillations. However, upon introduction of another artificial pressure disturbance, the oscillation disappeared.

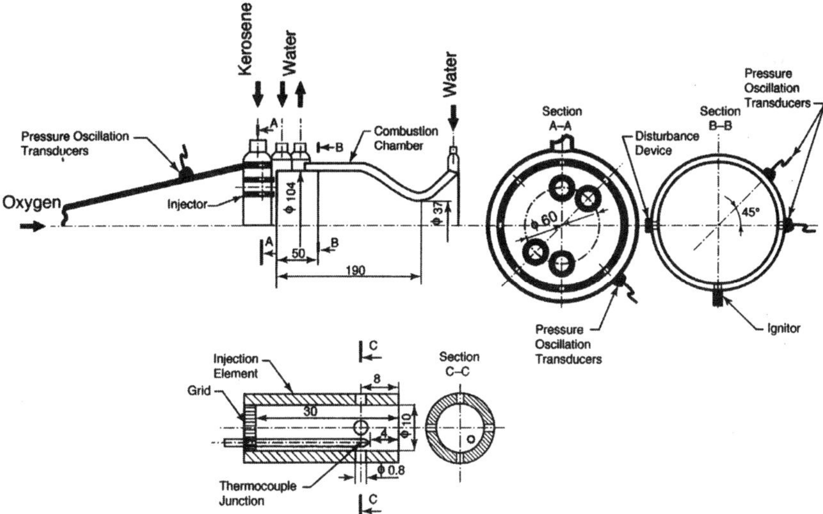

Fig. 4.7 Setup for combustion instability study: DD, disturbance device; I, ignitor; and ΛX-608, ΛX-604, pressure pulsation transducers.

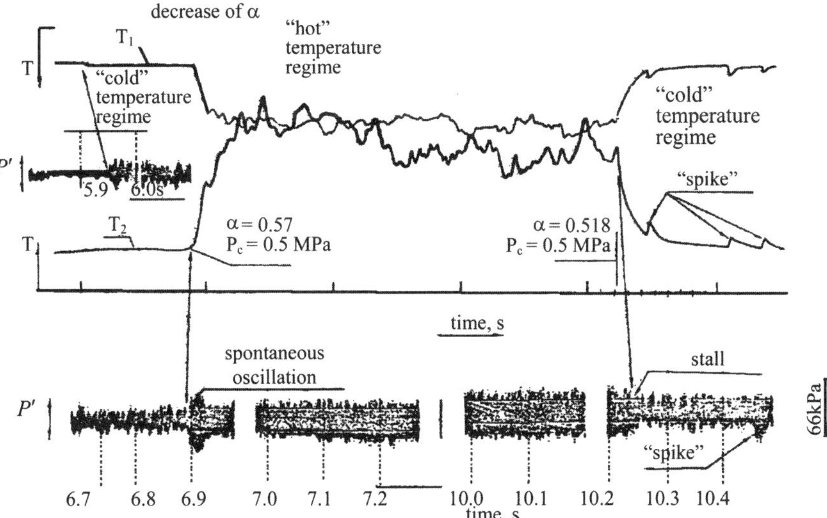

Fig. 4.8 Spontaneous oscillation and stall in the course of mode change, dependence on the reciprocal of the equivalence ratio.

B. Second Setup

Experiments were carried out on a free flame setup to observe flame stabilization, as shown in Fig. 4.9. The same injector head as in the combustion chamber tests (see Fig. 4.10) was used in this study. Instead of kerosene, gaseous methane was fed into the gaseous oxygen flow neutralized by some amount of air. Flame stabilization was varied by changing the neutralizing air flow (that is, the nitrogen concentration in the oxygen and air mixture). The level of noise emitted during combustion was recorded by a piezoelectric pressure transducer.

Two flame stabilization modes were observed:

1) The first stabilization mode, wherein the flame (Fig. 4.10) embraces the entire fuel jet root around the periphery, has a clearly defined structure emitting a low level of broadband noise.

2) The second stabilization mode in which the flame is partially separated from the orifice of the fuel jet, which is stabilized in the recirclating flow behind the jet, has no clear shape and causes an elevated level of noise emission.

The high-level noise has a self-oscillatory nature. The main difference between the excited oscillations and oscillations commonly observed in a combustion chamber is their wide-band spectrum. It has been found in numerous observations that excitation of oscillations is preceded by partial separation of the flame. The stabilization mode changes from the first to the second on at least two opposite jets in one of four injectors.

Restructuring of the flame stabilization on all four injectors is followed by a sharp increase (by an order of magnitude) of the level of emitted noise. This combustion mode is characterized by a considerable hysteresis zone within the

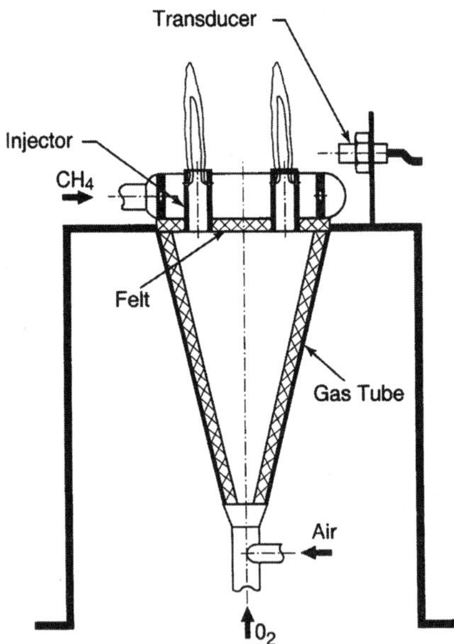

Fig. 4.9 Experimental setup of "gas burner."

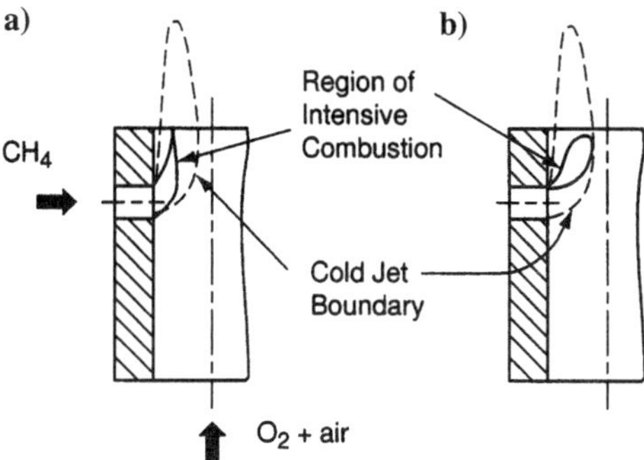

Fig. 4.10 Modes of flame stabilization in injection element: a) first mode and b) second mode.

values of the governing parameters (flow rates of oxygen, air, and methane). When the injectors are operated in the first flame stabilization mode, transition to oscillations is observed after some delay time, after reaching a stationary value of the governing parameter, in this case the flow rate of neutralizing air. With reduction in the airflow, the average value and dispersion of the delay time increased.

Therefore, during operation in the hysteresis zone, the probabilistic nature of the transition from the first flame stabilization mode to the second one was observed.

It is indicated in Fig. 2.1 that during studies and engine testing other cases of the unsteadiness of propellant conversion to combustion products were observed. Thus, for example, there can be uncertainty in the reversals of reversed flows formed in the combustion chamber depending on the scatter of characteristics during engine starting. Another type of unsteadiness is caused by the combustion of colliding jets of propellant components in emulsion injectors. Experimental results obtained during the development of combustion chambers and gas generators called for additional requirements when selecting stability characteristics. The constancy of stability characteristics must be checked by estimation of pressure oscillation decrements (δT) in the combustion chamber before and after introducing an artificial disturbance. The change of decrement before and after introducing the disturbance should not exceed the accuracy of its estimation.

A wide scatter of pressure-oscillation decrements in repeated testing of the same engine is evidence of uncertainty in the operating process.

Chapter 5

Studies of Operating Process Stability at Various Stages of Combustor Development

ANALYSIS of LRE evolution indicates that high-frequency instabilities in the combustion chambers and gas generators are one of the main structural and functional factors limiting the LRE reliability. On the basis of long-standing experience in achieving LRE combustion stability and a great volume of studies, technologies have been developed to provide required operating process stability margins within the limits of engine specifications. The technology involves application of a number of solutions providing operating process stability with respect to high-frequency pressure oscillations at all stages of engine development.

The technology includes the following stages: 1) design stage; 2) laboratory tests of hydraulic, acoustic, and combusting models of combustion chambers and gas generators; 3) stand tests of combustion chambers and gas generators in actual engine operating regimes, and tests of those combustors as part of the engine assembly; and 4) checking the consistency of combustion-chamber and gas-generator stability characteristics during the manufacturing of LRE for practical use. These stages are generally combined to save time for design and technology development for each engine project.

I. Design Stage

Prior experience with the development of combustion chamber and gas-generator mixing systems is used in the design stage based on solutions providing operating process stability. Those solutions have some unique features, depending on propellant thermodynamic states (liquid–liquid, gas–liquid), propellants used, and performance (e.g., thrust) and control ranges. All solutions providing operating process stability should take into account the requirements of attaining high combustion efficiency in the chamber and temperature uniformity in the gas generator. In some cases these two requirements for combustion chambers become contradictory. At present, an open circuit (liquid–liquid) with either pump or pressure feed of propellants is used only for engines with a thrust less than 10 tf. In the 1960s and 1970s, however, open-circuit engines with a much higher vacuum thrust of one chamber were developed, such as the RD-216, which produced 44.35 tf (435 kN), and the RD-250, which developed 44.95 tf (440.5 kN). In the same period, the LRE F-1 with a great thrust for the Saturn V launch vehicle and engines for the first and second stages of the Titan III rocket were developed in the United States. In developing the F-1 engine, operating process stability had to be provided at the stage of flight tests, at the expense of tedious and time-consuming work [10].

Nowadays, engines having thrust greater than 10 tf are generally developed with the use of a closed circuit (gas–liquid) with pump feed of propellants. The closed circuit (i.e., staged combustion cycle) enables the solution to a number of problems: 1) increase in specific thrust by attaining a higher pressure in the combustion chamber; 2) decrease in dimensions, which is very important for naval and mine rockets; and 3) in a closed circuit, the amount of energy released in the chamber decreases as a result of some energy release in the gas generator. However, there exist some exceptions, such as the open-circuit oxygen–hydrogen engine Vulcain with the vacuum thrust of one chamber of 1075 kN. The engine was designed in France for the Ariane 5 rocket.

In general, one- or two-component centrifugal injectors are mainly used for open-circuit engines in Russia. A screw or component injection through tangential holes is used to generate swirling flows in the injectors [16, and17]. In the United States, one- or two-component injectors with impinging jets were mainly used in the United States [10]. Two-component impinging-jet injectors provide rapid mixing of liquids, such that the combustion zone is generally located close to the injector face plate. Owing to these characteristics, the two-component injectors are frequently used in very small-size engines, which require very fast propellant combustion to provide good combustion efficiency.

One of the techniques providing the operation process stability is creation of distributed combustion throughout the chamber volume, with the maximum energy release moved to the region with a reduced pressure oscillation level (see Chapter 7). The distribution of propellant burning throughout the chamber can be achieved by the following methods: 1) through the application of proper design of the mixing head, such as the three-level mixing system used in the RD-216 engine as described in Sec. I of Chapter 4; 2) through modification of the size of atomization by changing the number of injectors and their characteristics (e.g., diameter of tangential holes, screw groove area, and the pressure differential across an injector); 3) by changing the propellant mixture composition for engines running on kerosene fuel, which is especially efficient for kerosene-nitric acid propellants (NA-27 I and NT); and 4) by using injectors with a high propellant flow rate for a closed circuit; for example, the chamber mixing heads of the engines RD-0208, RD-0209, RD-0213 (of the second and third stages of the Proton rocket) have only 61 umbrella-type injectors. The main fuel is introduced into the oxidizing gas through four holes in an injector with a diameter of 2.54 mm.

With the use of vibration baffles, it is undesirable to complete the preparatory processes beyond the zone of the baffles' action [18–22], because these processes are most sensitive to pressure oscillations.

As evident from the following sections of the book, the flow rating and its variation across the head radius substantially affect stability characteristics. The flow rating q is the ratio of the flow rate through the chamber head-end in grams per second, to the product of the injector head area in square centimeters and the chamber pressure in bars. It is not advisable that a value of q exceed 0.8 g/s/(cm^2 · bar) for open-circuit engines. It is advisable that a value of q be equal to about two for closed-circuit engine chambers and gas generators.

In the design of chamber heads to improve the operating process stability with respect to acoustic oscillations of the first tangential mode, it is advisable to reduce the flow rating from the center to the periphery of the chamber [10], such as in the chamber C5.3.0100 of engine RD-216 [23]. Each designer used different injector head designs for closed-circuit engines. For self-igniting (hypergolic) propellant components (see Chapters 13 and 14), they used 1) injectors with introduction of a jet into the gas flow and with injection of some fuel through centrifugal injectors; 2) umbrella types of injectors in which the major portion of the fuel is introduced into the oxidizing gas flow and about 10% of the fuel is injected to the chamber through the centrifugal cascade around the main flow; 3) injectors with fuel-injection centrifugal cascade; and 4) injectors with fuel-injection centrifugal cascade adjusted to one of the chamber oscillation frequencies.

For non-self-igniting (nonhypergolic) propellant components designers often used 1) slotted injector head in which the oxidizing gas is fed through annular slots and fuel jets are introduced into it; 2) injector head in which the oxidizing gas is fed through cylindrical sleeves and fuel is fed through centrifugal injectors; 3) injector head consisting of injectors with centrifugal fuel injection into the oxidizer gas flow; and 4) chamber head consisting of a sufficient number of small-flow injectors with centrifugal fuel injection to the oxidizer gas flow and adjusted to one of the chamber oscillation frequencies.

Injectors having coaxial injection of hydrogen around an oxygen jet were mainly used for oxygen–hydrogen propellants. Part of the hydrogen was used for porous bottom cooling (i.e., transpiration cooling).

The following were used in closed-circuit engines with different mixing circuits to improve the operating process stability: 1) change in chamber head permeability (defined as the ratio of the injector flow area to the head area); 2) change in gas temperature (i.e., change in the gas velocity through injectors); 3) change in the ratio of the liquid-to-gas velocity head (i.e., the depth of fuel penetration into the oxidizer flow); and 4) change in the number of injection holes for fuel injection. In all countries where LREs are developed, pulsation baffles of different designs, acoustic absorbers, and well-tuned injectors for closed-circuit engines are commonly used to absorb the oscillatory energy in combustion chambers and gas generators as effective means for improving the stability margin (see Chapters 15 and 16).

As an example, the chamber head of the LE-5 engine is shown in Fig. 5.1. The engine is designed for the second stage of Japan's H-2 launch vehicle. The propellants are oxygen and hydrogen. The vacuum thrust P_v is 10.5 tf, and the specific impulse J_v is 448 s. Injectors with coaxial mixing of propellants are mounted on the chamber head. The head has a porous bottom; 24 acoustic absorbers tuned to a frequency of about 6000 Hz are installed around the chamber mixing head to absorb the acoustic energy.

The design of the gas generator of the Vulcain engine is shown in Fig. 5.2. The engine was designed for the Ariane-5 rocket in France. This open-circuit engine

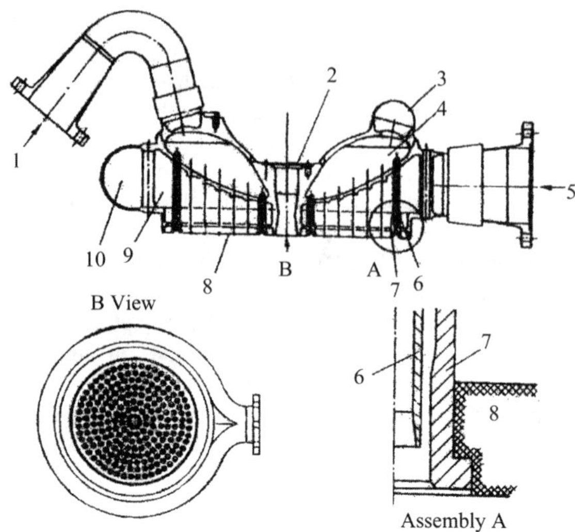

Fig. 5.1 Mixing head of the LE-5 engine combustion chamber.

runs on oxygen–hydrogen propellants. The vacuum thrust P_v is 1075 kN; the sea-level thrust P_g is 815 kN; the specific impulse J_v is 431 s; and the chamber pressure P_{ch} is 105 bar. The gas generator has three-blade vibration baffles and acoustic absorbers.

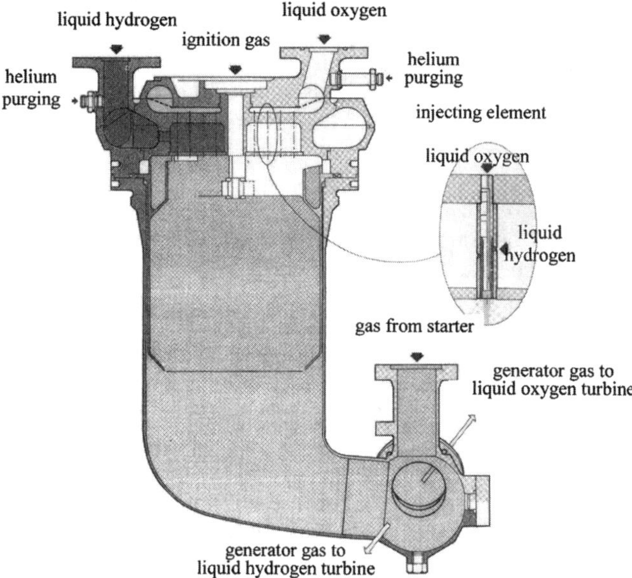

Fig. 5.2 Gas generator of Vulcain engine.

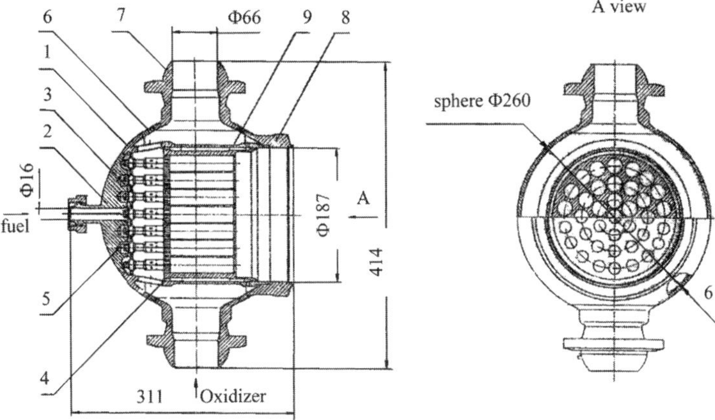

Fig. 5.3 Modular gas generator.

Acoustic absorbers were not used in practice in Russian engines, but vibration baffles were successfully used. Adjusted (well-tuned) injectors were used to advantage, and a part of them were extended to the combustion chamber and formed vibration baffles.

The gas generators of modular design have been developed for a number of engines such as RD-120, RD-170, RD-180, etc., as shown in Fig. 5. 3. The modular design of gas generators makes it possible to optimize the injector gas generator assembly and thereafter to equip the gas generator with the required number of injectors. A small diameter of an injector gas-generator assembly virtually rules out the occurrence of high-frequency oscillations.

Structurally, the gas generator has a spherical high-strength external shell covering the cylindrical combustion chamber (see Fig. 5.3). A space between the spherical shell and an external wall of the combustion chamber serves as a distribution cavity (i.e., manifold) for oxidizer feed to the mixing head. This design of the gas generator is convenient for engine arrangement, optimal from the standpoint of dimension, weight, and strength characteristics of the unit. The gas generator comprises a mixing head with a fuel inlet pipe and a high-strength spherical shell with two oxidizer inlet pipes. The mixing head comprises a ring 1, an outer bottom 2, a middle bottom 3, and an inner bottom 4. The middle and inner bottoms are interconnected with injectors 5. The inner bottom is secured in the ring 1 and the cylindrical shell 9.

The spherical high-strength shell comprises a spherical bottom 6, two oxidizer inlet pipes 7, and a locking ring 8. Fuel is fed to the gas generator through the inlet pipe connected to the mixing head outer bottom and from the distribution cavity between the outer and middle bottoms to the injectors. To reduce the volume of the fuel cavity, the outer bottom is made with a thickened wall, which serves as a filler. The oxidizer is fed to the gas generator through the two inlet pipes attached to the spherical high-strength shell and enters the mixing head through the ports in ring 1 and a hole of shell 9. The head is equipped with two-component injectors arranged around concentric circles with a spacing of 23.5 mm (37 injectors).

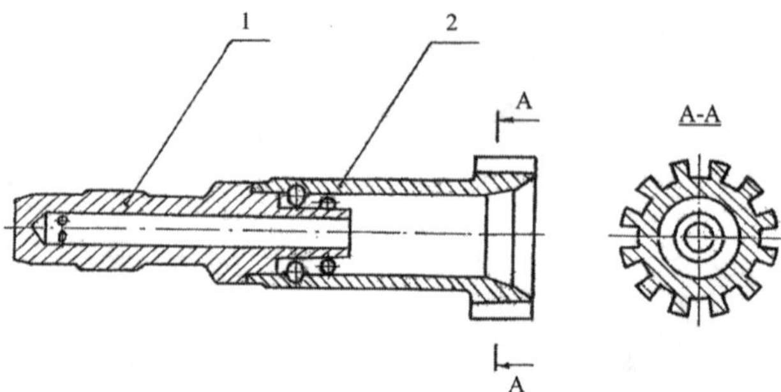

Fig. 5.4 Injector of modular gas generator.

Each injector consists of a body 1 and a sleeve 2 connected by welding, as shown in Fig. 5. 4. Tangential holes are made in the body for fuel injection. Eight tangential holes (in two rows with four holes in each of them) and 12 interfin passages of oxidizer feed jet cascade are made in the sleeve. The flow rates through the jet and tangential holes are selected so that at the total mixture ratio of propellants $\mathcal{K}m_{gg}$ is 56.7. The ratio of propellants fed through the tangential holes to the injector internal cavity (prechamber) $\mathcal{K}_i$ is 12. The inner bottom thickness is 90 mm. Thirty-seven cylindrical passages (neutralizing chambers) with a diameter equal to the diameter of the injector outlet including the jet cascade fins are made in the inner bottom. Each neutralizing chamber has an annular collar at the bottom outlet, which serves as a turbulence stimulator and promotes a uniform temperature field of the combustion products at the head outlet.

A gas temperature of ~2000°C in the prechamber provides practically complete combustion of fuel coming into the gas generator. The oxidizer fed through the jet cascade neutralizes the high-temperature gas to the specified temperature values. The selected mixing circuit with a high proportion of jet injection of excess propellant results in an essential increase in the head permeability and hence reduction of dimensions and improvement of weight characteristics of the gas generator. The selected mixing circuit provides high margins of operation process stability with respect to high-frequency pressure oscillations.

In the design and manufacturing of combustion chambers and gas generators to achieve a steady operating process, rather strict tolerances and manufacturing control systems are especially of concern for the head mixing elements. To provide the mixing system stability, a number of measures are taken. For example, an arrangement of injectors having a specific tolerance should be made according to a certain scheme as part of the overall manufacturing tolerance (for example, for a flow rate at a specified pressure differential) for the chamber heads. Such a method of mounting injectors on the heads of combustion chambers and gas generators reduces the effect of manufacturing tolerances and provides less difference between different engines in terms of operation process stability.

In engines with afterburning (i.e., staged combustion) of generator gases, the holes for fuel injection into the generator gas flow on the combustion chamber head have the same orientation.

At the engine design stage, the procedure of optimizing operating process stability is treated as a basic problem of providing the engine reliability. The purpose is to develop a product with the required reliability at minimum costs. The scope of tests required for confirming the required operating process reliability and their characteristics are defined by a number of factors. The advisable working and guaranteed limits of variations of engine operating conditions and regimes during the development and check tests are given in Table 5.1. The working limits are fixed by engine specifications; the limits in development are defined by the engine designer. The scope of tests depends on the operating process stability margin. It is evident that the higher the operating process stability margin, and the better the consistency of the results obtained are, the less the number of tests required.

Table 5.1 Working and guaranteed variation limits for engine operating conditions and modes during development and check tests

Ref. No.	Test modes and conditions	Working limits[a]	Guaranteed limits[a]	Ranges in development[a]
1	Operation time	$\tau = \tau_{check} + \tau_{flight\ test}$ s	$(2\text{–}3)\tau$	$(2\text{–}3)\tau$
2	Number of switches	$n = n_{check} + n_{flight\ test}$	$(1\text{–}3)n$	$(1\text{–}3)n$
3	Pressure in chamber and gas generator	$P_{ch,min}$, (atm abs) $P_{ch,max}$	$(0.75\text{–}0.9)P_{ch,min}$ $(1.10\text{–}1.25)\ P_{ch,max}$	$(0.75\text{–}0.9)\ P_{ch,min}$ $(1.10\text{–}1.25)\ P_{ch,max}$
4	Mixture ratio of propellants	$K_{m,min}$ $K_{m,max}$	$(0.90\text{–}0.95)\ K_{m,min}$ $(1.05\text{–}1.10)\ K_{m,max}$	$(0.90\text{–}0.95)\ K_{m,min}$ $(1.05\text{–}1.10)\ K_{m,max}$
5	Propellant temperature at engine inlet	T_{min}, °C T_{max}, °C	$T_{min} - 5$°C $T_{max} + 5$°C	$T_{min} - 5$°C[b] $T_{max} + 5$°C[b]
6	Structure temperature	T_{min}, °C T_{max}, °C	$T_{min} - (5\text{–}10)$°C $T_{max} + (5\text{–}10)$°C	$T_{min} - (5\text{–}10)$°C $T_{max} + (5\text{–}10)$°C
7	Propellant pressure at engine inlet	$P_{in,min}$, (atm abs) $P_{in,max}$, (atm abs)	$P_{in,min} - \Delta h\gamma$[c,d] $P_{in,max} + \Delta P$[e]	$< P_{in,min} - \Delta h\gamma$[c,d] $> P_{in,max} + \Delta P$[e]
8	Maximum gas saturation of propellants[g]	C (g/l) $C = \chi$ (g/l•atm abs)[f] P(atm abs)[h]	$(1.10\text{–}1.20)\ C$	(1.10–1.20) C

[a]Working limits are specified by engine specifications; guaranteed limits are specified by customers; and limits in development are specified by engine designers.
[b]With account of preheating during pressurization.
[c]$\Delta h = 0.5\text{–}1$ m for engines with booster pumps, and $\Delta h = 2\text{–}4$ m for engines without booster pumps.
[d]γ is the specific weight of propellant components.
[e]$\Delta P = 2\text{–}3$ atm.
[f]Henry coefficient χ depends on the temperatures and types of liquid and pressurizing gas [26].
[g]Maximum amount of driving gas that can be dissolved in a unit volume of liquid propellant at given pressure and temperature during a fixed time duration.
[h]P is the absolute pressure of the driving gas displacing the propellant from a tank to an engine.

The tests depend on the method of reliability evaluation: statistical or parametric. In selecting the optimal design, the parametric method is most effective. The proposed method makes it possible to select the designs of the combustion-chamber and gas-generator injector heads having the highest stability margins or to select the optimal version taking into account not only the operating process stability, but also other requirements (combustion efficiency, cooling, temperature uniformity in a gas generator, and so on) imposed on LRE combustion chambers and gas generators.

The parametric method is effective in evaluating the influence of ultimate operating modes on the operating process stability (see Table 5.1) and in evaluating the units manufactured with ultimate tolerance values. The effectiveness of the parametric method depends on 1) the accuracy of the selected procedure for evaluating the operating process stability; 2) the information obtained and used; 3) the accuracy of the information obtained; and 4) consistency of characteristics (tolerances) in production at all stages of engine manufacturing. When using the parametric method, the scope of tests is defined by the following considerations. 1) selection of the optimal designs of combustion chambers and gas generators; 2) evaluation of the measures to improve reliability; and 3) evaluation of the admissibility of the accepted manufacturing tolerances defining the operating process stability. As known from experience, to solve these problems it is necessary to carry out two to five tests with one to three specimens of combustion-chamber or gas-generator units. The testing procedure and criteria for quantitative assessment of the stability margins with respect to soft and hard excitations of high-frequency pressure oscillations are described in the following chapters of this book.

Engine optimization with respect to operating process stability in combustion chambers and gas generators should be included in a complex plan of engine experimental optimization (CPEO). The scope of testing depends on the novelty of the engine being developed. Engine optimization should provide, to a maximum possible extent, concurrent optimization of each structural element. When conducting complex tests, defects are detected in sequence. In separate tests, defects are detected concurrently.

An engine reliability control system used for commercial products is defined at the design stage. Two control systems are used, depending on rocket designation. The IST-SVT system is used for military rockets with expendable engines. This control system has also been used for all engine modifications of the second and third stages of the Proton rocket (see Chapter 17). IST stands for inspection-sampling tests, confirming the reliability of the manufactured lot of engines, and SVT denotes special verification tests. The tests are conducted in the regimes and under external conditions such that most frequent failures were observed during engine optimization. The number of engines in a lot is defined by the engine reliability attained, based on the test results and a prespecified rate of reliability improvement. A lot of engines are accepted as commercial products, if the engine characteristics include quantitative stability characteristics that comply with the engineering specification requirements.

The ITT-IST-SVT control system is used for engines of space rockets (reusable engines) and heavy ballistic rockets. ITT stands for inspection technological tests used as a means of rejection of unreliable engines, including those exhibiting operating process instability. However, based on experience, the fact of passing the ITT does not guarantee that the operation process instability will not occur during repeated tests because the operating process instability has a probabilistic nature and can take place in case of inadequate quantitative stability characteristics. ITT is carried out in two different ways: without overhaul and with top or complete overhaul.

A specific system for measuring the parameters is provided in the engine design. It comprises the checking of operating process stability. The measurement system to estimate the operating process stability in a chamber generally comprises five transducers: three vibration transducers (accelerometers) and two pressure transducers measuring pulsations in the fuel and oxidizer cavities. After the quantitative characteristics of stability in response to artificial disturbances were estimated and initial tests of engines were carried out, a transducer for direct measurement of pulsation in the chamber was installed. The acceptable locations of transducers during engineering development of an engine should be kept unchanged during ITT, IST, and SVT in order to control the production stability.

It follows from the aforementioned that a great deal of experience in developing combustion chambers and gas generators featuring stable operation process has been gained. However, without using quantitative stability characteristics, it is not possible to evaluate the effectiveness of design modifications and even much less possible to evaluate their optimality according to other requirements imposed on the design, such as combustion efficiency for combustion chambers, temperature uniformity for gas generators, and so on.

II. Laboratory Tests of Hydraulic, Acoustic, and Combustion Models of Combustion Chambers and Gas Generators

These studies are carried out primarily at the stage of research work (RW). Hydraulic characteristics of injectors are studied, and the following are determined: 1) compliance of the mixing system with design requirements and its stability; 2) atomization characteristics; and 3) dynamic characteristics of atomization (i.e., the influence of pressure variations at the injector inlet and outlet on atomization characteristics), etc.

For this purpose, the dependence of flow rate on pressure differential is determined experimentally for all injectors under the conditions that will be realized at the factory during serial production. In addition, special studies of atomization in a medium with an increased pressure are carried out. Special pulsators are used to determine relevant dynamic characteristics of injectors.

The next stage is aimed at a detailed study of the acoustic characteristics of combustion chambers and gas generators (see Chapter 7). The following characteristics are determined: 1) frequencies for a number of different acoustic modes; 2) optimality of vibration baffles (solid baffles, injectors extended into the chamber and other designs); 3) adjustment of resonance absorbers; and 4) efficiency of absorption by the porous bottom of a combustion chamber, etc. This stage is also aimed at obtaining promptly preliminary information about the operating process stability with various versions of actual injectors (or mixing heads) based on the results of testing under subscale laboratory conditions: low pressure levels, small mass flow rates, and application of model components instead of actual ones, should the need arise. The acquired stability data are used for selecting the most promising versions of mixers and for developing and introducing modifications into the original designs to increase the stability. The processing of model test data and extension of the results to actual conditions are made using the parametric criteria obtained from the theoretical and numerical studies that characterize the energy conditions of oscillations.

Rating of operating process stability during model firing tests is generally made by estimating the departure of simulated actual regimes from the boundary within which oscillations take place. A comparative efficiency evaluation of different injector versions is performed by changing the operating points in the instability and hysteresis regions. The injector evaluation is also based on the criteria accepted for evaluating the operating process stability under actual conditions.

Simplified combusting models are still used in full-scale stand tests to find out promptly (if necessary) in which way stability will be affected by structural modifications to be introduced into the selected versions of injectors and mixing heads under actual operating conditions.

Apart from the main problem of determining the oscillation regions for different versions of injectors, studies on the effects of the length of injector gas passage on stability characteristics for gas–liquid engines, the effects of harmonic oscillations in the feed system on the dynamics of the combustion zone, and other effects are carried out using combustion models.

The optimization of modular gas-generator elements is successfully performed at the research stage. Tests are carried out with different gas-generator versions assembled with one and three injector units. At this stage, tests are carried out on experimental units to obtain the required data for designing fundamentally new units. These include, for example, tests of engines using new components, which were not used earlier, and tests for selecting a mixing system for three-component chambers [24 and 25].

III. Tests of Combustion Chambers and Gas Generators Under Actual Regimes and as Part of Engine Assembly

At the third stage, the design and development work (DDW) is basically completed. The object of this stage is to select from several promising mixer versions developed according to recommendations from the first and second stages. The most optimal version in terms of sufficient operating process stability margins under practical conditions will be taken into account. Considerations will also be given to the influences of various disturbances that can occur during operation of actual engine assemblies and units.

The stability characteristics in actual operating modes are determined from the properties of the amplitude-frequency spectra of pressure oscillations that were obtained under conditions with and without externally imposed pulse disturbances.

As a stability characteristic determined from noise inherent in a combustion chamber, the decrement of pressure oscillations at natural acoustic frequencies in the chamber is used. As a stability characteristic determined from the operating-process response to artificially imposed pulse disturbances, the relaxation time for a certain pressure pulse disturbance is used (see Chapter 6).

Operating process stability is rated by comparing experimental values of stability characteristics with the corresponding values obtained from practical experience acquired by using the stability estimation procedure.

During tests at the design and development stages, the combustion chamber and gas generator are equipped with transducers measuring pressure pulsations and vibrations (see Sec. IV of Chapter 5). For further examination of operation process stability, when the engines are supplied as commercial products, five transducers at fixed spatial locations are installed in each combustion chamber and gas generator. The transducers for measuring fuel and oxidizer pulsations are mounted upstream of the injectors, and three vibration transducers are mounted in three directions: along the chamber axis, perpendicular to the axis, and along the tangent to the chamber shell. It is advisable to mount transducers in the area of maximum oscillations for low acoustic frequencies.

Statistical data obtained from the decrements and amplitudes of pressure oscillations during the tests are used for control of production stability, when the engines are supplied for commercial applications.

IV. Test Organization, Measurement Requirements, and Processing of Rapidly Varying Parameters

A. LRE Preparation for Firing Tests

Pressure transducers mounted on a combustion chamber (or a gas generator) provide a source of information about pressure oscillations. They are used for direct measurements in the gas volume, pre-injector cavities, cooling circuit, and feed pipes. Vibration transducers mounted on the chamber surface are also sources of information. For a deeper investigation of operating processes, the following optical methods were used under laboratory conditions: motion-picture recording, schlieren and shadowgraph methods, holography, electro-optical method, etc. However, only pressure and vibration transducers were used for actual engines. Thus, quantitative evaluation of operating process stability is based on the information obtained from measurements of pressure and vibration oscillations. The number and mounting locations of sensors for measuring pressure oscillations and vibrational acceleration of structural elements are dependent on the method of analysis that provides quantitative evaluation of stability margins.

Let us consider the installation of pressure transducers on a combustion chamber or a gas generator. To measure pressure oscillations, it is advisable to install at least two transducers on a combustion chamber and two transducers on a gas generator at an angle of 135 ± 10 deg on a plane perpendicular to the chamber axis and located as close as possible to the inner bottom of the mixing head. For a gas generator, the pressure transducers should not be isolated from the reaction volume by structural elements (grates, washers, etc). If the chamber is equipped with vibration baffles of various designs, it is advisable to arrange the pressure transducers at the edge of the baffles. For stability evaluation by pulse disturbances, one of the pressure transducers should be placed at an angle of 180 ± 10 deg with respect to the point of disturbance introduction to the chamber. To measure the pressure oscillations in pre-injector cavities, at least one transducer should be mounted in each cavity of the combustion chamber and gas generator. The transducers are arranged on the generating lines, with a circular deviation no more than ± 30 deg from the generating line on which the transducer for measuring pressure oscillations in the reaction volume is mounted. In the case of stability evaluation by pulse disturbances, the transducers for measuring pressure oscillations in pre-injector cavities should be oriented at 180 ± 30 deg with respect to the disturbance introduction point.

When high-frequency operating process instability is studied, very strict requirements are imposed on a pressure transducer. As the transducer is exposed to high heat fluxes (10–13 MW/m^2 under normal combustion conditions and up to 60 MW/m^2 at conditions with self-oscillations) and to large acceleration forces (up to 1000 g), it is desirable to use a small-size transducer and diaphragm (with a diameter of about 6 mm) for measurements during engine firing tests to reduce losses of the shell integrity and disturbances in the cooling flow. The transducer must endure a high heat flux without losing its sensitivity or zero drift. It should have a low sensitivity to overloads, or the sensitivity should be accurately allowed for. A high output signal sensitivity of transducers is also desirable because this allows one to distinguish small-amplitude pressure pulses from system noise. Transducers should have a high stability so that they can be calibrated with the help of an electrical simulator at the test site. Transducers should have high natural frequencies. With a small diaphragm (of 6 mm diam or less), a high natural frequency up to 100 kHz or higher can be obtained. It is clear that transducers with a constant frequency response up to 20 kHz should be used to ensure accurate recording of high-frequency pressure oscillations. To provide good frequency response and high sensitivity, transducers should be flush mounted on the inner wall of the chamber. If there is a passage between the transducer diaphragm and the chamber wall, it should not distort the measurement data. Under these conditions, transducers endure the most severe conditions in the chamber, including high heat fluxes and fragments from disturbing devices. To reduce the probability of fragment impact on the transducer diaphragm, the transducer seat should be slightly displaced from the disturbing device axis. A number of measures have been worked out to protect transducers from overheating: double diaphragms with a water flow between them, transducers with coolant discharge into the chamber, heat-resistant coating, etc. Each of them is effective to certain degree and can be applied in suitable conditions.

Vibrational acceleration transducers on combustion chambers and gas generators are commonly employed, depending on the specific problem to be solved. At least two transducers are mounted on the mixing head body to measure vibrational accelerations in the radial and axial directions. It is advisable to arrange transducers near one of the transducers for measuring pressure oscillations in pre-injector cavities. At the early stage of LRE firing tests in the course of design and development, it is advisable to increase the number of pressure transducers and accelerometers for better detection of oscillation sources and for identification of forms of oscillations. When conducting LRE firing tests during manufacturing of engines to be supplied for use, the number of transducers can be reduced. In particular, transducers for measuring pressure oscillations in the reaction volumes of combustion chambers and gas generators are generally not installed.

B. Installation of Disturbance-Generating Devices on Combustion Chambers and Gas Generators

Devices for generating multiple-impulse disturbances with explosive charges are located outside the reaction volume of the chamber and are used for creating pressure pulse disturbances in the chamber. These devices and their characteristics are described in detail in Chapter 9. Let us consider some general issues of equipping combustion chambers or gas generators with disturbance-generating devices. A pulse is fed from such a device to the chamber reaction volume through a connecting channel with a diameter of 7–10 mm and a length of not more than 50 mm (from the chamber fire surface to the orifice of the disturbance-generating device). The connecting channel is positioned radially. It is advisable that the distance from the channel axis to the inner bottom of the mixing head be less than 50 mm.

If vibration baffles are used in the chamber, the connecting channel is positioned 1–2 mm downstream of the trailing edge of baffles. In a gas generator, the connecting channel is located so that its exit area and mixing head are not separated by gas-generator structural elements (e.g., grates, washers, etc). If it is necessary to increase the strength of the pulse disturbance without increasing the explosive-charge size, the connecting channel can be positioned tangentially to the circumference with a radius equal to 0.6–0.7 of the inner radius of the chamber.

C. Requirements for Measurement Systems and Processing of Rapidly Varying Parameters

On the basis of long-standing experience with the evaluation of operating process stability, specific requirements are established for systems of measuring rapidly varying parameters and their constituent parts (such as recording and processing units), in order to achieve the highest possible accuracy under conditions with and without introduction of pulse disturbances.

Pressure transducers and signal conversion and recording devices should provide simultaneous operation of two recording channels (with low and high sensitivities) from one transducer. These two channels should have different amplitude and frequency responses. The amplitude ranges of calibration signals are specified, depending on the average pressure in the volume being studied: 1) for high-sensitivity channels within limits of 2–6% and 2) for low-sensitivity channels within limits of 10–30%.The ranges of amplitudes are defined more exactly in the course of LRE development. The specified amplitude ranges must correspond to the working range of transducers.

The frequency ranges of pressure measurements should satisfy the following requirements:

1) The first is low-pass filters with a cutoff frequency f_{low} = 0.2–1.0 kHz and high-pass filters with a cutoff frequency f_{high} = 20.0 kHz (depending on values of recorded frequencies and characteristics of recording devices employed). It is not advisable to use filters with a steep front of frequency characteristic as it severely distorts pulse signals.

2) Filters are not installed on a low-sensitivity channel.

General requirements for vibration measurements are described in relevant literature. Here only the following requirements are noted. The amplitude ranges of calibration signals should lie within the working range of accelerometers and in general should afford the measurement of vibrational acceleration up to $3 \cdot 10^4$ m/s^2. The frequency range of measurements should be 0.2–20.0 kHz for high-sensitivity channels and 0.01–20.00 kHz for low-sensitivity channels.

D. General Requirements for High-Speed Signal Processing Devices

The signal processing systems should provide the capability of continuous (without skips) processing of rapidly varying parameter (RVP) signals throughout the test using contemporary high-performance computer facilities. The system should be able to process signals in a frequency range of 0.2 to 20.0 kHz. In the case of digital processing, the following should be provided: 1) number of digitization levels of no less than 128; and 2) sampling frequency $\Omega > 4f_m$, where f_m is the maximum frequency in the spectra of rapidly varying parameters. An absolute limit of reproduction error in a current-time unified scale should be 0.05 ms. The averaging time interval in obtaining natural noise parameters under steady-state conditions of combustion chambers, and under nonsteady conditions with slow variation of parameters (i.e., parameter variation in more than 1 s), should not be less than 1 s. The number of parameter realizations should not be less than eight. The average time intervals in obtaining natural noise parameters under rapid transient conditions (i.e., variation of parameters in less than 1 s) should not be less than 1) 0.02 s when obtaining the values of total signals of rapidly varying parameters and 2) 0.04 s when obtaining the parameters of amplitude-frequency spectra.

The number of parameter realizations throughout the transient regime should be no less than five. It is advisable to increase the number of realizations through obtaining the parameters of amplitude-frequency spectra by processing signals with a displacement by 0.01–0.02 s. The system should obtain the average and dispersion of parameters determined from natural noises in the chamber within testing time intervals specified for the analysis. The signal processing should also provide the averages, instantaneous values, and dispersions of parameters determined from noises in the chamber in form of tables and plots as functions of time and frequency.

The signal processing techniques for obtaining the quantitative stability characteristics from noises in the chamber are discussed in detail in Chapter 8. The processing methods provide the necessary data under both stationary and transient conditions. On the basis of experience, the following requirements for errors resulting from the processing of RVP have been set forth.

The RVP processing techniques for steady-state operation regimes normally have the following relative errors of processing, including the error of the signal processing method itself, when determining the following parameters: 1) maximum and peak values of pressure oscillation signals, ~15%; 2) average values of total signals, ~5%; 3) principal maximum of the signal amplitude spectrum, ~10%; 4) frequency of the amplitude-frequency spectrum, ~3%; 5) average values of filtered signals in the prescribed frequency band, ~10%; and 6) oscillation decrements with their values being higher or equal to 0.05, ~20%.

The RVP processing technique for rapid transient operation regimes normally has the following relative processing errors, including the error of the signal processing method itself, when determining the following parameters: 1) maximum and peak values of pressure oscillation signals, ~30%; 2) average values of total signals, ~10%; 3) frequency of the dispersed amplitude-frequency spectrum, ~6%; 4) principal maximum of the signal amplitude spectrum, ~20%; and 5) oscillation decrements with their values being higher or equal to 0.05, ~40%. The errors of oscillation-decrement determination at discrete frequencies in the amplitude-frequency spectra of rapidly varying parameters increase in the following cases: 1) the ratio of the amplitude at a discrete frequency to background noise in the vicinity of that frequency is less than two, and 2) there are close frequencies with similar amplitudes. To improve the accuracy of oscillation decrement in such cases, it is necessary to implement noise-subtraction and frequency-filtering methods during processing. With preliminary filtering of RVP signals, an additional error of other parameters should not exceed 5%.

E. Test Checks of Natural-Noise Processing System

To provide the required accuracy, it is necessary to check the signal processing errors by digital and/or analog simulations of typical random processes using a test-magnetogram (or other information media). The signal from a noise generator passing through resonance circuits with known oscillation decrements is recorded on an information medium according to the following scheme:

The oscillation decrement in the resonance circuit is calculated using the formula

$$\delta T = \pi \cdot \frac{\Delta f}{f_r}$$

where Δf is the frequency width in the amplitude-frequency spectrum at the level of 0.707 of its maximum value and f_r is the resonant frequency. In this case, it is advisable to carry out checks for decrements δT with values from 0 to 0.35–0.40 with intervals of 0.05. The resonant frequency should be within a range $f_r = 2.0$–6.0 kHz. A signal with oscillation decrement equal to zero should represent a sinusoid with a constant amplitude. It is desirable to record on a test-magnetogram the sinusoids with frequencies $f_r \sim 0.5$, 1.0, 2.0, 4.0, 6.0, and 8.0 kHz.

The data-processing system error is determined by multiple (not less than 20 times) processing of test-magnetograms. It is necessary to obtain the absolute error not exceeding 0.01 for oscillation decrement that equals zero at all frequencies. If a preliminary filter is used in the RVP processing system, the test-magnetograms also should be processed with the use of the same filter. It is necessary to smooth the amplitude (or energy) spectra and correlation functions obtained as a result of processing by averaging their instantaneous values in order to reduce false signals and measurement errors.

V. Control of Consistency of Stability Characteristics in Serial Production

As mentioned in Sec. I, an engine reliability control system for commercial products is defined at the design stage. Because operating process stability is one of the most important constituents of engine reliability, the results of measuring pulsations and vibrations during ITT, IST, and SVT of engines are used for control of the operating process stability and consistency of quantitative stability characteristics. The IST-SVT control system, gives less information as compared to the ITT-IST-SVT system and is generally used for expendable engines, for example, for all-welded "drowned" engines (e.g., engines located in rocket propellant tanks). The check tests for each engine allow one to determine its quantitative stability characteristics, and in case of their deviation beyond tolerance limits, to reject such an engine, and to find the causes of production instability. Similar decisions are also taken in case of inspection of one engine from a lot (IST-SVT). But in this case, the whole lot of engines is rejected, and the causes of production instability are examined (see Chapter 17).

As mentioned earlier, in serial production, five transducers are generally used to monitor combustion chamber stability and to keep the same stability characteristics during manufacturing. Two transducers for measuring pulsations in the fuel and oxidizer cavities upstream of injectors and three transducers for measuring vibrations along three axes are used. The mounting locations of transducers and their characteristics during design tests must remain unchanged throughout the period of engine supply for commercial use.

The pressure-oscillation decrement and amplitude characteristics in combustion chambers and pre-injector cavities, as well as vibrational accelerations of combustion-chamber and gas-generator structural elements, must lie within the limits stipulated in engineering specifications. Limiting values are defined from a set of statistical data for a particular engine. The production quality control is also performed during other checking operations: determination of pressure differential dependence of flow rate through injectors and through the cooling circuit of a combustion chamber; determination of pump and turbine characteristics; etc. Changes in the characteristics of combustion-chamber elements and elements of other engine units in the course of manufacturing can result in changes of the quantitative characteristics of operation process stability.

Experience with investigation of production stability concerning the control over stability characteristics of operation processes in a combustion chamber and in a gas generator indicates the necessity to retain pulsation and vibration records obtained during firing tests and other checks on engine characteristics throughout the period of commercial supplies (see Chapter 17).

In case of significant degradation of quantitative stability characteristics, and impossibility to restore the manufacturing technique to an initial condition, it is necessary to introduce modifications to the combustion-chamber (gas-generator) mixing system. When introducing such modifications to the mixing system, the availability of manufactured engines should be taken into account. Thus, designers should make efforts to improve stability characteristics by minimizing modifications of the design. The effectiveness of modifications introduced into a combustion chamber design is evaluated on two to three engine specimens in accordance with the procedure described in this book.

Chapter 6

Quantitative Characteristics for Estimating Stability of LRE Combustion Chambers and Gas Generators

I. Definition of Problems

THE oscillatory processes in a LRE combustion chamber and gas generator have a complicated physical nature. Under certain conditions in the combustion chamber, the transition takes place from combustion with small values of pressure oscillations ("noise") to a pressure oscillation with large-amplitude, vibrating combustion. An instability of the combustion processes, a transition from some pressure oscillation level to another, is defined by interactions of the combustion processes with oscillations of the medium in the reaction volume.

Complete mathematical description of all processes proceeding in the combustion chamber is practically an intractable problem. However, a number of authors have tried to describe the unstable combustion process by essentially simplifying the phenomena described, as shown schematically in Fig. 6.1.

In theoretical analysis of the stability of a LRE, the latter is represented in the form of the simplest structural diagram: a closed circuit consisting of three simple members—a supply system, the combustion process, and the combustion chamber as an acoustic oscillatory system interconnected by direct and feedback couplings.

Direct coupling corresponds to transformation of the reactants (fuel and oxidizer) to combustion products; feedback couplings include the sensitivity of the supply system and the combustion processes to pressure oscillations in the combustion chamber. However, even in the linear approximation, plotting the boundaries of stable operation of this closed circuit is a challenging task. The combustion process is the most complicated part for description. Two approaches are used: a phenomenological approach, wherein the combustion process is described by a burnout curve; and an approach based on specific combustion process models. The main disadvantage of the phenomenological approach lies in the fact that it does not involve direct consideration of features of the mixing system design, operating parameters of the engine, and physical and chemical properties of propellants. The influence of all of the preceding factors is indirectly taken into account by specifying a representative time of combustion, relevant amplification factors, or similar quantities. Dependence of these values on the

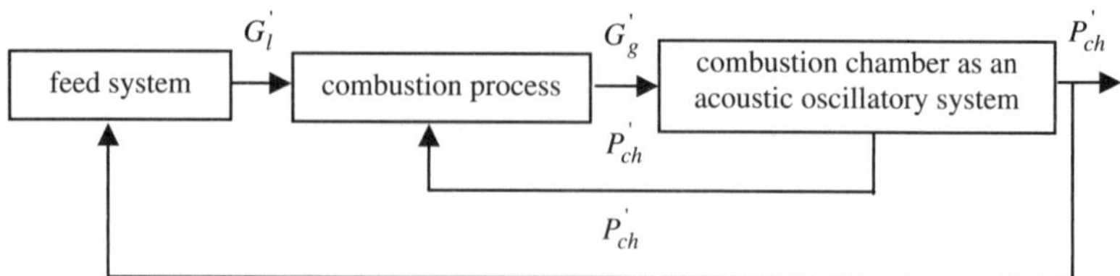

Fig. 6.1 Simple block diagram of LRE used in theoretical analysis of stability. G_l' = fluctuations of liquid (fuel and/or oxidizer) flow rate; G_g' = fluctuations of gas flowrate in the process of converting propellants to combustion products; and P_{ch}' = pressure oscillations in the combustion chamber.

engine operating and design parameters is not considered in the theory, but is determined on the basis of qualitative considerations. In a number of cases, this strategy does not allow one to make unambiguous conclusions.

The transition from a phenomenological description of the combustion processes to a theory applying specific combustion models allows one to obtain direct information concerning the influence of operating and design parameters upon the occurrence of soft modes of oscillation. This approach, however, has several drawbacks. First, there is a lack of fundamental knowledge of the physical phenomena involved in the process being studied. It is necessary to consider that the operating processes in a LRE from an essentially nonlinear oscillatory system. Finally, the existence of a multiplicity of simultaneous couplings between the combustion processes and the oscillatory system preclude practical data in this line of studies.

Practical solutions to the problem of diagnosing stability have been found both in Russia and in the United States by using empirical methods of stability evaluation with application of all existing knowledge of the observed phenomena. In early experiments, the basic criterion of stability margin was based on departure of operation from the boundary of transition to vibrating combustion. The disadvantage of this method consists in the fact that there is not always a possibility of changing the operating parameters up to the instability boundary. Moreover, there is a considerable scatter of the data obtained, caused by physical phenomena occurring near the self-excitation boundary. Examples of such phenomena include nonlinearities and random excitations of the instability (Sec. III of Chapter 3).

Reaching the stability boundary under conditions far from normal operation can give wrong information about the stability margin within the limits of engine operation. In addition, estimates based on the boundaries of spontaneous oscillations is incomplete because of the essential nonlinearity of processes in a combustion chamber. The procedure developed for estimation of the stability margin fundamentally differs from existing analogs. Practical realization of the method of quantitative estimate of the stability margin for a LRE called for the solution of a great number of scientific and technical problems, including 1) choosing the optimal characteristics of the stability of operation in combustion chambers and gas generators; 2) development of the equipment, initial data-processing methods, and relevant devices for determining numerical values of stability characteristics; 3) study of the relation between stability characteristics and the engine design and operating parameters; 4) determination of numerical values of stability characteristics required for providing sufficient stability margins for different types of engines; and 5) a set of statistical data for introducing the methods developed.

II. Stability Characteristics

For choosing optimum stability characteristics, let us represent a running engine, or a combustion chamber tested separately, by a complete set of nonlinear equations. We introduce into this set of equations disturbances that are consistent with the equations. Let us represent a set of responses to the disturbance as a sum of expressions of the form:

$$y = Ay_0 \exp\left(-\int_0^t \delta \,\mathrm{d}t\right) \cdot \sin\left(\int_0^t \omega \,\mathrm{d}t + \varphi_0\right) \tag{6.1}$$

where Ay_0 is the initial value of the deviation of the parameter from its mean value. This value is conditionally referred to as the initial amplitude of the disturbance; δ and ω are the damping factor and oscillation frequency; and φ_0 is the initial phase of oscillation.

Thus, the oscillations experimentally obtained in LRE combustion chambers and gas generators can be analyzed in the form of Eq. (6.1). Values of δ and ω depend in general on time, $\delta = \delta(t)$ and $\omega = \omega(t)$. They characterize completely the oscillations in LRE combustion chambers (gas generators) at any given moment of time. That is why quite definite values of the quantities correspond to every characteristic state of the oscillatory system under consideration. This allows one to use these quantities as experimentally determined characteristics of the oscillatory system being studied, in particular, its stability characteristics.

On the basis of available experimental data for the stationary mode, it has been found that the frequencies of pressure oscillations in a combustion chamber vary only slightly. This result allows one to assume that the oscillation frequency is constant, or to introduce a correction for the frequency variation, and to use a single stability characteristic, that is, the damping factor δ. The damping factor has a definite physical meaning. Its value is equal to one-half the ratio of the difference between oscillatory energy flux dissipated by the system E_2 and oscillatory energy flux entering the system E_1 in unit time, to the total oscillatory energy E_Σ in the system:

$$\delta = \frac{(E_2 - E_1)}{2E_\Sigma}$$

Therefore, the analysis of stability characteristics can be coupled with analysis of the energy in the oscillatory system and the oscillatory energy fluxes entering the system and dissipated within the system.

A LRE combustion chamber or gas generator, being nonlinear systems, can be represented as two members, an active and a passive one, with direct and feedback coupling between them. The active member is a source of oscillatory energy, and the passive member is a sink. Nonlinear properties of these members show up as the dependence of the oscillatory energy fluxes entering the system $E_1(A)$ and dissipated by the system $E_2(A)$ on the oscillation amplitude A. The relation among these fluxes of the oscillatory energy in a system defines its stability characteristics.

The combustion chamber, being a nonlinear system with distributed parameters, has some features making it different from nonlinear systems with lumped parameters. Some of these distinctive features are 1) a great number of oscillation modes, many of which are realized simultaneously with distinct amplitudes; 2) the possibility of interaction among oscillations of different modes; and 3) fluctuations of parameters in the chamber volume.These, and perhaps other, peculiarities necessitate preliminary study of acoustic oscillation modes, which should be followed by defining the quantitative stability characteristics for these oscillation modes (see Sec. I of Chapter 7).

The parameter fluctuations result in some scattering of oscillatory energy flux and values of the stability characteristics. However, for a clearer presentation of ideas and methodological approaches to stability estimates in this section, the oscillations in combustion chambers will be treated from the deterministic standpoint. For this purpose, only mean values of energy fluxes and stability characteristics will be used.

Let us consider some peculiarities of the dependence of the energy fluxes $E_1(A)$ and $E_2(A)$ on the amplitude of oscillation. The change pattern of the energy flux entering the oscillatory system, depending on the amplitude $E_1(A)$, is affected by the propellant transformation conditions. It should therefore differ markedly for gas generators and combustion chambers with different thermodynamic states of propellants (e.g., gas–gas, gas–liquid, and liquid–liquid). The maximum level of the oscillatory energy flux is determined at least by the oscillation period, by the amount of energy stored in the propellants entering the system during this period, and by the completeness of their transformation to combustion products.

For qualitative analysis, let us plot the energy fluxes entering and dissipated by the oscillatory system against the oscillation amplitude, as shown in Fig. 6.2. Oscillatory energy losses in the combustion chamber include energy losses through the injector head, the nozzle, and the chamber side surface, as well as losses caused by medium heterogeneity in the combustion chamber and flow irregularity. In addition, special dampers such as vibration baffles and acoustic dampers are installed in the combustion chamber for oscillatory energy losses. Some processes connected with the interaction between disturbances and the active member of the oscillatory system also cause oscillation energy dissipation (for example, atomization, droplet vaporization, and others). Figure 6.1 shows three possible variants of the positions of amplitude dependence of the oscillatory energy fluxes. The energy characteristics representing the amplitude dependence of excess energy fluxes ΔE in the systems are also shown. The values $\Delta E(A) = E_2(A) - E_1(A)$ define the amplitude dependence of the damping factor. The value of $2E_\Sigma$ in Eq. (6.1) is always positive, and so the sign of the damping factor is defined by the sign of the difference $\Delta E(A)$.

The curve shown in Fig. 6.2a corresponds to a stable oscillatory system. The curve shown in Fig. 6.2b corresponds to a system in which high-frequency pressure oscillations are excited spontaneously and the amplitude reaches the value of A_2. Figure 6.2c corresponds to the most common case when self-oscillations are excited by an initial disturbance exceeding the critical amplitude A_{cr}. Each of the plots shown in Fig. 6.2 corresponds to certain operating and design parameters of the

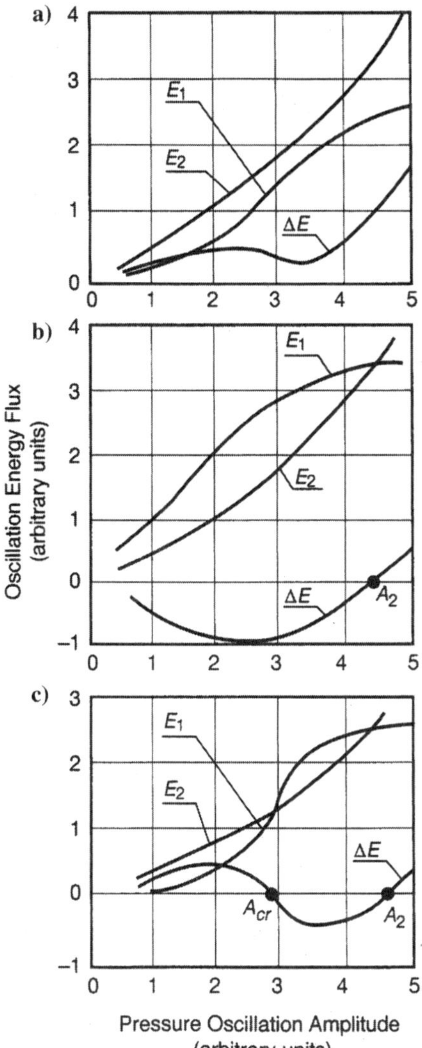

Fig. 6.2 Fluxes of energy entering the oscillatory system E_1, dissipated by the system E_2, and their difference $\Delta E = E_2 - E_1$ vs oscillation amplitude A.

system. As is seen from Fig. 6.2c, besides the damping factor, there is one more characteristic, which defines the system stability. This is the relation between the amplitude of an initial disturbance influencing the oscillatory system and the critical amplitude A_{cr}. Variations in the operating mode of the combustion chamber affect the relation between the fluxes of the oscillatory energy entering and dissipated by the system. In turn, these changes cause changes in the relation $\delta(A)$.

Figure 6.3 shows five possible variants of the curve $\delta = F(A)$, from an initial position shown in Fig. 6.3a to the position in Fig. 6.3d characterizing a completely stable system. In the case shown in Fig. 6.3b the change in the position of the curve relative to its position in Fig. 6.3a indicates that the damping factor becomes

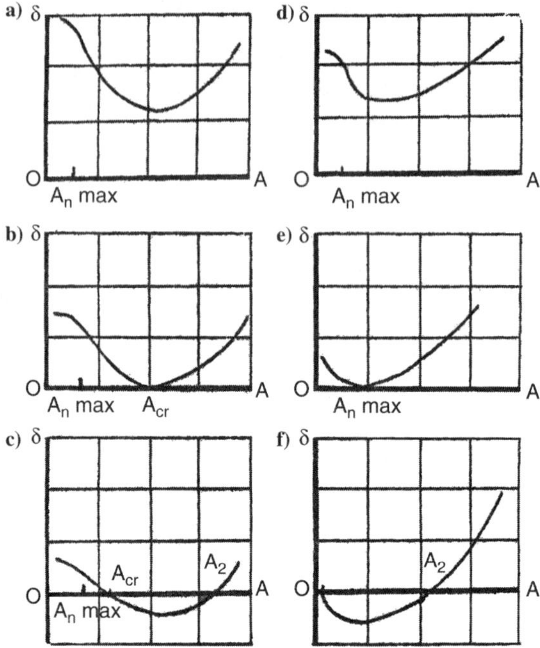

Fig. 6.3 Possible changes of $\delta = F(A)$ curve on approaching the instability boundaries.

zero at amplitudes exceeding a noise level in the combustion chamber. With the mode changing further, the range of amplitudes at which the value of the damping factor is equal to or less than zero can be extended (Fig. 6.3c). Another behavior of the amplitude dependence of the oscillation damping factor with change of operating mode can occur; it differs in that its most essential changes take place in the range of amplitudes belonging to the noise region (Figs. 6.3e and 6.3f). It is seen from Fig. 6.3 that excitation of oscillations is possible in two cases:

1) The first case is when the value of the damping factor equal to zero is achieved at amplitudes corresponding to the noise level (Figs. 6.3 and f). This case of oscillation excitation is referred to as a *soft excitation mode* (see Sec. I of Chapter 3).

2) The second case is when the amplitude of a random disturbance exceeding the value of A_{cr} is necessary for excitation of high-frequency pressure oscillations (Fig. 6.3c). This case is referred to as a *hard excitation mode*, as accepted in most studies (see Sec. II of Chapter 36).

In the case shown in Fig. 6.3c. (when the disturbance exceeds A_{cr}) and Fig. 6.3d, the oscillation amplitudes will change until finally they will reach steady state at a level of A_2 corresponding to stable oscillations.

In practice, it is convenient to normalize the damping factor by the inverse of the oscillation period T, that is, the oscillation decrement δT.

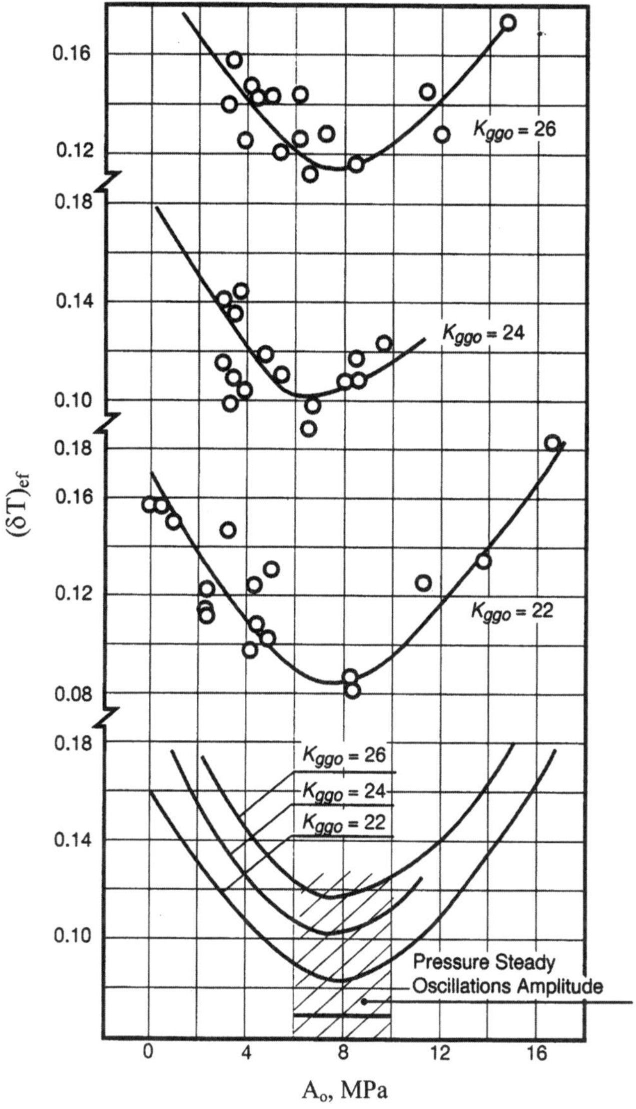

Fig. 6.4 Variation of oscillation decrement against initial pressure pulse A_0 for injector heads characterized by the relative flow area of oxidizing gas injectors.

As an example confirming the behavior of the amplitude dependence of the decrement (the damping factor) shown in Figs. 6.2 and 6.3, experimental results are shown in Fig. 6.4. The studies were carried out in the gas–liquid unit shown in Fig. 6.5 using the following propellant components: NT as the oxidizer and UDMH as the fuel. The unit consisted of a gas generator running with excess oxidizer

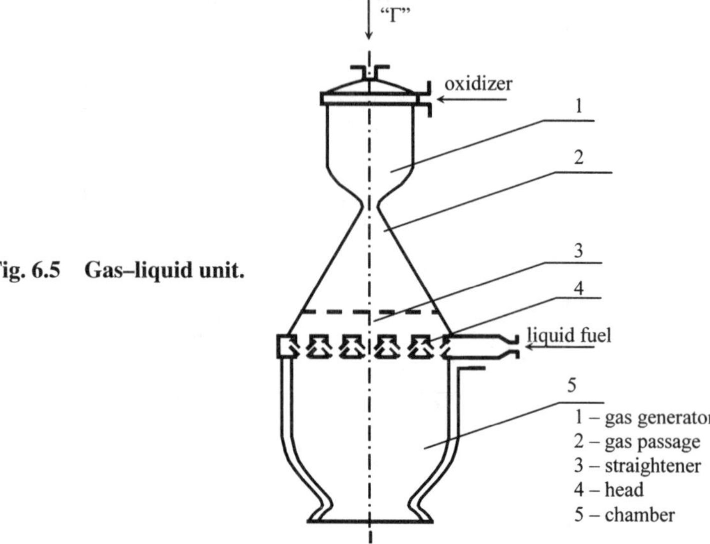

Fig. 6.5 Gas–liquid unit.

(in the experiments shown in Fig. 6.4, $K_{ggo} = 26$), a gas passage with a flow straightener, a set of interchangeable mixing heads, and a combustion chamber with a water-cooled shortened nozzle. The chamber diameter was 260 mm.

The characteristics of the three injector heads under study are shown in Fig. 6.6 and Table 6.1. Each head had 91 injectors with a gas duct having different diameters

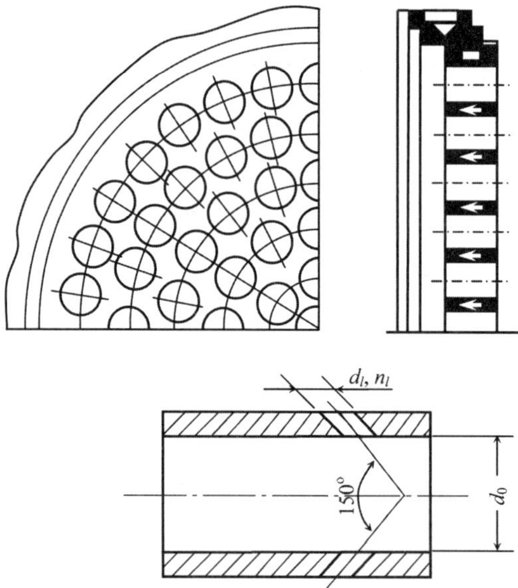

Fig. 6.6 Design parameters of injector heads.

Table 6.1 Injector head design parameters

No.	Legend	n_i	d_o, mm	n_1,	d_1, mm	d_{ch}, mm
1	91-18-8-1.2	91	18	8	1.2	260
2	91-14.9-8-1.2	91	14.9	8	1.2	260
3	91-12.2-8-1.2	91	12.2	8	1.2	260

(18; 14.9; 12.2 mm). The injector had eight holes with diameters of 1.2 mm for feeding the fuel. The oscillation decrement was determined by introducing disturbances of the prescribed amplitude by a shock-tube device (see Sec. IV in Chapter 9). The decrement was calculated from the relationship of $(\delta T_i)_{\text{ef}} = \ell n(A_i/A_i)$ with relevant amplitude averaging—the effective decrement during one period of attenuating oscillations from A_i to A_{i+1}. The three injector heads had a relative permeability:

$$\bar{f}_{\text{r}} = F_i/F_{\text{ch}} = n_i \cdot d_0^2/D_{\text{ch}}^2$$

where

F_i = area of all gas ducts of the injectors
F_{ch} = area of injector head combustion bottom
n_i = umber of injector on the head
d_0 = diameter of injector gas duct
D_{ch} = chamber diameter

As seen from Fig. 6.4, depending on the disturbance amplitude $\delta T_{\text{ef}}(A_o)$, the curve has a pattern similar to those shown in Figs. 6.2 and 6.3, and a minimum value at amplitudes where stable self-oscillations are realized at the instability boundary ($\delta T = 0$). The parameter K_{ggo} is defined as the ratio of the oxidizer flow rate G_{ogg} to the fuel flow rate G_{fgg} in the gas generator. With the decrease in head relative permeability, the decrement-amplitude curve moves down, which is indicative of the process approaching the unstable region. The causes of such a relationship between δT_{ef} and d_o are analyzed in Sec. IV of Chapter 14.

It can be seen from Figs. 6.2 and 6.3 that besides δT, the value of A_{cr} defines the system stability. A_{cr} is the amplitude of minimum disturbance, which causes the process instability. The random-parameter (for example, a flow of propellant components through the combustion chamber, ratio of components, flow rating,...) dependence of amplitude A, that is, a bifurcation diagram, is shown in Fig. 6. 7.

Three zones can be seen: 1) a zone where the combustion chamber or gas generator is absolutely stable; 2) a zone where the process is potentially stable; and 3) a zone of operation process instability at the given high-frequency oscillation mode. All of these boundaries are quite scattered, as is a level of noise in the chamber as a result of combustion fluctuations. However, for convenience of analysis, let us take them to be stationary. The zone of absolute stability of the process corresponds to the energy flux curve in Fig. 6.2a, the zone of potential stability corresponds to Fig. 6.2c, and the zone of process instability corresponds to Fig. 6.2b.

The relation A_{cr}/A_n characterizes the departure from the lower hysteresis boundary. To determine A_{cr}, artificial pressure disturbances should be introduced to the combustion chamber. The requirement for the characteristics of

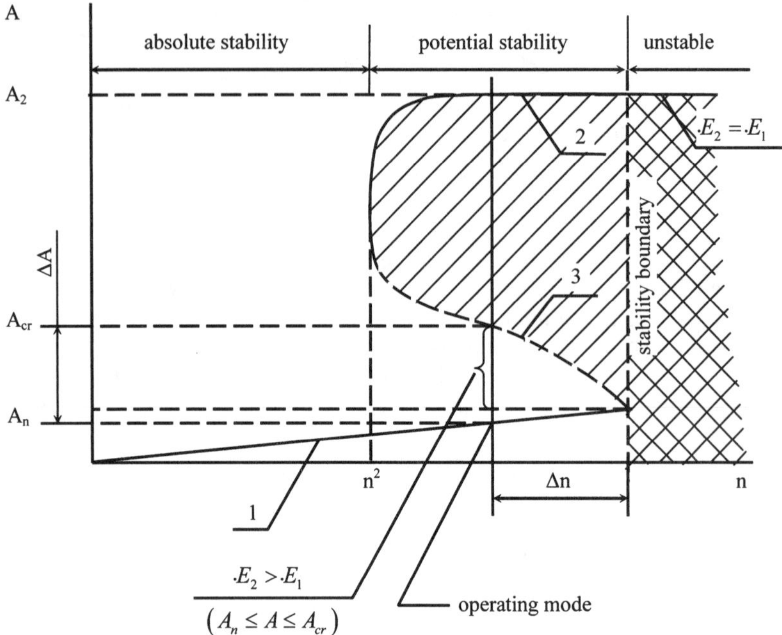

Fig. 6.7 Pressure oscillation amplitude *A* in combustion chamber against the parameter *n* defining stability (bifurcation diagram).

the artificial disturbance and of the devices generating the disturbances are described in detail in Chapter 9. In practice, serious problems were faced when determining A_{cr}. Introducing very considerable disturbances distorts the operation. The whole set of equations describing it is disturbed, which can result in an incorrect view of the tendency to hard excitation. In addition, introducing the disturbance above A_{cr} results in engine failure, which is inadvisable. Thus it is acceptable that the stability margin for hard excitation is sufficient if, after applying the pulse disturbance, the pressure oscillations damp quite rapidly and

$$N > \frac{A_0}{2A_{nef}} = n$$

Here, A_0 is the initial peak value of the pressure deviation from the average at the pulse disturbance; A_{nef} is the effective value of the high-frequency signal just before the pulse pressure disturbance, and N is the number determined experimentally.

The bifurcation diagram obtained on unit D495 by the artificial disturbance method is shown in Fig. 6.8. In unit D495 (Fig. 6.9) NT was used as an oxidizer and UDMH as a fuel. The unit comprised gas-generator feeding gas with an excess oxidizer, a gas passage, a set of interchangeable mixing heads, and a water-cooled combustion chamber with an adjustable critical section. The regulating cone was

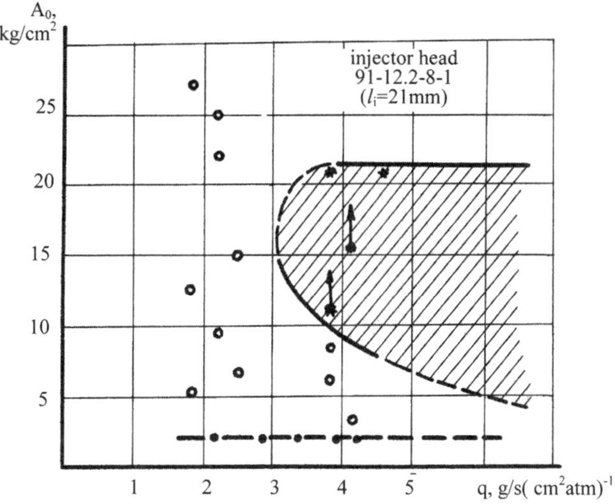

Fig. 6.8 Bifurcation diagram: •, a level of natural noise in chamber; o, initial pressure pulse, which induced damped oscillations; and *, initial pressure pulse exciting undamped oscillations.

water cooled and was moved by hydraulic cylinders. The chamber diameter was 267 mm. During experiments, whose results are shown in Fig. 6.8, the injector head was equipped with 91 pneumatic injectors with internal mixing. The injector had a gas duct diameter of 12.2 mm and eight holes with a diameter of 1.0 mm for feeding the fuel. More detailed data concerning the tests carried out in this unit with different head versions are presented in Chapter 13.

The dependence of stability in the coordinates *A*-*q* (bifurcation diagram) is shown in Fig. 6.8. A constant flow through the chamber ~35 kg/s and a constant ratio of components in the chamber $K = 3$ and in the gas generator were

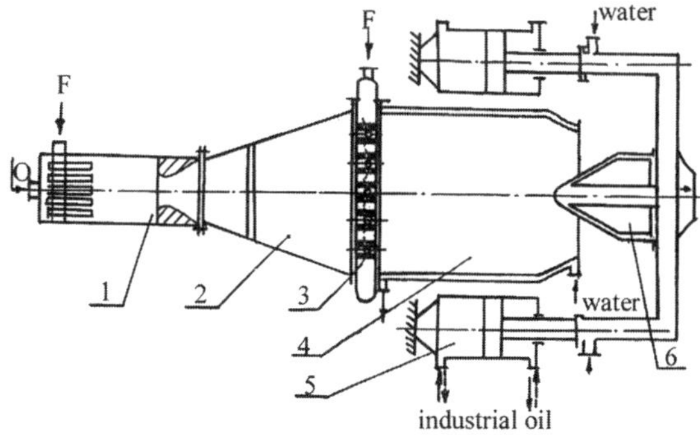

Fig. 6.9 Units D-495: 1, gas generator; 2, gas passage; 3, mixing head; 4, combustion chamber; 5, hydraulic cylinders; and 6, regulating cone.

maintained during the experiments. When the cone was extended, the critical section (nozzle throat) of the chamber increased, and the pressure changed from 40 to 15 bar. The artificial pulse disturbances produced by a shock-tube device were applied to the chamber for estimating stability. Four shock tubes were mounted on the chamber.

When constructing the bifurcation diagram, the oscillation amplitudes corresponding to the upper limit cycle were assumed equal to the amplitude of the developed oscillations occurring after applying an artificial pulse. It follows from Fig. 6.8 that for this head design the absolutely stable zone exists up to the flow rating of 2.5–3.0. Thereafter with an increase in the flow rate, at disturbances above A_0~10 bar there occur high-frequency pressure oscillations with amplitudes above 20 bar and a frequency of ~2400 Hz.

In further studies, the method of estimating stability with respect to finite disturbances was essentially improved as regards the design of the disturbing devices, the processing methods, and analysis of the obtained results (Chapter 9).

One of the quantities that might characterize the rate of damping of pressure oscillations generated by the pulse disturbance is the time of relaxation of these oscillations, that is, the time of their decrease by a factor of e after an initial transient region with duration τ_1. When estimating stability with a relaxation time, the maximum value of pressure deviation from the average value caused by a pulse disturbance should markedly differ from combustion-chamber pressure peaks, but not so much as to distort the process characteristics.

A quantity inversely proportional to the relaxation time is directly proportional to the effective (amplitude-averaged) oscillation decrement δT_{ef}. Therefore the applied stability characteristics determined from natural noise δT and from pulse disturbances τ_r have identical physical meaning based on the damping rate of pressure disturbances of different nature (natural or artificial).

As a result of long-term studies, it was found that a recommended value of τ_r should be below $\tau_r = 0.15$ s as a value defining a sufficient margin of stability to hard excitation of pressure oscillations. This result holds if pressure disturbances applied to the chamber or to the gas generator are $15 < A_0/A_{\mathrm{mr}} < 25$, where A_0 is the value of initial disturbance, and

$$A_{mr} = \frac{1}{n} \cdot \sum_{i=1}^{n} |Y_i|$$

is the mean rectified value of the amplitude of oscillation for a certain averaging range prior to applying the disturbance. The absolute value of parameter deviation from the average value recorded by the computer is $[Y_i]$.

Sufficiency of the selected stability characteristics and their actual values has been confirmed by numerous data obtained in experimental and operational LREs.

III. Relationship Between Amplitude and Decrement of Pressure Oscillations

Operation of LRE combustion chambers and gas generators is always accompanied by occasional random pressure pulsations (combustion noise), which are mainly caused by pulsations of heat-release rate during propellant burning. The turbopump unit makes a certain contribution to pulsations in the combustion chamber and gas generator. A level of the pressure pulsation amplitudes is determined by peculiarities of the mixing system and by the pressure in the units. An increased level of pressure pulsations causes prohibitive vibration overloads, which is especially important for reusable engines. Thus the design of combustion chambers and gas generators with a reduced level of pressure pulsations represents a separate problem. However, in comparing different mixing heads, increased level of pulsations does not mean their stability margin is reduced.

Let us determine the decrement dependence of the pattern of oscillation amplitude variation with time. Assuming a random process at the system inlet to be stationary, for the rms value of the amplitude variation within a time interval τ, we get,

$$\overline{[A(t+\tau)-A(t)]^2}=\overline{A^2(t+\tau)+A^2(t)}-\overline{2A(t+\tau)A(t)}$$

$$=\overline{2A^2(t)}-2B_A(\tau)=2A^2(t)[1-R_A(\tau)] \tag{6.2}$$

where $A^2(t)$ is the rms value of the amplitude; $B_A(\tau)$ is the correlation function of the envelope function $A(t)$; and $R_A=B_A(\tau)/B_A(0)$ is the correlation factor for the function $A(t)$.

It is known from [27] that the correlation factor $R_A(\tau)$ of the envelope of a random process at the outlet of a simplest oscillatory system has the form

$$R_A(\tau)=\frac{4}{5}\left(1+\frac{1}{4}\,l^{-2\delta T}\right) \tag{6.3}$$

Substitution of Eq. (6.3) into Eq. (6.2) gives

$$\frac{\overline{[A\,(t+\tau)-A(t)]^2}}{\overline{A^2(t)}}=\frac{2}{5}\,(1-e^{-2\delta T}) \tag{6.4}$$

It can be seen from Eq. (6.4) that with a decrease in the value of the damping factor δ, the relative rms variation of the amplitude within a fixed time interval decreases; that is, the time dependence of the oscillation amplitude becomes smoother.

Let us consider the relationship between the pressure oscillation amplitude and the decrement. Let the broadband random signal $X(t)$, which for simplicity is assumed to be stationary white noise, be applied to the inlet of the oscillatory system with a transient pulse function

$$h(t)=\begin{cases} e^{-\delta T}\sin\omega t & t>0 \\ 0 & t<0 \end{cases} \tag{6.5}$$

Then for the signal $Y(t)$ at the outlet of the system we get

$$\mathrm{Y}(t)=\int_{-\infty}^{\infty}\mathrm{h}(t-\tau)\mathrm{X}(\tau)\,d\tau$$

$$\mathrm{Y}^2(t)=\int_{-\infty}^{\infty}\int_{-\infty}^{\infty}\mathrm{h}(t-\tau)\mathrm{h}(t-\tau')X(\tau)X(\tau')\,d\tau\,d\tau'$$

or, taking into account that the function for white noise is

$$\overline{X(\tau)X(\tau')} = \delta(\tau-\tau')\overline{X^2(t)}$$

where $\delta(\tau-\tau')$ is Dirac's delta function, we get

$$\overline{Y^2}(t) = \overline{X^2(t)} \int_0^{\infty} h^2(\tau)\, d\tau \tag{6.6}$$

Substitution of Eq. (6.5) into Eq. (6.6), we obtain

$$\overline{Y^2(t)} = \overline{X^2(t)} \int_0^{\infty} l^{-2\delta T} \sin \omega\tau\, d\tau = \frac{1}{2}\overline{X^2(t)} \int_0^{\infty} l^{-2\delta T} (1-\cos 2\omega\tau)\, d\tau = \overline{X^2(t)}/4\delta \tag{6.7}$$

The lower integration limit is taken from zero as the influence on the system starts with $\tau = 0$. Representing the random oscillatory process $Y(t)$ in the form

$$Y(t) = A(t)\sin[\omega t + \varphi(t)]$$

we get

$$Y^2(t) = A^2(t)\sin^2[\omega t + \varphi(t)] = \frac{1}{2}A^2(t) \tag{6.8}$$

Substitution of Eq. (6.7) into the preceding equation, we find

$$\frac{A^2(t) \cong X^2(t)}{2\delta}$$

or

$$\frac{\sqrt{\overline{A^2(t)}} \cong \sqrt{\overline{X^2(t)}}}{\sqrt{2\overline{\delta}}} \tag{6.9}$$

Therefore, the rms value of the oscillation amplitude at the oscillation system outlet is inversely proportional to the square root of the damping factor δ of this system when a broadband random signal is applied to its inlet. When the combustion chamber process changes on approaching a soft disturbance mode (that is, upon approaching instability), a decrease in the oscillation decrement (damping factor) is accompanied by a growth in the pressure oscillation amplitude of the corresponding acoustic mode, as follows from Eq. (6.9).

As the system approaches the instability mode, besides the change of the oscillation decrement, a change of energy spectrum parameters also takes place. The oscillations become more regular: the energy at the natural acoustic frequencies-corresponding to a high quality factor of the system, increases.

Restructuring of the oscillation energy spectrum is characterized by a number of parameters. All of them can be used in separate cases as the characteristics evaluating the approach of the system to the instability region, but they appear to be related to the quantity δT.

The use of the absolute value of the pressure oscillation amplitude for stability evaluation is essential because the amplitude depends on the location of a pressure transducer in the combustion chamber or in the gas generator.

The quantitative value of the oscillation decrement does not depend on transducer location. However, the position of a transducer in the combustion chamber can affect the accuracy of determination of the decrement for different shapes of acoustic oscillations.

IV. Stability Studies for a Model Chamber by Shutting off the Nozzle (D-45)

One of the first studies of combustion chamber sensitivity to finite disturbances was carried out on the unit D-45 [28]. The unit was a self-contained system with a supply system. The combustion chamber had an uncooled cylindrical part with a diameter of 50 mm and a length including a nozzle of 390 mm and a water-cooled nozzle with a throat diameter of 20 mm made without an expansion section. The combustion-chamber head had 13 screw injectors for fuel and six oxidizer injectors arranged around the circle between the central and peripheral fuel injectors. The propellant components were UDMH and NA27I or TG-02 and NA27I.

Longitudinal oscillations in the combustion chamber were periodically excited by shutting off the nozzle with a shutter of rectangular shape having a width equal to a nozzle diameter rotated by a pneumatic turbine. The experimental unit is shown in Fig. 6.10. The shutter configuration provides a shape of the excited pulse quite close to sinusoidal. The magnitude of the pulse is defined by a gap between the nozzle exit and the shutter, as well as by the ratio between the diameters of the combustion chamber and nozzle.

To determine the relation between the pulse magnitude and the initial amplitude of the damped high-frequency oscillations, let us consider the equations of longitudinal oscillations in a LRE combustion chamber:

$$\frac{\partial p'}{\partial t} + \bar{u}\frac{\partial p'}{\partial x} + \frac{\partial u'}{\partial x} = 0$$

$$\frac{\partial u'}{\partial t} + \bar{u}\frac{\partial u'}{\partial x} + \frac{\partial p'}{\partial x} = 0$$

$$p' = \gamma p' \tag{6.10}$$

Solutions have the form of traveling waves:

$$u' = \varphi_1(t - a_1 x) + \varphi_2(t + a_2 x) \tag{6.11}$$

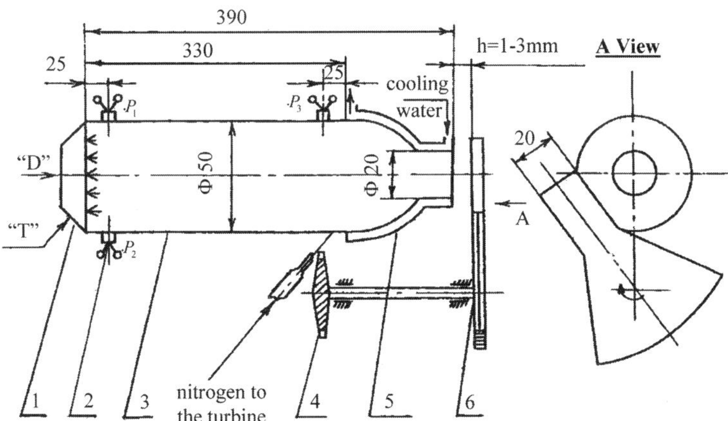

Fig. 6.10 Schematic diagram of experimental unit D45: 1, head; 2, pressure transducer; 3, chamber; 4, turbine, 5, nozzle; and 6, shutter.

$$\rho' = \frac{1}{\gamma} p' = \varphi_I(t - a_I x) - \varphi_2(t + a_2 x)$$

$$u' = \frac{\nu - \overline{\nu}}{c_0}; \quad \overline{u} = \frac{\overline{\nu}}{c_0}$$

$$p' = \frac{p - p'}{p}; \quad \frac{1}{a_1} = 1 + \overline{u}$$

$$p' = \frac{p - p'}{p}; \quad \frac{1}{a_2} = 1 - \overline{u} \tag{6.12}$$

where ν is the velocity of combustion products in the chamber; p is the pressure; c_0 is the sound velocity; γ is the adiabatic exponent; ρ is the density; t, x are the dimensionless time and axial coordinates, respectively; and $\varphi_1(t - a_1 x)$, $\varphi_2\ (t + a_2 x)$ are arbitrary functions corresponding to the waves running downstream and upstream, respectively. A bar above a letter indicates the average value of a quantity.

Let us find a relation between the dimensionless pressure oscillation p' and the weight flow

$$G' = \frac{G - \overline{G}}{G} \tag{6.13}$$

in a traveling wave. The weight flow is

$$G = g\rho\nu F \tag{6.14}$$

where g is the gravitational acceleration and F is the cross-sectional area of the cylindrical section of the combustion chamber.

From Eqs. (6.14) and (6.13), substituting the variables in Eq. (6.12), and ignoring second-order quantities, we have

$$G' = \frac{P' + u'}{u} \tag{6.15}$$

From Eqs. (6.15) and (6.11) and assuming $\varphi_2(t + a_2 x) = 0$, we get

$$G' = \frac{1}{\gamma}\left[1 + \frac{1}{\overline{u}}\right] P' = \left[1 + \frac{1}{\overline{u}}\right] \varphi_I(t - a_I x) \tag{6.16}$$

Assuming $\varphi_1(t - a_1 x) = 0$, we get

$$G' = \frac{1}{\gamma}\left[-1 + \frac{1}{\overline{u}}\right] P' = \left[-1 + \frac{1}{\overline{u}}\right] \varphi_2(t + a_2 x) \tag{6.17}$$

Taking into account that

$$\overline{u} = \frac{\overline{\nu}}{c_0} = \left(\frac{2}{\gamma+1}\right)^{\frac{\gamma+1}{2(\gamma-1)}} \frac{F_{cr}}{F}$$

where F_{cr} is the area of the nozzle throat. For $\gamma = 1.2$, from Eq. (6.4.7) we get

$$G' = 0.83\left[1.7 - \frac{F_{ch}}{F_{cr}} + 1\right] P' \tag{6.18}$$

and from Eq. (6.4.8)

$$G' = -\ 0.83\left[1.7 - \frac{F_{ch}}{F_{cr}} - 1\right] P' \tag{6.19}$$

Equation (6.18) gives a relation between the fluctuating component of the flow and pressure in the wave traveling with the flow, that is, on a traveling-wave excitation at the combustion-chamber head side. Equation (6.19) gives the same relation for a wave traveling against the flow, that is, on traveling-wave excitation at the nozzle throat. Let us evaluate the order of the disturbances by a flow.

Taking $F_{ch}/F_{cr} = 6.25$ and $G' = -1$ in Eq. (6.19) corresponding to a complete cutoff of flow through the nozzle, we get $p' \sim 0.125$, that is, the fluctuating pressure will be ~12.5% of the average pressure in the combustion chamber. The presence of gaps between the shutter and the nozzle reduces the value of a disturbance by ($1-4\ h/d_{cr}$), where h is the gap and d_{cr} is the diameter of the nozzle throat. For a gap of 1–1.5 mm, $p'_{ef} = p'_x$ (0.8–0.7). That is, it makes up 10–9% of the average pressure in the combustion chamber.

It is seen from Eqs. (6.18) and (6.19) that the shape of a traveling wave in the combustion chamber is defined by the law of flow oscillation. Hence the formation of a sinusoidal wave requires sinusoidal variation of the pulse, which can be obtained, for example, by shutting off the nozzle of the combustion chamber, without a supersonic section, for a short time with the help of a rotating shutter (Fig. 6.10). The nozzle or the shutter is shaped as required.

Suppose we specify at the nozzle side a pulse as a portion of a sinusoid with a period and duration equal to the time of twice the path along the combustion chamber. That is, it is equal to the period of the first longitudinal mode of gas-volume free oscillations. The time of the pulse termination is set interference of a direct traveling wave and of a traveling wave reflected from the head end of the chamber. A standing wave is formed with frequency of the first longitudinal mode. More precisely, such a wave should be considered a standing wave only approximately because the oscillations are decaying.

When a period or duration of the pulse does not coincide with the period of natural oscillations, as well as when the pulse has nonsinusoidal shape, the damped oscillations will have a complex composition. With small deviations from a required pulse shape and duration, the oscillation amplitude of the mode to be excited will prevail.

Tests were carried out in different modes, under a constant pressure in the combustion chamber, and with a constant mixture ratio of components, as well as with variable values of $0.5 < \kappa_m < 6$ and $20 < P_{ch} < 45$ bar. In most runs, the pulse duration defined by the nozzle shutoff time was assumed equal to the period of the first longitudinal mode, 1/1300 s.

To save equipment, the duration of every test was approximately 4.0–5.0 s. This period was sufficient to obtain 200–250 records of damped pressure oscillations during every test. Before starting, the turbine was driven by air to a required rotational speed (~50 rev/min) measured by a frequency counter; the signal to the counter came from a speed transducer.

Calculation of the damping factor is not a problem when only one of the oscillation modes is excited. In this case the damped pressure oscillation curve is

$$p = \bar{p} = A_e \sin(\omega t + \varphi)$$

The damping decrement and oscillation frequency f are easily calculated from the formulas

$$\delta T = \mathcal{L}\mathcal{N}a; \quad f = \frac{1}{T}; \quad a = \frac{h_i}{h_i + 1} = \sqrt[k]{\frac{h_i}{h_i + k}}$$

where T is the period of oscillation and a is the damping factor determined from the relation of the adjacent amplitudes (Fig. 6.11). Processing of the damped oscillation records starting with measurement of h_1 should be made from the moment when the action of the exciting pulse terminates, and the oscillations become free.

Let us discuss some results of the studies. The disturbances introduced exceeded the amplitude occurring with high-frequency oscillations, and so calculating the oscillation decrement in the stable region of chamber operation by the formula

$$\delta T = \mathcal{L}\mathcal{N} \cdot \frac{h_i}{h_{i+1}}$$

gives the amplitude dependence of decrement similar to that shown in Fig. 6.4. The minimum value of decrement ~1 bar to the amplitude of the developed longitudinal high-frequency oscillations. The average decrement calculated by the following formula

$$\delta T = \mathcal{L}\mathcal{N} \cdot \sqrt[k]{\frac{h_i}{h_i + k}}$$

gives an overestimated value. This explains an insignificant change of decrement on approaching the instability boundary and its difference from zero in Fig. 6.12. The oscillation decrement in the K_m and P_{ch} diagram is shown in Fig. 6.13. With P_{ch}~28 bar and K_m~6 the values of oscillation decrement of the first longitudinal mode were

$$\delta T \sim 0.33\text{-}0.40$$

With K_m decreasing to the values close to ~1, the decrement decreased to 0 and high-frequency oscillations were excited.

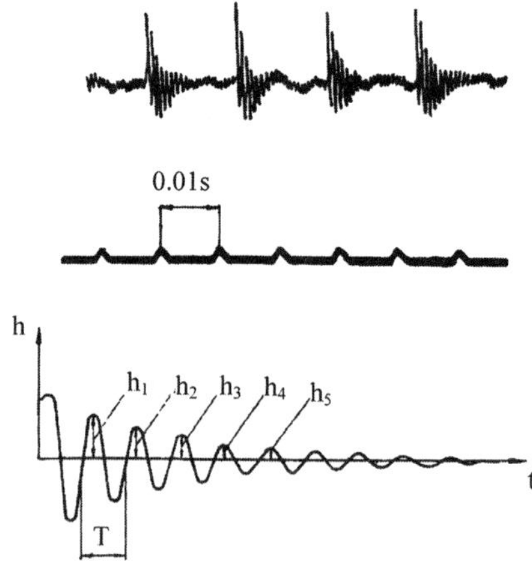

Fig. 6.11 Experimental record and history of damped pressure oscillations.

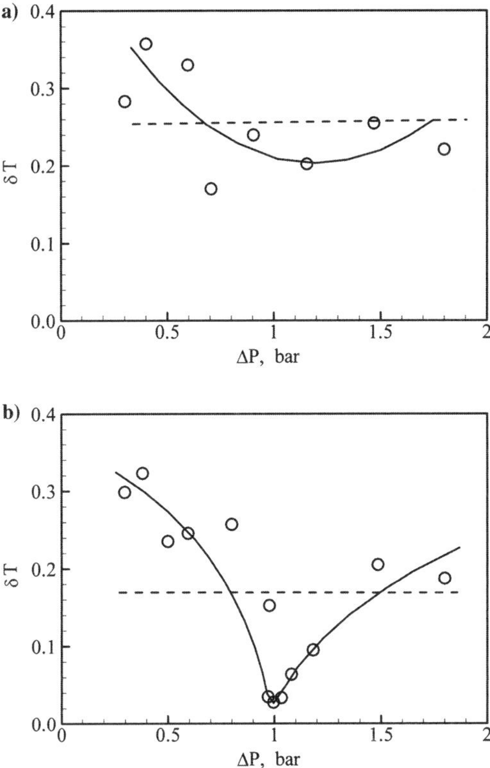

Fig. 6.12 Dependence of decrement amplitude on damping of pressure oscillations induced by an external disturbance. The curves are represented for the experiment, and the results are shown in Figs. 6.1–6.4. a) Curve corresponds to a testing time of about 1 s, and b) curve corresponds to a testing time of about 1.5 s in Fig. 6.14.

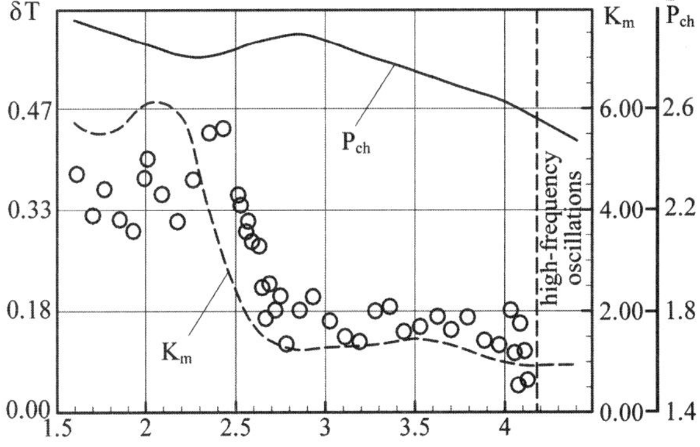

Fig. 6.13 Damping factor variations with respect to chamber pressure P_{ch} and mixture ratio between components K_m; o, averaged decrement.

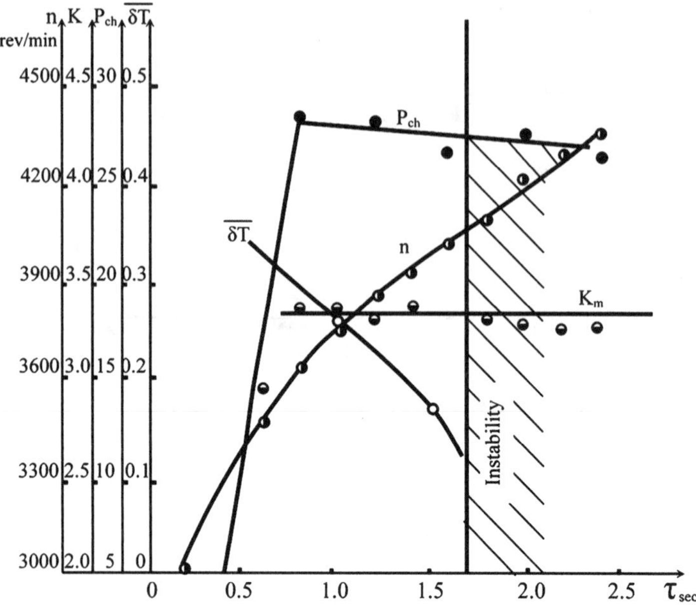

Fig. 6.14 Variation of parameters in time (combustion chamber D-45).

The results of the experiment to study the dependence of the decrement on characteristics of the disturbance are shown in Fig. 6.14. With practically constant values of P_{ch} and K_m, the turbine rotational speed was increased. It is seen from Fig. 6.14 that this resulted in a decrease of the decrement and excitation of high-frequency oscillations.

The process was affected by the change of the disturbance characteristics (pulse spectral distribution). The problem of the equivalence of the action of different forms of the disturbance is discussed in detail in Sec. III of Chapter 9.

Chapter 7

Acoustic Study of Combustion Chamber Stability Characteristics

COMBUSTION instability in the form of self-excited pressure oscillations develops at resonant frequencies in the combustion-chamber and gas-generator volumes. To each resonant frequency, there corresponds its own field of standing waves within the chamber, the so-called oscillation mode. The oscillation mode excited by an external source is damped after removal of the source because of radiation and friction losses. The damping rate of self-excited natural pressure oscillations is characterized by an oscillation decrement.

The acoustic study of combustion-chamber characteristics is aimed at determining expected resonance frequencies, modes, and decrements of oscillations.

I. Identification of Natural Modes of Pressure Oscillations

Identification of pressure oscillations in combustion chambers and gas generators amounts to determination of oscillation processes of the natural modes. The identification must be performed at every stage of engine development. At the design stage, it is desirable to use numerical methods or analytical models to determine natural frequencies of oscillations in combustion chambers, and pressure pulsations downstream of pumps, in order to identify the main engine operating modes. In engine firing tests, the identification of frequencies in the amplitude spectrum and vibrations is performed, based on the spectral analysis of recorded pressure pulsations.

In identification by numerical and experimental methods, the natural frequencies of acoustic oscillations in gas and liquid cavities and the frequency spectrum of vibrations of combustion-chamber (and gas-generator) structural elements are determined. The measured pressure oscillation frequencies and vibrations are examined and assigned to the specific systems from which the oscillations originate. Analysis is also carried out to determine the means by which pressure oscillations and vibrations transmit from one oscillatory system to another.

Individual harmonics in the amplitude spectrum of recorded pressure pulsations and vibrations can be excited by sources external to combustion chambers and gas generators. Such sources include pressure pulsations and vibrations generated by pumps and turbines of an engine turbo-pump unit.

The frequencies f_p of pressure pulsations produced by pumps and turbines depend on the rotor rotating speed N (rev/s); the number of rotor blades Z_r; the number of straightener blades (for a pump); the number of distributor blades (for a turbine); and the harmonic number n ($n = 1, 2, \ldots$; an integer number):

$$f_p = nZ_rN$$

If there exists a nonuniform azimuthal flow in the pump cavity caused by the superposed screw, local cavitation, or alternation of impeller blades having different length and spacing, well-defined peaks can appear in the amplitude spectrum with frequencies

$$f_n = \frac{1}{n}Z_rN$$

When the traces of rotor blades intersect those of stator blades, there occur pressure pulses with frequencies equal to the product of the rotating speed N and the least common multiple of the rotor-blade number Z_r and stator-blade number Z_{st}, m

$$f_n = mN$$

For example, for the oxidizer pump of the RD-170 engine, $Z_r = 14$, $Z_{\mathrm{st}} = 12$, and $N = 240$ rev/s. The least common multiple for the given numbers of blades is m = 84. The expected repetition rate of pressure pulse becomes $f_p = 84 \times 240 = 20{,}160$ Hz. Experimental vibration transducers with a frequency response of up to 60,000 Hz confirmed the existence of such a frequency in the amplitude spectrum of pump-casing vibrations.

The transmission of a disturbance from the turbopump unit to the combustion chamber (or gas generator) depends on the natural frequencies of fuel delivery pipes and resonance properties of structural elements.

The natural frequencies of the gas and liquid cavities in a combustion chamber and a gas generator are approximately calculated as for a cylindrical volume of a diameter D and a length L with prescribed boundary conditions on the walls.

The natural frequencies depend not only on the geometrical dimensions of combustion chambers and boundary conditions on the walls, but also on the speed of sound c in the medium filling the cavity as well as on the oscillation mode of concern.

The speed of sound in a gaseous medium can be calculated by the following formula [29]

$$c = \sqrt{\gamma R T_0} \tag{7.1}$$

where γ is the adiabatic exponent (i.e., the ratio of specific heats), R is the gas constant in J/kg·K, and T_0 is the temperature in K. Their values are calculated by thermodynamic analysis of combustion products [30]. The sound speed in combustion products can also be estimated from the measured specific impulse β [31]

$$c \approx 0.708\beta \tag{7.2}$$

where $\beta \approx 1570$ m/s for nitric acid propellants and $\beta \approx 1770$ m/s for oxygen-hydrocarbon propellants [52]. The data on the speed of sound in liquid propellants can be found in reference books on physiochemical properties of rocket propellants [26 and 32].

Sound propagation in a cylindrical pipe can be described by a special function Φ_{mn}, commonly referred to as velocity potential [30 and 33].

$$\Phi_{mn}(r, \varphi, x, t) = AJ_m(\nu_{mn} r)\cos(m\varphi - \varphi_0)\exp i\left(\omega t \pm \sqrt{k^2 - \nu_{mn}^2}\, x\right) \tag{7.3}$$

where r, φ, x are the cylindrical coordinates in the radial, azimuthal, and axial directions, respectively; t the time; $k = \omega/c$ the wave number of the acoustic wave; ω the circular frequency; c the sound speed; $\nu_{mn} = \omega_{mn}/c$ the wave number for transverse modes of oscillations in a pipe of unit radius; $m, n = 0, 1, 2, 3\ldots$ are prime integer numbers; and $J_m(\nu_{mn} r)$ the Bessel function of the mth number.

The velocity potential Φ_{mn} is a solution to the wave equation. The oscillating pressure P' and velocity V' are expressed in terms of the partial derivatives of the velocity potential Φ_{mn} with respect to the spatial coordinates and time [30]:

$$V' = -\frac{\partial \Phi_{mn}}{\partial x}, \quad P' = \rho \frac{\partial \Phi_{mn}}{\partial t} \tag{7.4}$$

where ρ is the density of the medium. If $m = 0$, $n = 0$, the velocity potential (7.3) describes planar waves in a pipe

$$\Phi_{00} = Ae^{i(\omega t - kx)} \tag{7.5}$$

Using Eqs. (7.4) and (7.5), we get the ratio of the oscillating pressure P' to its velocity counterpart V′:

$$\frac{P'}{V'} = \frac{i\rho\omega\Phi_{00}}{ik\Phi_{00}} = \frac{\rho\omega}{\omega/c} = \rho c \tag{7.6}$$

The product of density and sound velocity ρc is called the acoustic-wave resistance of the medium.

Boundary conditions on the surfaces defining the volume of a combustion chamber (or gas generator) are assigned by the ratio of the pulsating pressure P' to the pulsating velocity V':

$$Z = \frac{P'}{V'} = R + iY = \rho c(\overline{R} + i\overline{Y}) \tag{7.7}$$

where Z is the surface impedance, R the real part of impedance, Y the imaginary part of impedance, and $\overline{R}$ and $\overline{Z}$ the dimensionless components of impedance.

If the acoustic resistance of a medium changes during the propagation of an acoustic wave, partial- or full-wave reflection occurs. The magnitude r_p and phase 2δ of the reflection coefficient in terms of pressure oscillation are calculated by the following formulas [33]:

$$r_p = \left|\sqrt{\frac{(\overline{R}-1)^2+\overline{Y}^2}{(\overline{R}+1)^2+\overline{Y}^2}}\right|; \qquad \cos 2\delta = \frac{\overline{R}^2 + \overline{Y}^2 - 1}{\left|\sqrt{(\overline{R}^2 + \overline{Y}^2 - 1)^2 + r\overline{Y}^2}\right|}$$

$$\sin 2\delta = \frac{2Y}{\left|\sqrt{(\overline{R}^2 + \overline{Y}^2 - 1)^2 + 4\overline{Y}^2}\right|} \tag{7.8}$$

The impedance Z_l at the end of the pipe with a length L with mild wave attenuation over the length L can be transposed to the pipe by the following formula [30]:

$$Z_{00} = \rho c \frac{i\tan kL + \bar{Z}_l}{i\bar{Z}_l \tan kL + 1} \tag{7.9}$$

where the impedance Z_{00} is called the transposed impedance.

Three different modes of natural oscillations in a cylindrical pipe are distinguished [33–35]:

1) The first mode is purely longitudinal oscillations, that is, $m = 0, n = 0$, and $\nu_{mn} = 0$. The velocity potential takes the form $\Phi_{00} = Ae^{i(\omega t - kx)}$ with $0 \le k \le \infty$;

2) The second is purely transverse oscillations, $m \ne 0$, or $n \ne 0$, or $m \ne 0, n \ne 0$, $k = V_{mn}$. The velocity potential takes the form: $\Phi_{mn} = AJ_m(\nu_{mn} r)\cos(m\varphi - \varphi_0 e^{i\omega t})$; and

3) The third is combined transverse-longitudinal oscillations, $m \ne 0$, or $n \ne 0$, or $m \ne 0$, $n \ne 0$. The velocity potential takes the form: $\Phi_{\mathrm{mn}} = AJ_m\,(\nu_{mn}\, r)$ $\cos(m\varphi - \varphi_0) \exp i(\omega t - \sqrt{k^2 - v_{mn}^2}\, x)$, with $k \ge \nu_{mn}$.

Equation (7.3) for the velocity potential describes all three types of oscillation modes in the pipe.

The wave numbers ν_{mn} for purely transverse modes of oscillations are calculated from the boundary conditions on the walls in terms of the wall impedance Z_w. If $Z_w \to \infty$ (i.e., rigid impermeable wall), the oscillating velocity component normal to the wall is equal to zero. Using Eqs. (7.3) and (7.4), we get an equation for calculating the wave numbers for purely transverse modes of pressure oscillations:

$$\frac{\partial J_m(\nu_{mn} r)}{\partial r} = 0 \tag{7.10}$$

The Bessel functions are almost periodical. Thus, Eq. (7.10) has a set of solutions [30 and 36], which give the values of wave numbers for transverse modes of oscillations. The wave number ν_{mn} for the transverse mode with indices m, n is the nth root of Eq. (7.10). The wave numbers for several low oscillation modes are given in Table 7.1.

The natural frequencies of transverse modes in a cylindrical pipe of an arbitrary radius r can be calculated as follows.

Table 7.1 Wave numbers ν_{mn} for transverse modes of oscillations in a cylindrical pipe of unit radius with rigid walls

	n			
m	0	1	2	3
00	0	3.832	7.016	10.172
11	1.841	5.331	8.537	11.705
22	3.054	6.707	9.068	13.169
33	4.200	8.014	11.344	14.743

The wave number by definition [29 and 30]

$$k_{mn} = \frac{\omega}{c} = \frac{2\pi f}{c} = \nu_{mn}\frac{1}{r}$$

Hence,

$$f_{mn} = \frac{\nu_{mn} c}{2\pi r} = \frac{\alpha_{mn}\, c}{D}\ \text{Hz} \tag{7.11}$$

where $D = 2r$ is the pipe diameter in meters and α_{mn} is defined as

$$\alpha_{mn} = \frac{\nu_{mn}}{\pi}$$

If the medium in a pipe flows at a velocity U, a correction in the form of a multiplier $\sqrt{1-M^2}$ is introduced to Eq. (7.11) in calculating the frequencies [4, 33, and 37]:

$$f_{mn} = \frac{\alpha_{mn} c}{D}\sqrt{1-M^2} \tag{7.12}$$

where M is the Mach number, defined as $M = U/c$. The Mach number in a combustion chamber depends on the extent of combustor opening (i.e., the ratio of the chamber diameter D_{ch} to the nozzle throat diameter D_{cr}) and on the adiabatic exponent of combustion products γ [31]:

$$M = \left(\frac{2}{\gamma+1}\right)^{\frac{\gamma+1}{2(\gamma-1)}}\left(\frac{D_{cr}}{D_{ch}}\right)^2 \tag{7.13}$$

The diabatic exponent for hydrocarbon-oxygen propellants with oxygen is in the range $\gamma = 1.12$–1.15, and for nitric acid propellants $\gamma = 1.17$–1.20 [31]. The mean velocity correction for transverse oscillation frequencies as a function of the ratio of D_{ch}/D_{cr} is shown in Fig. 7.1, whereas the Mach number vs this ratio is given in Fig. 7.2.

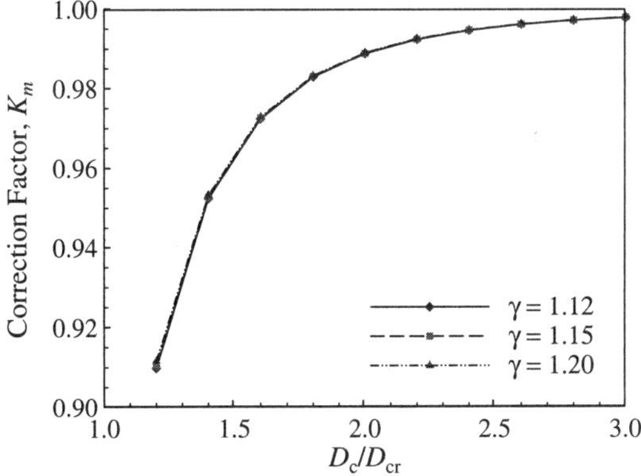

Fig. 7.1 Correction factor for transverse mode frequencies vs the ratio of combustion-chamber to nozzle-throat diameter.

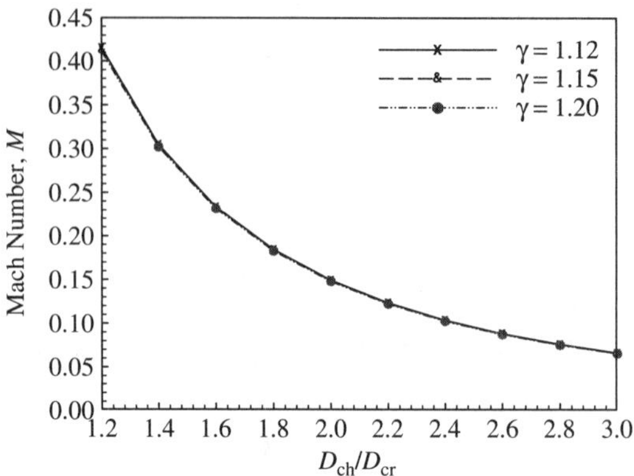

Fig. 7.2 Mach number vs the ratio of combustion-chamber to nozzle-throat diameter.

The natural frequencies of longitudinal oscillations depend on the sound velocity c, the combustion chamber length L, and the impedance at the chamber ends Z_o and Z_l, whereas the frequencies of mixed longitudinal-transverse modes depend on the diameter D as well. The natural frequencies of these modes are determined by the condition that the phase between the waves traveling upstream and downstream, compensated by the phase changes of $2\delta_0$ and $2\delta_l$ because of the end impedances, must be exactly equal to $2\pi p$, where p is an integer, $p = 0, 1, 2,\ldots$[30]. This condition applied for pure longitudinal oscillations is written as

$$2\delta_0 - 2kL + 2\delta_l = 2\pi p \tag{7.14}$$

with $\sin 2kL = \sin(2\delta_0 + 2\delta_l)$ and $\cos 2kL = \cos(2\delta_0 + 2\delta_l)$, where cosines and sines of angles $2\delta_0$ and $2\delta_l$ are calculated by Eq. (7.8).

For mixed longitudinal-transverse modes of oscillations, Eq. (7.14) is of the same form, but instead of the wave number k, the expression $\sqrt{k^2 - \nu_{mn}^2}$ is substituted:

$$2\delta_0 - 2\sqrt{k^2 - \nu_{mn}^2}\, L + 2\delta_l = 2\pi p$$

with

$$\begin{aligned} \sin(2\sqrt{k^2 - \nu_{mn}^2}\, L) &= \sin(2\delta_0 + 2\delta_l) \\ \cos(2\sqrt{k^2 - \nu_{mn}^2}\, L) &= \cos(2\delta_0 + 2\delta_l) \end{aligned} \tag{7.15}$$

The natural frequencies of pure longitudinal and mixed longitudinal-transverse oscillations can also be determined on the condition that in resonant modes; the imaginary part of the transposed impedance in a reference pipe section with coordinate x on the Z_0 side is equal to that on the Z_l side taken with an opposite sign. The transposed impedance is calculated by Eq. (7.9). This condition can be written as

$$iY(x) = iY(L - x) \tag{7.16}$$

where $Y(x)$ and is the imaginary part of the transposed impedance as a function of the axial coordinate x.

The natural frequencies of mixed longitudinal-transverse oscillations in gas cavities of combustion chambers and gas generators are approximately calculated as for a cylindrical pipe of a diameter D and a length L with acoustically closed ends ($Z_0 \to \infty$, $Z_l \to \infty$). From Eq. (7.15) along with Eqs. (7.8) and (7.9), we get

$$f_{pmn} = \frac{\alpha_{mn} c}{D} \sqrt{1 + \left(\frac{p}{2\alpha_{mn} \overline{L}} \right)^2} \tag{7.17}$$

where $\overline{L} = L/D$ is the relative length of the combustion chamber. The preceding equation can be corrected for the presence of mean flow with a Mach number M as

$$f_{pmn} = \frac{\alpha_{mn} c}{D} (1 - M^2) \sqrt{\frac{1}{1 - M^2} + \left(\frac{p}{2\alpha_{mn} \overline{L}} \right)^2} \tag{7.18}$$

Another extreme case is that the combustion chamber is acoustically open at the exit ($Z_l = 0$). Under this condition, Eq. (7.18) assumes the form

$$f_{pmn} = \frac{\alpha_{mn} c}{D} (1 - M^2) \sqrt{\frac{1}{1 - M^2} + \left(\frac{2p - 1}{4\alpha_{mn} \overline{L}} \right)^2} \tag{7.19}$$

Figure 7.3 shows the relative frequency of the mixed first tangential-longitudinal mode f_{110}/f_{010} as a function of the relative length of the chamber L/D, calculated by Eq. (7.18) for several different Mach numbers. The same relationship calculated by formula (7.19) is shown in Fig. 7.4. The resonant frequency of the first tangential-longitudinal mode decreases with increasing chamber length and mean-flow

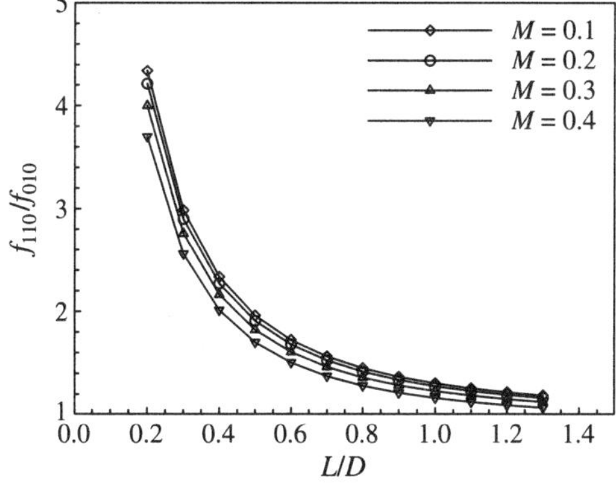

Fig. 7.3 Relative frequency of first tangential-longitudinal mode vs relative length of combustion chamber with acoustically closed boundaries at both ends.

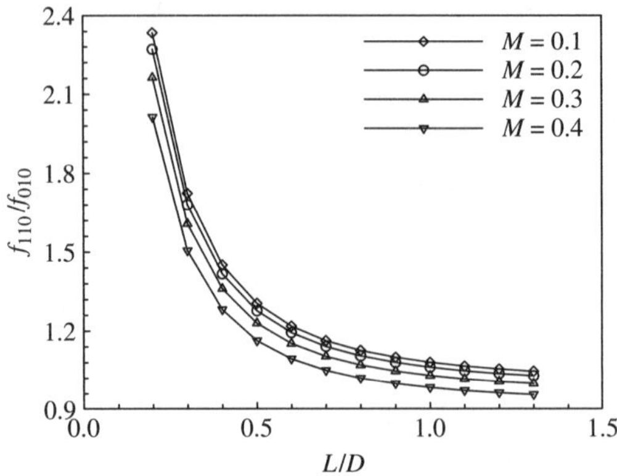

Fig. 7.4 Relative frequency of first tangential-longitudinal mode vs relative length of combustion chamber with acoustically open boundary on the nozzle side.

Mach numbers. Results calculated by Eq. (7.19) are in better agreement with experimental data. Thus, it is advisable to use the condition of acoustically open end at the combustor exit for estimating the frequencies of low longitudinal-transverse modes.

In calculations by Eqs. (7.17) and (7.19), the sum of the length of the cylindrical part of the chamber L_c and two-thirds of the length of the subcritical part of the nozzle L_n, $L = L_c + 2/3\ L_n$, is generally taken as the effective length of the combustion chamber.

The pre-injector oxidizer and fuel cavities of the mixing heads are very diverse and intricate. So the method of exposing such cavities to acoustic oscillations under simulated conditions with subsequent recalculation for actual conditions is the most reliable method of determining the natural frequencies of oscillations in such cavities. The frequencies of the pre-injector cavities of the mixing heads can be approximately estimated by Eq. (7.11), assuming them as short cylinders with a diameter D and acoustically closed boundaries at both ends. Baffles in the cavities, attached volumes (for example, a cooling jacket), and cavities encumbering with injector bodies can substantially affect the natural frequencies.

As an example, the resonant frequencies in a hollow pipe of $D = 150$ mm and $L = 400$ mm with acoustically closed boundaries at both ends and those in the same pipe fitted with 16 bars of $d = 21$ mm uniformly arranged in groups of eight pieces on diameters of 64 and 120 mm are calculated. The results are given in Table 7.2.

Table 7.2 Resonant frequencies in acoustically closed hollow pipe $D = 150$ mm, $L = 400$ mm with acoustically closed ends and in the same pipe uniformly fitted with 16 bars of $d = 21$ mm

Experimental conditions	Resonant frequencies in increasing order, Hz					
Hollow pipe	1329	1439	1720	2131	2198	2265
Pipe fitted with bars	1028	1174	1532	1695	1790	2040

It is apparent from Table 7.2 that fitting a pipe with bars results in a reduction of resonant frequencies. Thus, frequency of the first tangential oscillation mode decreased from 1329 to 1028 Hz, or by 23%. The reduction of resonant frequencies of transverse oscillation modes also can be expected in the mixing-head cavity containing with injector bodies.

II. Identification of Natural Modes of Pressure Oscillations Using Vibration Measurements

Identification of pressure pulsations based on vibration measurements at specified frequencies is performed in two steps: 1) calculation (or experimental determination), of natural frequencies and shapes of combustion-chamber shells, preparation of data on vibration measurements (i.e., obtaining the amplitude spectra and spectral analysis of vibration measurements), and extraction of information from vibration amplitudes; and 2) observation of a pressure pulsation and identification of its mode shape at the frequency under consideration.

Vibrations are recorded in accordance with the work scope and specifications. One of the parameters in making vibration measurements is the minimum number of transducers shown in Fig. 7.5.

When considering the reactions of the combustion chamber shell at acoustic frequencies f_{res} recorded by vibration transducers installed on the shell surface, one must take into account that the level of vibration recorded by transducers is

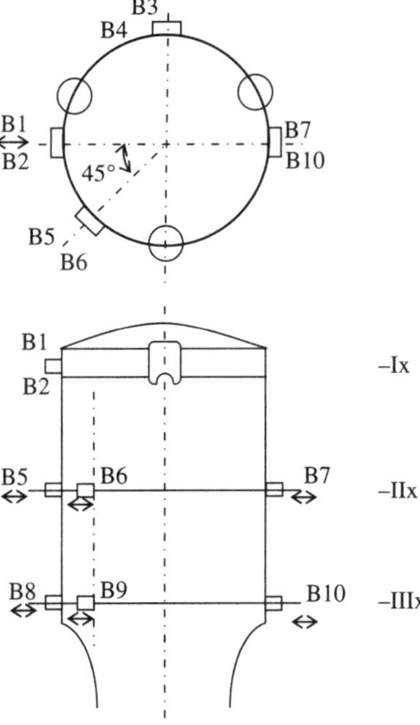

Fig. 7.5 Arrangement of vibration transducers on an LRE combustion chamber shell.

defined by the forced oscillations of the shell due to intrachamber pressure pulsations. For all combustion chambers studied in practice, the frequency spectra of the shell vibration and acoustic oscillations within the chamber do not coincide at the minimum wave numbers on the shell.

The natural frequencies of shell oscillations of a cylindrical combustion chamber are estimated by the following formula:

$$f = \frac{1}{2\pi}\sqrt{\frac{E}{Q}}\frac{1}{R}\left[\frac{\alpha^4}{m^2(m^2+1)L_0^4} + \frac{h^2m^2(m^2-1)^2}{3(1-V^2)(m^2+1)}\right]^{\frac{1}{2}} \tag{7.20}$$

where E is Young's modulus, N/m^2; Q is the density of the shell material, kg/m^3; R is the radius of the cylindrical section (i.e., median surface), m; α is a coefficient depending on the shell sealing ($\alpha = \pi/4$ or $3\pi/4$); m is the number of waves around the periphery, $m = 1, 2, 3, \ldots$; $L_0 = L_c/R$ is the relative length of the cylindrical section; L_c is the length of the cylindrical section of the combustion chamber, m; $h = h_{shell}/2R$ is the relative half-thickness of the shell; and ν is Poisson's coefficient.

The recorded vibration field contains vibrations of the shell itself recorded by transducers B5–B10, and vibrations are measured by B1–B4 characterizing the rigidity of attachment.

The pressure pulsations of the natural modes of the chamber volume can be classified into two main types:

1) Transverse oscillations of the gas volume whose action upon the shell is recorded by vibration transducers located on the shell surface at natural acoustic frequencies (e.g., f_{1T}, f_{1T1L}, f_{1T2L}, etc.). In addition, the moment and force generated by the attachments are recorded by the transducers on the mixing head (B1–B4).

2) Longitudinal oscillations of the gas volume (e.g., f_{1L}, f_{3L}, etc.) that produce significant forces on the mixing head and the convergent section of the exhaust nozzle. This type of oscillation can be recorded by the vibration transducers located along the chamber wall, B1 and B3. For ideal mode shapes of the longitudinal pressure pulsations, the shell vibration levels measured in sections IIx and IIIx should be lower than other measured values because axisymmetric shell oscillations are low in amplitude even in case of resonance.

Longitudinal oscillations in the chamber volume at the frequency of f_{2L} are self-compensating, and in general they are difficult to determine on the basis of vibration measurements.

When the acoustic mode frequencies in the combustion chamber and the shell vibration frequencies coincide (i.e., $2f_m^{\text{shell}} = f_L$, with $m \geq 2$ being the wave number at the phase angle φ), parametric resonance occurs, which makes it possible to estimate and identify longitudinal oscillations in the gas volume based on the measurements made on the shell surface at the B1 and B3 locations.

Reconstruction of the mode shapes of pressure pulsations based on the vibration-field measurements is carried out by decomposition of the shell vibration field into its own shape using measurements on the shell. At elevated levels of pressure pulsations on one of the natural modes in the combustion chamber, the shell deformation as a linear system reproduces the pressure mode shape in accordance with the vibration component normal to the shell surface. In the case of high-frequency pressure oscillations in the combustion chamber, the shell displacement approaches the mode shape of the shell oscillation that is closest to the mode shape of pressure pulsation.

The level of pressure pulsation can be estimated using the following relation:

$$A_{\text{trans}}^{\text{vibr}} = A_{\text{trans}}^{\text{puls}} \Big/ \frac{\rho}{E}\left[\omega^2 - \frac{1}{1-ij}(\omega_{\text{res}}^{m})^2\right] \tag{7.21}$$

where $A_{\text{trans}}^{\text{vibr}}$ is a recorded amplitude of shell vibration; $A_{\text{trans}}^{\text{puls}} = [P(\xi,\gamma,\varphi), g(\xi, \cos m\varphi)]$ the amplitude of pressure pulsations scalar-wise projected on one of natural shell mode shapes; ρ the shell density; γ the coefficient of shell internal friction; ω_{res} the cyclic resonant frequency of the shell natural mode; and ω the pulsation frequency of the chamber volume.

At elevated noise levels imposed on pressure pulsations of acoustic modes of concern, it is necessary to approximate the vibration field with the help of a small set from the shell oscillations of natural shapes in two mutually orthogonal directions.

1) From the angular coordinate φ in sections IIx and IIIx, respectively, the displacement of the shell surface can be represented as

$$W_\xi(\varphi) = A_1\cos(\varphi + \Psi_1) + A_2\cos(2\varphi + \Psi_2) + A_3\cos(3\varphi + \Psi_3) + \cdots, \tag{7.22}$$

where

$$A_k = \sqrt{(A_k^c)^2 + (A_k^s)^2} \quad k = 1, 2, 3 \ldots$$

and

$$\varphi_k = \sin^{-1} A_k^s \Big/ \sqrt{(A_k^c)^2 + (A_k^s)^2}$$

Substitution of the spatial coordinates of vibration transducers into the preceding expression and the vibrations $W_\xi(\varphi_i)$ into the left-hand side of Eq. (7.22), we get two sets of linear equations in terms of cosine and sine functions, respectively, as follows:

$$W_\xi(\varphi_i) = A_m^c \cos m\varphi_i \qquad \text{and} \qquad W_\xi(\varphi_i) = A_m^s \sin m\varphi_i \tag{7.23}$$

where i is an integer and $i = 1, 2, 3 \ldots n$.

Having solved each set of the preceding equations, we obtained the contribution from the acoustic field to the natural mode shapes of the shell vibrations in each of the sections. Calculations of such type can be easily made by means of standard programs.

2) With the values of vibration amplitudes in each of sections IIx and IIIx of the shell surface, we can plot the envelope along the flow-passage direction of the combustion shell. In this case, the transducers located along the normal direction record the vibratory displacements of the shell as a whole on fastening girder B1–B4 and on girder and shell deflections B5–B10.

Separation of vibrations is made possible with the use of n measurements of the vibration field along the fixed flow-passage direction on the shell and the solution of the resultant equations:

$$W(\xi) = C_{n-1} + C_n\xi_i + \sum_{n=1}^{n-2} C_n\omega_n\,(\xi_i) \qquad (7.24)$$

where $C_i (i = 1, 2, \ldots, n-2)$ are the coefficients corresponding to the amplitudes of vibrations with the natural mode shapes of the shell $\omega_i(\xi_i)$; C_{n-1} and C_n the coefficients characterizing the displacement and turning of the shell for the measured signals; and $W(\xi_i)$, $i = 1, 2, \ldots n$ the vibratory displacements recorded by vibration transducers at the frequency under consideration at points ξ_i.

Because in practice acoustic oscillations occur with a limited number of modes (f_{1T}, f_{1T1L}, f_{1L}, etc.), identification of pressure pulsations in a combustion chamber can be simplified accordingly. To analyze the shell vibration field along the flow-passage direction, it is normally sufficient to install vibration transducers in three sections (see Fig. 7.5). Ix transducers record vibrations of combustor attachments and mixing head in two mutually orthogonal directions. IIx and IIIx transducers record the changes of the vibration field along the flow-passage direction and at different circumferential locations. In the latter case, it is sufficient to use three vibration transducers in each section to reconstruct the pressure pulsation field with the frequency f_{2T}.

The preceding procedure, along with the use of combustor-shell dynamic characteristics and Eqs. (7.21–7.24), allows identification of the mode shapes of pressure pulsations on the basis of vibration measurements.

III. Methods for Simulating Acoustic Pressure Oscillations in Combustion Chambers

Acoustic simulation in the study of high-frequency oscillations in LREs is used for experimental determination of resonant frequencies and oscillation decrements, as well as the field of standing waves in a combustor, by conducting measurements on a physical model.

The acoustic simulation involves the following steps:

1) Study the design features of combustion-chamber (or gas-generator) and thermodynamic properties of combustion products. The following data are found in the design documentation: a) engine diagram, which can have an open loop (liquid–liquid system) or closed loop with second burning of generator gas (gas–liquid system); b) types of oxidizer and fuel; c) mass flow rates of oxidizer G_o and fuel G_f; d) ratio of oxidizer and fuel flow rates, $K_m = G_o/G_f$; e) nominal pressure in the combustion chamber P_{ch} and P_{gg}; f) operating ranges for pressure and propellant mixture ratio; g) design of injectors and their pressure differentials; h) geometrical dimensions of combustors, with diameter D_{ch}, length of the cylindrical part L_c, diameter of nozzle throat D_{cr}, length of the subsonic part of nozzle L_n, and nozzle converging angle $2\alpha_n$; and i) presence and design features of pulsation baffles and resonant acoustic absorbers.

The thermal and physical properties of combustion products, such as density ρ and adiabatic exponent γ, can be determined using thermodynamic analyses. The speed of sound c is calculated using either Eq. (7.1) or (7.2). The Mach number of the mean flow M is estimated using Eq. (7.13).

2) Develop a mathematical model qualitatively describing the dependence of resonance frequencies on geometrical dimensions of an actual combustion chamber and on thermodynamic properties of combustion products. Similarity coefficients for combustion chambers of the same kind should also be established.

The construction of a mathematical model is preceded by the definition of the simulation task. This can include, for example, determination of the following parameters and effects: a) resonant frequency and standing wave field for a particular oscillation mode; b) resonant frequency spectrum and standing wave fields for a specified oscillation frequency range; c) effects of pulsation-baffle and acoustic-absorber design features on resonant frequencies and oscillation decrements; and d) effects of injector flow characteristics on the spectral properties of the unsteady pressure amplitude and other related quantities.

The boundary conditions are formulated using the impedance function $Z = P'/V'$ on the surfaces enclosing the chamber volume.

Using Eq. (7.3) for calculating the velocity potential and applying Eqs. (7.14) and (7.15), we obtain the equations for resonant frequencies as functions of linear dimensions of the combustion chamber, sound velocity in the combustion products, and the boundary conditions.

Generally, a model combustion chamber (gas generator) differs from actual ones by dimensions. The gaseous medium in a model combustor is as a rule air under normal conditions ($P_m = 0.1$ MPa and $T_m = 290$ K). It is seen from the structure of Eqs. (7.14) and (7.15) that the ratio

$$\alpha_L = \frac{c_a L_m}{c_m L_a} \tag{7.25}$$

can be taken as a scaling factor for longitudinal oscillation modes, and

$$\alpha_R = \frac{c_a D_m}{c_m D_a} \tag{7.26}$$

for transverse modes, where c_a, L_a, and D_a are the sound velocity, the chamber length, and diameter of an actual combustor, respectively. Here c_m, L_m, and D_m are the corresponding quantities for a model chamber.

3) Construct a physical model simulating the acoustic behavior of an actual combustion chamber.

The model should, as a rule, be geometrically similar to an actual combustor and have the same acoustic properties of the walls. The dimensions of a physical model are chosen for convenience of making experimental measurements. If it is necessary to reproduce the actual oscillation frequencies on the model, the geometrical dimensions are so chosen that the frequencies in the model and in the actual unit are equal, that is, $f_m \approx f_a$. Holes are made on the walls of the physical model for installing microphones and a source of oscillation excitation in the model cavity.

4) Make experimental measurements on a physical model of the combustion chamber.

Experimental measurements on physical models supplement and sometimes replace numerical and analytical predictions of the acoustic characteristics of an actual combustion chamber. The purpose of measurements is three-fold: a) acquisition

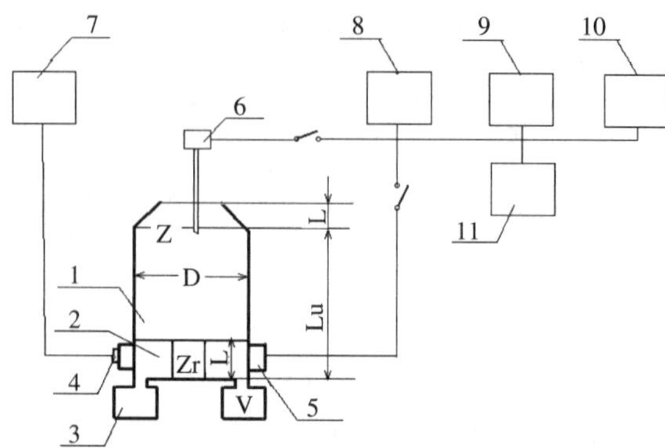

Fig. 7.6 Diagram of the combustion-chamber model and measurement system: 1, combustion chamber model; 2, vibration baffles; 3, acoustic absorber; 4, source of excitation (loudspeaker); 5, stationary microphone; 6, microphone with a wave guide; 7, source of sine voltage; 8, voltmeter; 9, oscillograph; 10, frequency meter; and 11, personal computer.

of preliminary information about the resonance frequencies and oscillation decrements of different modes at the design stage of a combustion chamber; b) dentifying resonant (natural) frequencies of the combustion chamber from the pressure-pulsation spectrum obtained during fire tests; and c) identifying the physical nature of pressure oscillations at different frequencies during recorded fire tests.

These objectives are achieved by measuring resonance frequencies and pressure oscillation decrements, as well as the acoustic fields of standing waves, in the model cavity.

One possible version of a physical model of a combustion chamber and the associated measurement system is shown in Fig. 7.6. The resonant frequencies of the model are obtained by exciting oscillations from an external source with varying frequency ω. When the excitation frequency ω coincides with one of the resonant frequencies of the model cavity ω_r, the amplitude of excited oscillation reaches the highest value.

After noting the frequency ω_r, at which the amplitude of pressure oscillations attained a maximum, the distribution of pressure pulsations in the model cavity is determined with the help of a microphone with a wave guide (see Fig. 7.6). By comparing the measured distribution of acoustic oscillation with the velocity potential described by Eq. (7.3), the mode of excited oscillations is identified. The oscillation decrement is determined from the amplitude characteristic (see Fig. 7.7) or from the measurement of damped pressure oscillations after the loudspeaker has been switched off (see Fig. 7.8).

In the first case illustrated in Fig. 7.7, the oscillation decrement is calculated from the formula

$$\delta T = \pi \frac{\omega_2 - \omega_2}{\omega_r} \tag{7.27}$$

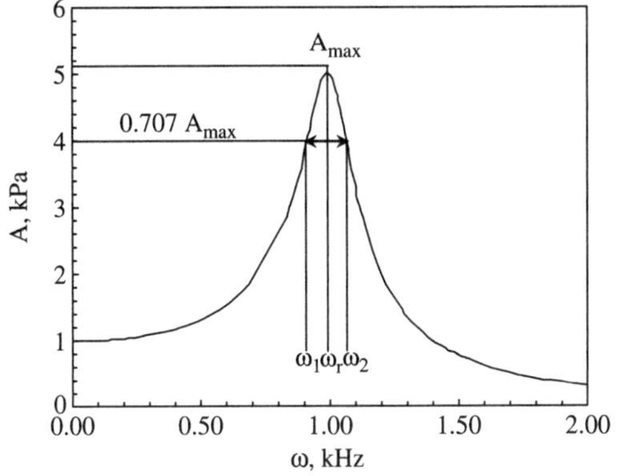

Fig. 7.7 Determination of the oscillation decrement from amplitude characteristic.

In the second case illustrated in Fig. 7.8, the oscillation decrement is calculated from the formula

$$\delta T = \frac{1}{n-1} \ln \frac{A_1}{A_n} \tag{7.28}$$

5) Analyze of measurement data obtained on a physical model of combustion chamber and its scaling to actual conditions.

The analysis of measurement data consists in plotting frequencies, amplitudes, and decrements of pressure oscillations as functions of the parameters of concern, for example, the length of pulsation baffles.

Recalculation of the resonance frequencies measured on a model in terms of actual conditions is made using Eqs. (7.25) and (7.26), depending on the specific oscillation mode of interest. The oscillation modes characterized by a certain field of standing waves in a model cavity are identical for a model and an actual unit.

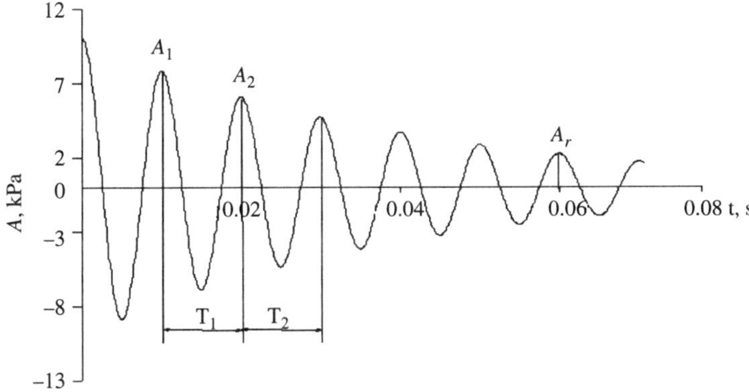

Fig. 7.8 Determination of oscillation decrement from the recording of decaying natural oscillations.

IV. Effects of Combustion-Chamber and Nozzle Configurations on Stability Characteristics

The most commonly used combustion chamber has a cylindrical shape smoothly transiting to a conical shape in the subsonic part of the nozzle. The changes of combustion-chamber and nozzle shapes can be reflected on the acoustic characteristics of the chamber in the following areas: 1) frequencies of natural oscillation modes, 2) distributions of standing waves for the same oscillation mode, 3) oscillation energy loss caused by radiation from the nozzle, 4) energy stored in the standing wave field, and 5) Mach number of the mean flow.

These changes usually take place simultaneously and can exert different effects on combustion stability. In some cases, the collective effect of the changes improves the stability, but in other cases an opposite trend occurs [10]. As an example, let us consider the measurement data obtained on two model chambers with excitation of the first tangential-longitudinal mode. The first chamber is cylindrical and has the dimensions of $D_{ch} = 115$ mm, $L_c = 148$ mm, $L_n = 50$ mm, $D_{cr} = 57.5$ mm, and $L_\Sigma = 198$ mm. The second chamber was conical with an angle of 15 deg. The dimensions are $D_{ch} = 115$ mm at the head end, $L_{ch} = 162$ mm, $L_n = 36$ mm, $D_{cr} = 57.5$ mm, and $L_\Sigma = 198$ mm.

The measured resonance frequency in the cylindrical chamber was 1802 Hz. It became 1911 Hz in the conical chamber with a 6% increase. The decrement of oscillations in the conical chamber also increased from 0.01 (for the cylindrical chamber) to 0.03. Such an enhanced decrement is caused by the increased radiation loss and by the reduced energy stored in the standing-wave field. Figure 7.9 shows the distributions of the pressure oscillation amplitudes along the axes of both chambers in millivolts.

As seen from Fig. 7.9, the oscillation amplitude at the exit of the conical combustion chamber is lower than that in the cylindrical one. Thus, the radiation loss in the conical chamber should be lower than its counterpart in the cylindrical chamber. However, because the oscillation decrement in the conical chamber is higher than that in the cylindrical one, the observed phenomenon is apparently concerned with the reduction in the energy stored in the standing wave field; the

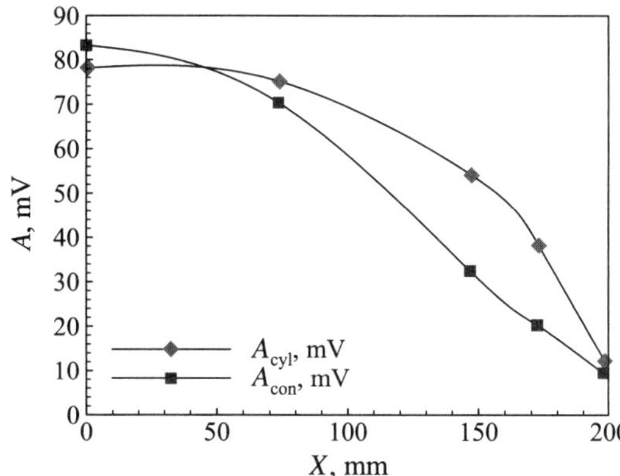

Fig. 7.9 Amplitude variation along the length of cylindrical and 15-deg conical chambers.

oscillation amplitude declines more steeply toward the nozzle in the conical chamber.

The net effect of applying a conical configuration can only be estimated by properly accounting for the mixing and combustion process sensitivities to the amplitudes and frequencies of pressure oscillations, as well as to the Mach number of the mean flow. When adopting a conical shape for the combustion chamber, the change of resonance frequency will also affect the stability characteristics, if the combustion sensitivity is frequency dependent.

Results of firing tests of a combustion chamber with a conical section near the injector head that expands to the cross-sectional area of the nozzle entrance ran are reported in [10]. Two different cone angles, 30 and 60 deg, were tested. The chamber was operated on oxygen and hydrogen propellants. The stability threshold was assessed by the hydrogen temperature. The higher the hydrogen temperature is at the stability boundary, the more stable is the system. It has been found that a conical section diverging from the injector head improves the combustion stability. This phenomenon can be attributed to the following two factors: decreased pressure oscillation phenomenon at the injector head and reduced energy stored in the standing wave field.

The decrease in the nozzle convergence angle with a fixed throat diameter improves the stability of longitudinal oscillations, whereas the effect of the convergence angle on mixed longitudinal-transverse modes is uncertain [10]. The increase in the nozzle throat diameter with a fixed combustor diameter results in an increased Mach number and improves the stability [10].

V. Acoustic Characteristics of Combustion Chambers with Vibration Baffles

Vibration baffles proved to be an effective means for preventing hard excitation of pressure oscillations in combustion chambers [10]. In addition, baffles limit the magnitudes of pressure oscillations in soft self-excitation of self-sustained oscillations.

Baffles are usually a set of plates mounted on the injector face plate of a combustion chamber. Tested baffles can have a wide variety of layouts [10]. Baffles in the form of six radial plates, which spread from the center of the injector head or from a circular ring with a diameter of about 40% of the chamber diameter, have the widest application.

Vibration baffles have long been used [1], but a well-rounded explanation of the mechanisms for their stabilizing effects has not been found and probably does not exist.

The studies of physical models of combustion chambers with vibration baffles have shown the following changes in the chamber acoustics: 1) resonant frequencies of mixed longitudinal-transverse modes, in particular when there occur resonant frequencies lower than those of pure transverse modes determined by Eq. (7.11); 2) fields of standing waves and hence the energy stored in oscillatory systems; 3) distributions of oscillatory velocities; 4) acoustic energy losses caused by radiation from the nozzle and viscous losses caused by the flow turning around the baffle edges and the gas flow between baffle compartments; and 5) presence of sources of self-excited oscillations of aerodynamic nature when the mean flow is convected downstream around the baffle plate edges (i.e., vortex shedding). Apparently, in some particular cases, one or several of the preceding factors can prevail. There is thus most probably no unambiguous explanation of the effects of vibration baffles on combustion stability.

Several studies [4 and 10] were performed to derive equations for calculating resonant frequencies lower than those of pure transverse modes calculated by Eq. (7.11). The following assumptions have been used when deriving these formulas: 1) the resonant frequency ω is lower than the frequency of purely transverse mode ω_{mn}; 2) purely longitudinal oscillations are realized in the compartments between baffle blades; 3) thermophysical properties of gases are the same in the baffle compartments and main combustion chamber; 4) baffle blades are infinitely thin; 5) impedance of the injector face plate is infinitely high (i.e., $Z_g \to \infty$); 6) there are no slots between the combustor head end, lateral walls, and baffle blades; and 7) the Mach number of the mean flow in baffle compartments is small (i.e., $M^2 \ll 1$).

As an example of using the acoustic equations (7.1–7.19) given in Sec. I, let us derive a design formula for calculating resonant frequencies with $\omega < \omega_{mn}$. The assumptions 3 and 4 are removed. Thus, 1) the gas density ρ_b and sound velocity c_b in baffle compartments are not equal to their counterparts of ρ_{ch} and C_{ch} in the main chamber; and 2) the thickness of baffle blades is finite, and the cross-sectional area of baffle compartments S_b is less than the cross-sectional area of the combustion chamber S_{ch}.

Using Eq. (7.3) for the velocity potential and Eq. (7.4) for the acoustic pressure and velocity, we get

$$P' = \rho \frac{\partial \Phi_{mn}}{\partial t} \propto i\rho\omega\Phi_{mn}$$

$$V' = \frac{\partial \Phi_{mn}}{\partial x} \propto \sqrt{V_{mn}^2 - k^2 \Phi_{mn}}$$

where $\propto$ is the sign of proportionality. The mechanical impedance of the combustion chamber section at the baffle edges is denoted by Z_{ch} [33]:

$$Z_{\text{ch}} = \frac{S_{\text{ch}} P'}{V'} = \frac{i\rho\omega S_{\text{ch}}}{\sqrt{V_{mn}^2 - k2}} = i\,\frac{S_{\text{ch}}\rho_{\text{ch}} C_{\text{ch}} \overline{\omega}_{mn}}{\sqrt{1 - \overline{\omega}_{mn}^2}}$$

where $\overline{\omega} = \omega/\omega_{mn}$ is the normalized frequency of mode (m, n).

Using assumption 5 and Eq. (7.9) for the transposed impedance, we get the impedance at the exit section of baffles

$$Z_b = -iS_b\rho_b C_b \cot\left(\frac{2v_{mn}\overline{L}_b\overline{\omega}}{\alpha_b}\right)$$

where $\overline{L}_b = L_b/D_{\text{ch}}$ is the normalized length of baffles and $\alpha_b = c_b/c_{\text{ch}}$ is the ratio of sound velocities in baffle compartments to that in the main chamber, $\alpha_b \leq 1$.

Transforming the combustion-chamber impedance from the chamber area S_{ch} to the area of baffle compartments S_b and using Eq. (7.16), we get the formula for calculating resonant frequencies:

$$\overline{L}_b = \frac{\alpha_{\text{b}}}{2V_{mn}\overline{\omega}_{mn}} \cot^{-1}\left[\frac{S_{\text{ch}}\rho_b c_b}{S_b \rho_{\text{ch}} c_{\text{ch}}} \frac{\sqrt{1 - \overline{\omega}_{mn}^2}}{\overline{\omega}_{mn}}\right] \tag{7.29}$$

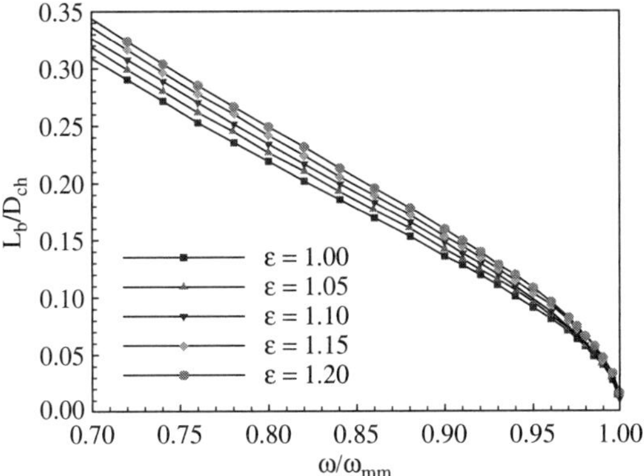

Fig. 7.10 Normalized baffle length L_b/D_{ch} vs frequency ω/ω_{mn} with different blockage of chamber section by baffles; first tangential-longitudinal mode.

Figure 7.10 shows the normalized baffle length L_b/D_b as a function of the normalized frequency $\overline{\omega} = \omega/\omega_{mn}$ for the mixed first tangential-longitudinal mode ($\overline{\omega} < 1$) with different blockages of the combustion-chamber area by baffle blades. The parameter ε is defined as $F_{ch}/(F_{ch} - F_b)$, with F_{ch} being the combustor cross-sectional area and F_b the total area of the baffles. The gaseous medium in baffle compartments and the main chamber is assumed to be identical, that is, $\alpha_b = 1$, and $\rho_b c_b = \rho_{ch} c_{ch}$. The figure indicates that with the baffle length fixed an increase in the blade thickness within the specified limits results in an insignificant increase of resonant frequencies.

The difference of sound velocity between the baffle compartments and the main chamber (or the gas generator), as evidenced from Eq. (7.29), results in a considerable difference between the $\overline{L}_b - \overline{\omega}$ relation. Figure 7.11 shows the

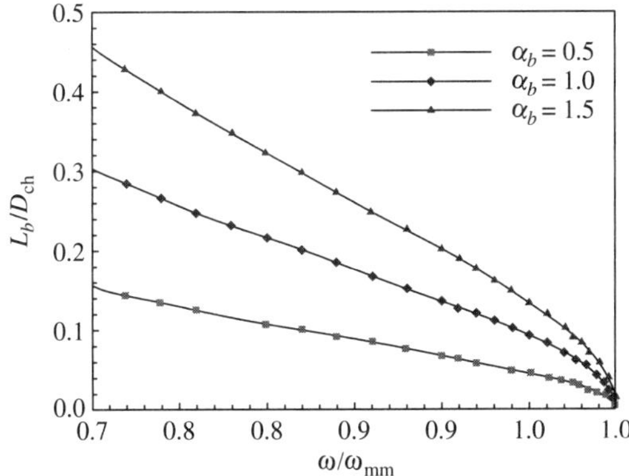

Fig. 7.11 Normalized baffle length vs frequency for the first mixed tangential-longitudinal mode with different ratios of sound velocities in baffle compartments and main chamber; thin baffle blades, $\varepsilon = 1$.

normalized frequency of the mixed first tangential-longitudinal mode for thin baffles with three different ratios of sound velocities $\alpha_b = c_b/c_{ch} = 0.5$, 1.0, and 1.5. The value of $\alpha_b < 1$ is typical for combustion chambers, whereas $\alpha_b > 1$ for gas generators with a two-zone combustion system.

It is seen from plots in Fig. 7.11 that the ratio of sound velocities in the baffle compartments and in the rest of the combustion chamber (gas generator) has a significant effect on the resonance frequencies of the tangential-longitudinal oscillation modes.

At the inlets of baffle compartments (i.e., the interface between the baffle compartments and main chamber), the amplitude of pressure oscillation varies across the radius and around the periphery of the combustion-chamber. The radial distribution of the pressure oscillation of the first tangential mode is shown in Fig. 7.12, whereas the azimuthal distribution around the combustion chamber periphery is shown in Fig. 7.13. This kind of nonuniformity, however, is reduced when a wave propagates deeply into the baffle compartments, because the natural frequencies of transverse modes inside baffle compartments are much higher than the incident wave frequency [see Eq. (7.3) for the velocity potential]. The measurements on a model combustion chamber confirm this conclusion, as evidenced in the data shown in Fig. 7.14. The gradient of the pressure pulsation amplitude dA is the difference of the amplitudes measured on the mixing head surface at two farthest points within the same baffle cell. In a physical sense, the pressure gradient is proportional to the potential energy of the gas mass and is equivalent to the velocity oscillation. A decrease in the pressure gradient corresponds to a reduction of the velocity pulsations along the head surface, thereby improving the combustion process stability in relation to high-frequency transverse pressure oscillations.

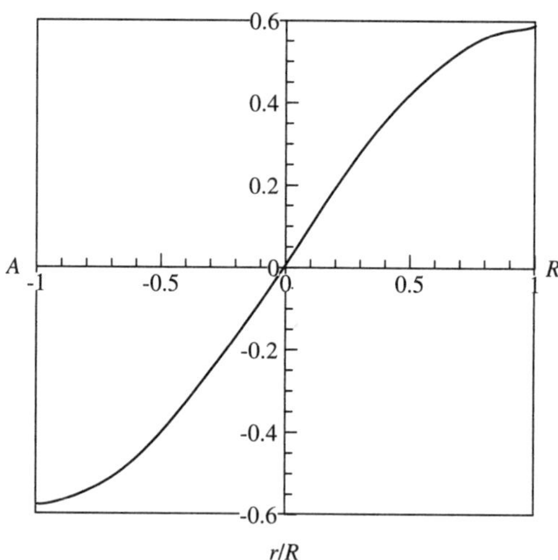

Fig. 7.12 Distribution of pressure oscillation amplitude of the first tangential mode across the combustion-chamber radius.

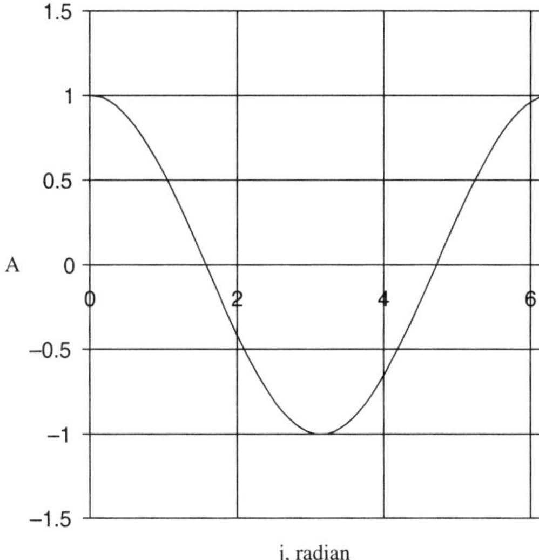

Fig. 7.13 Distribution of pressure oscillation amplitude of first tangential mode around the chamber periphery.

However, owing to the nonuniformity of pressure oscillations at the inlet of each separate compartments, even at the injector head bottom, there is a pressure gradient between separate compartments. Thus, additional damping of oscillations resulting from the gas flow between separate compartments can be obtained by making small slots between the compartments or by using porous plates for designing the baffles. The same mechanism of damping oscillations is realized when using the baffles formed by injectors extended into the combustion chamber.

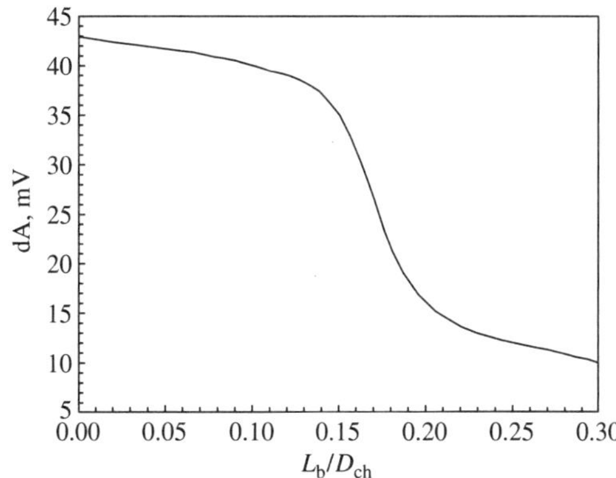

Fig. 7.14 Effect of baffle length on measured gradient of pressure oscillation amplitude of the first tangential mode at the bottom of baffle compartment.

The dimensions of slots and porosity of baffles are chosen experimentally because of the lack of reliable numerical methods.

Vibration baffles used in practice are not long, and baffle compartments communicate with each other through slots, such as in the baffles formed by extended injectors. Therefore, the reduction of resonance frequencies of mixed longitudinal-transverse modes with such baffles is not so great compared with situations without baffles.

Figure 7.15 shows the measured oscillation frequency and decrement as a function of the length of baffles formed by round pins with a diameter of 10 mm and a clearance of 0.5 mm between them. The frequencies calculated by Eq. (7.29) are also shown. The frequency of the mixed first tangential-longitudinal mode changed by a few percents, whereas the oscillation decrement increased by almost four times.

Equation (7.29) for calculating the resonance frequencies of mixed longitudinal-transverse modes, which are lower than the frequencies of pure transverse modes, was obtained under the assumption that the cross-sectional dimensions of baffle compartments are much smaller than the diameter of the combustion chamber, that is, the baffles are thick. Table 7.3 shows the measured resonance frequencies of the mixed first tangential-longitudinal oscillation mode with different numbers of baffles having a relative length of 0.2. With an increase in the number of baffles, the measured data approach the calculations using the equation for thick baffles. The result implies that the assumption on the existence of pure longitudinal oscillations in baffle compartments becomes more valid with increasing number of blades.

Gaseous combustion products flow around the edges of vibration baffles, downstream of which shear layers form, along with steep velocity gradients. These shear layers are hydrodynamically unstable and, when interacting with the acoustic field in the chamber, can excite pressure oscillations with finite amplitudes at the frequency of the longitudinal-transverse mode. Such self-excited pressure

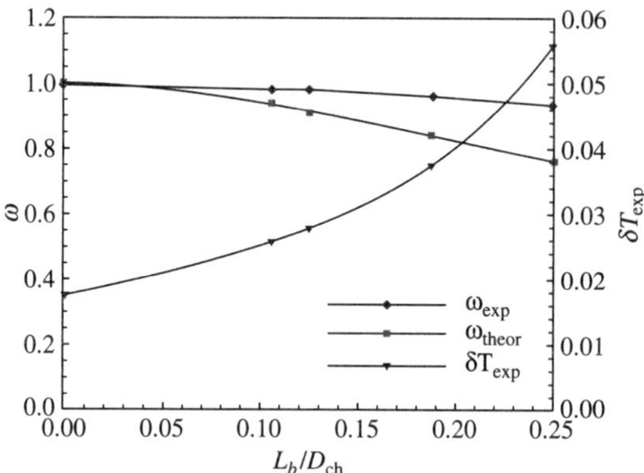

Fig. 7.15 Frequency and oscillation decrement as function of the length of baffles formed by round pins with a diameter of 10 mm and with a clearance between them of 0.5 mm.

Table 7.3 Normalized resonance frequency of first tangential-longitudinal mode in pipe with different number of baffles having a relative length of 0.2

Number of baffles, n	ω_{exp}	$\overline{\omega}_{theor}$	$\overline{\omega}_{exp}/\overline{\omega}_{theor}$
1	0.907	0.822	1.103
2	0.894	0.822	1.088
3	0.872	0.822	1.061
4	0.860	0.822	1.046
5	0.854	0.822	1.039
6	0.838	0.822	1.019
7	0.832	0.822	1.012

oscillations are observed at the Strouhal numbers of 0.12 … 0.25, defined as $Sr = f\delta/U$, where f is the frequency, Hz; δ the thickness of baffle plates, m; and U the gas flow velocity, m/s.

VI. Resonance Absorbers

A resonant absorber is represented by a cavity with a volume V communicating with a combustion chamber through a channel, as shown schematically in Fig. 7.16. The resonant frequency ω of an absorber is chosen to be 10–20% higher than the frequency of pressure oscillation in the combustion chamber. A set of uniformly arranged resonance absorbers is normally installed on the combustion chamber.

A cavity having cross-sectional area S_v and length L_v forms the volume V of a resonator. The walls are acoustically rigid. The resonator volume communicates with the combustion chamber through a channel with a diameter d_0, a length L_0, and a cross-sectional area S_0. In general, a gaseous medium filling the resonator volume and the channel can have different density and sound velocity. Let us denote the density and sound velocity in the resonator volume as ρ_v, c_v and those in the gas filling the channel (resonator throat) as ρ_0 and c_0, respectively.

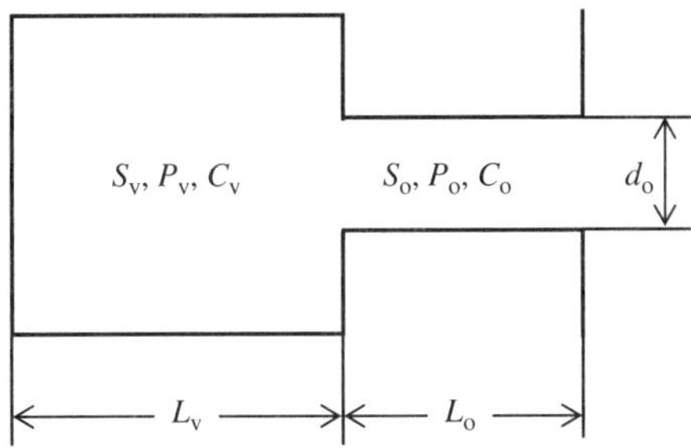

Fig. 7.16 Schematic diagram of a single resonance absorber.

Using Eq. (7.9) for the transposed impedance and transforming the throat impedance from area S_0 to area S_v, we equate the impedance immediately upstream and downstream of the interface, taken with the opposite sign,

$$i\frac{S_v^2}{S_o^2}S_o\rho_o c_o \tan\left(\frac{\omega}{c_o}L_o\right) = iS_v\rho_v c_v \cot\left(\frac{\omega}{c_v}L_v\right)$$

Thus, we get the equation for calculating the absorber resonant frequency with due account for the differences of thermophysical properties of gases in the cavity and the throat:

$$\tan\left(\frac{\omega}{c_o}L_{ef}\right)\tan\left(\frac{\omega}{c_v}L_v\right) = \frac{S_o}{S_v}\frac{\rho_v c_v}{\rho_o c_o} \tag{7.30}$$

where $L_{ef} = L_o + 0.8\,d_o$ is an effective length of the channel taking into account the added mass [30].

If the gaseous medium in a resonator is uniform (i.e., $\rho_o = \rho_v$ and $c_o = c_v$) and arguments of the tangent functions are small (i.e., $\omega L_o/c_o \ll 1$; $\omega L_v/c_v \ll 1$), the tangents can be replaced by their arguments. If these conditions are satisfied, Eq. (7.30) is transformed to the well-known Helmholtz formula

$$\omega = c\sqrt{\frac{S_o}{S_v L_v L_{ef}}} = c\sqrt{\frac{S_o}{VL_{ef}}} \tag{7.31}$$

Normally, this formula is used for calculating resonator frequencies without checking whether the just-mentioned conditions are met in a particular case. The oscillation frequency in hertz is connected with the relation $f = \omega/2\pi$ with the radian frequency calculated by Eq. (7.31).

If there were no losses in the resonator, pressure oscillations at the inlet of the resonator throat would cause unlimited growth of pressure oscillations in the resonator volume at resonant frequencies. Oscillation energy losses in a resonator consist of two parts: 1) internal loss by friction, vortex formation, and heat exchange; and 2) radiation loss from resonator throat.

In architectural acoustics, resonators are used wherein radiation losses predominate to provide good audibility in various areas of a big hall. If resonators are used for damping pressure oscillations in combustion chambers, dominance of the internal losses in resonators themselves is required. Such internal losses of oscillation energy primarily result from hydrodynamic resistance in the resonator throat. The procedure for calculating the optimum resistance at the resonator inlet is described in [30]. Hydrodynamic resistance is introduced into the analysis in the form of the real part of the transposed impedance at the inlet of a resonator $\overline{R}$.

With the aid of Eqs. (7.30) and (7.31), we can calculate, using Eq. (7.9), the transposed impedance normalized by $S_0\rho_0 c_0$ at the inlet of the resonator.

$$\overline{Z}_p = \overline{R} + i\left[\tan(k_0 L_{ef}) - \frac{S_o\rho_v c_v}{S_v\rho_o c_o}\cot(k_v L_v)\Big/1 - \frac{S_o\rho_v c_v}{S_v\rho_o c_o}\tan(k_o L_{ef})\cot(k_v L_v)\right] \tag{7.32}$$

If the gaseous medium is uniformly distributed in the resonator cavity and throat and the arguments of the tangent and cotangent functions are small (i.e., $kL << 1$), Eq. (7.32) is simplified to the following:

$$\overline{Z}_p = \overline{R} + i kL_{\text{ef}} - \frac{S_0}{S_v k L_v} \Big/ 1 - \frac{S_0 L_{\text{ef}}}{S_v L_v} \tag{7.33}$$

The absorption of a planar acoustic wave at normal incidence to a plane with uniformly distributed acoustic absorbers was studied in [30]. The transposed impedance of resonators can be expressed in terms of area S of the plane with just one resonator.

It has been found that the highest absorption coefficient at resonant frequency α_m is determined by a dimensionless active impedance $\overline{R}$:

$$\alpha_m = \frac{4\overline{R}}{(\overline{R} + 1)^2} \tag{7.34}$$

Two values of active impedance correspond to the highest absorption coefficient. They are calculated from the equation rearranged from Eq. (7.34):

$$\overline{R}^2 - 2\frac{2 - \alpha_m}{\alpha_m}\overline{R} + 1 = 0 \tag{7.35}$$

The two roots of this equation $\overline{R}_1$ and $\overline{R}_2$ are related by $\overline{R}_1\overline{R}_2 = 1$. One of them is greater than unity, and the other is smaller than unity. A flat distribution of the amplitude characteristic corresponds to the lower value of $\overline{R}$, whereas a steep amplitude characteristic corresponds to the higher value. A resonator damps oscillations over a wider frequency range in the first case than in the second case.

The operating conditions of resonance absorbers in an actual combustion chamber (or gas generator) significantly differ from those considered in [33], in which Eqs. (7.34) and (7.35) are derived, for the following reasons:

1) Resonant absorbers are part of the overall acoustic system that also include the combustion chamber (or gas generator). Boundary conditions by which the resonant frequency of the acoustic system are determined depend on the effects of absorbers.

2) Acoustic oscillations in a combustion chamber (or gas generator) with absorbers are represented by a system of acoustic (standing and/or spinning) waves incident to the interface with absorbers at random angles and not just in the normal direction.

3) The amplitude of pressure oscillations in combustion chambers (gas generators) is several orders of magnitude larger than the value within the validity of linear acoustics.

4) The gaseous medium in a resonator cavity can be remarkably nonuniform in thermal and physical properties.

5) Mean flows are convected through the throats of resonance absorbers. In some cases, there are even continuous plug flows for cooling.

The presence of a flow traveling through and around the throat not only changes the value of active impedance [4 and 10], but also causes excitation of pressure oscillations in resonators [10 and 37]. The howling of sounding soft toys when compressed are made based on this principle.

In [37], the condition for excitation of oscillations in a resonator by a gas stream flowing through the throat is formulated in terms of the Strouhal number.

$$Sr = \frac{f_n d}{Un} \tag{7.36}$$

where f_n is the frequency of the nth harmonic, Hz; d, the diameter of throat m; U, the flow velocity, m/s; and $n = 1, 2\ldots$ is the harmonic number. The largest amplitude of pressure pulsations in a resonator volume is attained on the first harmonic ($n = 1$, $Sr \approx 0.65$) [37].

So far, a great amount of information on the efficiency of a resonant absorber has been accumulated [4 and 10]. Results are summarized as follows:

1) The absorption coefficient α of a resonator depends on the amplitude of pressure oscillation [4]. If a resonator is adjusted to its maximum absorption (i.e., $\alpha = 1$) using linear acoustic approximation, the absorption coefficient decreases at large-amplitude oscillations. Figure 7.17 shows the coefficient of absorption α as a function of frequency at different levels of pressure-oscillation amplitudes, taken from [4]. As seen from the figure, for $\alpha < 1$ based on linear acoustics, the absorption coefficient α increases with increasing amplitude of oscillation. The trend is reversed after the maximum α is attained.

2) The absorption coefficient α also depends on the ratio of the throat area to the chamber head-end area, that is, the permeability ε. Figure 7.18 shows the real part of impedance at the resonance frequency as a function of sound intensity for different resonator permeability ε. With an increase in permeability, the real part of impedance and hence the absorption coefficient decrease, as indicated by Eq. (7.34).

3) Resonant absorbers should be installed at the pressure crests (i.e., acoustic pressure antinode points) of the oscillation mode to be damped [10]. For all modes, the pressure crests are located near the head end of the combustion chamber. Therefore, it is advisable to install resonators in the vicinity of the mixing head.

4) The change of resonant frequencies in combustion chambers after installation of resonant absorbers was noted both on cold-flow models and in firing tests of

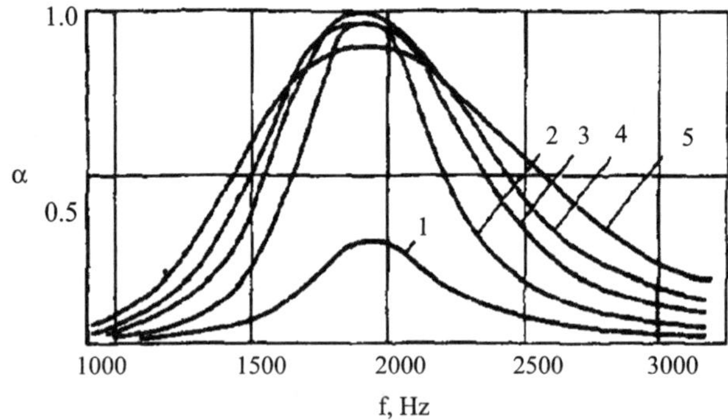

Fig. 7.17 Absorption coefficient vs frequency: 1, linear approximation; 2, $p'/\bar{p} = 0.01$; 3, $p'/\bar{p} = 0.03$; 4, $p'/\bar{p} = 0.05$; and 5, $p'/\bar{p} = 0.1$.

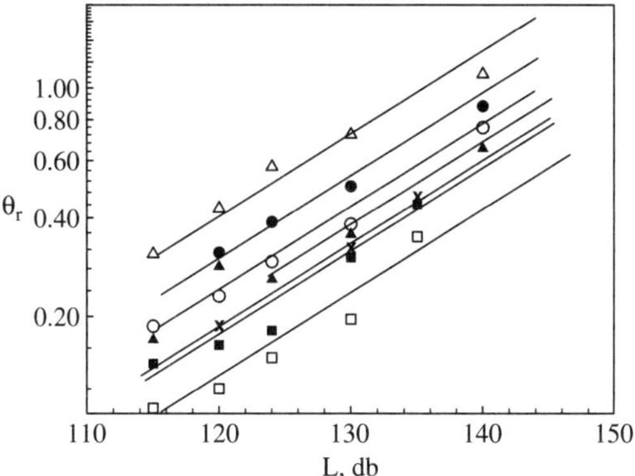

Fig. 7.18 Real part of impedance at resonance frequency vs amplitude of acoustic pressure with different permeability: o, $\varepsilon = 0.0379$; •, $\varepsilon = 0.0337$; ▲, $\varepsilon = 0.0526$; △, $\varepsilon = 0.032$; ×, $\varepsilon = 0.0672$; ■, $\varepsilon = 0.0673$ and □, $\varepsilon = 0.0865$.

combustion chambers [10]. In this situation, there exist additional resonant frequencies (so-called frequency bifurcation), and the acoustic field differs greatly from that in the original chamber without resonators.

Figure 7.19 shows the distributions of pressure amplitudes along a model combustor with different frequency adjustments of resonators for the first tangential-longitudinal mode. Installation of resonators resulted in bifurcation of the resonant frequency: one resonance occurs at a frequency higher than that of

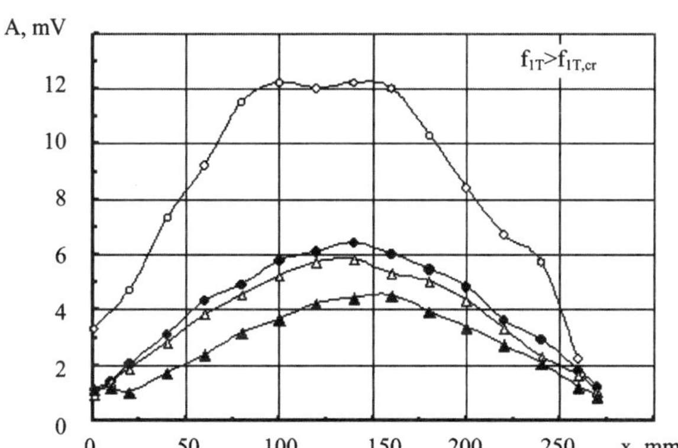

Resonance frequencies of absorber ($f_{abs} > f_{1T}$):			
o 1702 Hz $f_{101} = 1741$ Hz	• 1778 Hz $f_{101} = 1741$ Hz	△ 1865 Hz $f_{101} = 1755$ Hz	▲ 1966 Hz $f_{101} = 1756$ Hz
$\bar{f} = f_{abs}/f_{101} = 1702/1741 = 0.978$	$\bar{f} = 1.017$	$\bar{f} = 1.063$	$\bar{f} = 1.12$

Fig. 7.19a Distribution of pressure oscillation amplitude of the first tangential-longitudinal standing wave along a model combustion chamber.

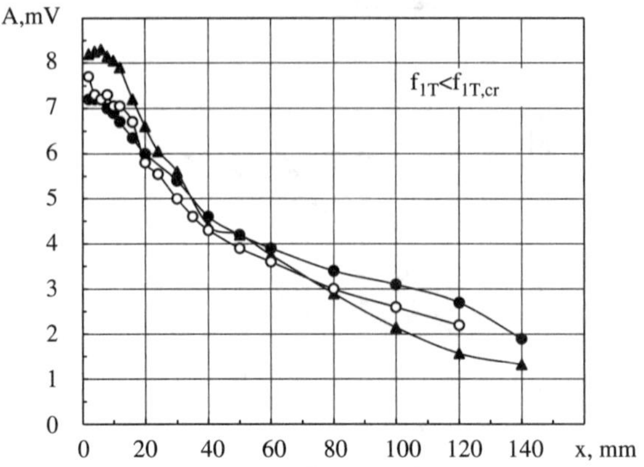

○ - 1702 Hz; V_{abs} = 7.17 cm^3	● - 1778 Hz; V_{abs} = 6.57 cm^3	▲ - 1966 Hz; V_{abs} = 5.37 cm^3
f_{100} = 1312.5 Hz	f_{100} = 1312.5 Hz	f_{100} = 1396.8 Hz
$\bar{f}=f_{abs}/f_{100}$ = 1.297	$\bar{f}$ = 1.331	$\bar{f}$ = 1.408

Fig. 7.19b Distribution of pressure oscillation amplitude of the first tangential-longitudinal standing wave along a model combustion chamber.

a pure tangential mode (see Fig. 7.19a), and the other occurs at a lower frequency (see Fig. 7.19b). The calculated critical frequency for a chamber of 121 mm in diameter at a temperature of 20°C is 1662 Hz:

$$f_{100}^{cr} = \frac{0.586 \times 343.4}{0.121} = 1662 \text{ Hz}$$

This agrees well with the experimental measurement of 1672 Hz for an acoustically closed chamber. Compared to the distribution of the pressure amplitude shown in Fig. 7.9 for the chamber without resonators, the acoustic field has changed greatly. Thus, it is impossible to identify unambiguously the mechanisms (such as additional damping only, or modifications of resonance frequencies and standing-wave fields) that are responsible for the effects of resonance absorbers.

5) Installation of resonant absorbers does not ensure absolute stability of operation process [10]. The application of absorbers can change the stability boundary for some governing parameters. It can change the level of pressure pulsations or might not give any positive effect, in particular, when the operation process has a tendency to hard excitation of pressure oscillations. Therefore it is advisable to use resonant absorbers in combination with vibration baffles.

VII. Adjusted Injectors

Tubes having a specific length and a contraction (i.e., a restriction) at the inlet are used as injectors for feeding gaseous propellants in combustion chambers for engines with a closed-loop system and secondary (staged) combustion of exhaust products from gas generators. Pressure oscillations in the combustion chamber excite certain oscillation modes in the gas duct upstream of the injectors. However, because the sound velocity in the combustion products from a gas generator is lower than that in the combustion chamber by a factor of two or even larger, the frequency of acoustic oscillation transmitted to the gas duct is considerably higher than the resonant frequencies of transverse modes in the gas duct. Therefore, the amplitude of pressure oscillations in the gas duct is much less than that in the combustion chamber [30]. The conditions at the inlet of injector channels can be considered acoustically open ($Z \approx 0$).

On the side of the combustion chamber, the total flow area of injectors is 0.1–0.3 of the cross-sectional area of the combustion chamber. In addition, the acoustic resistance ρc in the exhaust products from a gas generator is higher than that in the combustion chamber. Thus, the gas injector channels can be also considered to first approximation acoustically open at the interface with the combustion chamber. The gas-injector passages can be treated to first approximation as tubes with two acoustically open ends. According to Eq. (7.14), the resonance frequencies of such tubes depend on the sound velocity c and the effective length of injectors $L_{ef\,i}$.

$$f_{\mathrm{ni}} = \frac{nc}{2L_{\mathrm{efi}}} \tag{7.37}$$

where $L_{\mathrm{efi}} = L_i + 0.8\ d_i$ is the effective length of the injector, m, with L_i being the geometrical length of the injector, m, and d_i the injector diameter, m; c the sound velocity in the exhaust products from the gas generator, m/s; $n = 1, 2, \ldots,$ an integer number.

By choosing the length of injector channels based on the condition of equal frequency in the injector passage and the combustion chamber (i.e., $f_{ni} \approx f_{\mathrm{ch}}$), an additional damping of pressure oscillations in the combustion chamber can be obtained. In this case, injectors are considered as adjusted. The damping effect of adjusted injectors is attributed to energy losses resulting from the hydrodynamic resistances of a passage. Much of the hydrodynamic resistance is concentrated at the inlet (sudden contraction) and outlet (sudden expansion) of the passage, where the maxima of oscillating velocity are located.

The effect of injector passage length on the oscillation decrement for the first tangential-longitudinal mode is apparent even on a simple model combustion chamber without flow. The model represents a cylinder with a diameter of 160 mm. The cylinder is closed at the head end with a flat cover with vibration baffles made of pins (see Fig. 7.15) and also closed at the downstream end with a cover having a slot. A tube injector with a diameter of 12 mm was installed in a peripheral baffle compartment near the wall of the model chamber. The length of the tube was changed discretely. The tube was flush mounted on the cover, with the other end open to an open atmosphere at room conditions. The results of the measured oscillation decrement for the first tangential-longitudinal mode are shown in Fig. 7.20.

With an increase in the injector channel length from 94 to 124 mm, the effective length approaches the half-length of the acoustic wave, which in this case is 136 mm, and the oscillation decrement reaches its extreme value. Such a phenomenon

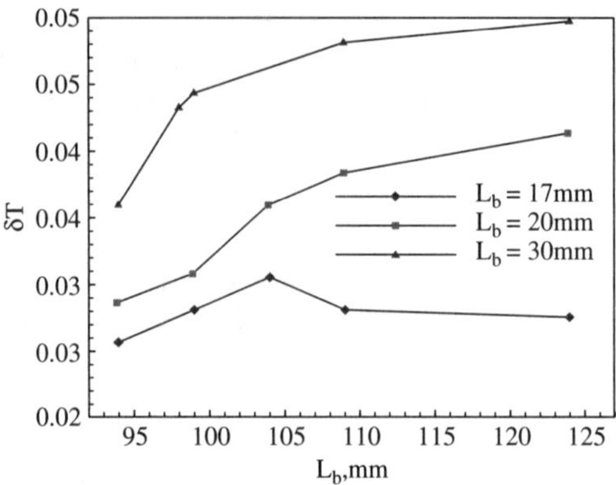

Fig. 7.20 Oscillation decrement on the first tangential-longitudinal mode vs the length of a single injector with different heights of baffles made of extended pins.

is observed with three variants of baffles. The oscillation decrement increases with increasing baffle length and injector passage length.

The damping effect of injector adjustment occurs in the case of external mixing of propellant components. If fuel is injected inside the injector passage, an opposite effect can be observed when the injection location is recessed from the exit section. The adjustment destabilizes the operation process.

VIII. Cold Zone

The propellant entering a combustion chamber is not burned instantaneously, and so a relatively cold zone is formed near the mixing head. Atomization, preheating, evaporation, and mixing of propellants take place in this zone. Let us denote the length of this zone as L_c and the density and sound velocity in this zone as ρ_c and c_c, respectively. We assume that the cold zone transits to a hot zone filled with combustion products (ρ_h, c_h) not smoothly but in a stepwise manner.

At the interface between the cold and hot zones, the normal oscillating velocity is continuous, and the tangential velocity can have a discontinuity. Using Eqs. (7.3), (7.7), (7.9), and (7.16), we obtain a relation between the length of the cold zone and the resonant frequency of the longitudinal-transverse mode, which is lower than the frequency of pure transverse mode in the hot zone.

$$L_c = \frac{1}{2\nu_{mn}} \frac{\rho_c c_c}{\rho_h c_h} \frac{\sqrt{1-\overline{\omega}_{mn}^2}}{\overline{\omega}_{mn}\sqrt{\overline{\omega}_{mn}^2/\alpha_v^2 - 1}} \tag{7.38}$$

where $\overline{L}_c = L_c/D$ is the relative length of the cold zone; ν_{mn} the nth root of Eq. (7.10); ρ_c, c_c, ρ_h, and c_h the density and sound velocity in the cold and hot zones, respectively; $\overline{\omega}_{mn} = \omega/\omega_{mn}$ the relative frequency; and $\alpha_v = c_c/c_h$ the ratio of the sound velocities in the cold and hot zones, respectively.

As an example, let us calculate the relationship between the relative resonance frequency $\overline{\omega}$ for the first tangential-longitudinal mode and the relative

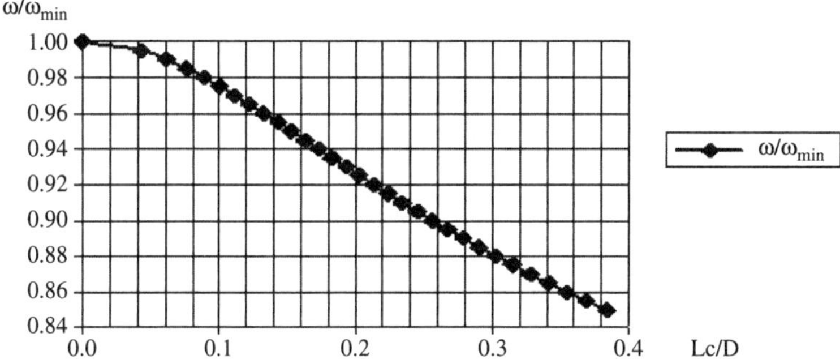

Fig. 7.21 Relative frequency vs relative length of cold zone for the first tangential mode.

length of the cold zone $\overline{L}_c$, using the following parameters for the cold and hot zones [10]: $\rho_c/\rho_h = 2$; $C_c/C_h = 0.7$; and $(\rho_c C_c)/(\rho_h C_h) = 1.4$. Figure 7.21 shows the calculated result. For a small length of the cold zone (less than 10% of the combustion-chamber diameter), the cold zone slightly affects the resonant frequency.

The length of a cold zone depends on the type and size of injectors, the spacing between injectors, and pressure differential across injectors. These parameters of the mixing head vary within a quite narrow range, for an operational combustion chamber.

In small-size combustion chambers, the relative length of the cold zone is larger than that in large combustion chambers, and its effect on resonant frequencies becomes more substantial.

For higher modes the value ν_{mn} increases (see Table 7.1). It is apparent from Eq. (7.38) that the length of cold zone has a greater effect on frequencies of higher oscillation modes.

In amplitude spectra of pressure pulsations recorded in firing tests, the frequency components are generally less than the frequencies calculated using the sound velocity determined from Eqs. (7.1) and (7.2). This can be attributed to the presence of the cold zone near the combustion-chamber mixing head.

IX. Absorption by Porous Bottom

For systems with hydrogen and oxygen as propellants, the combustion zone is arranged very close to the mixing head. For reliable cooling of the combustor head end, it was proposed to employ a porous bottom and feed some hydrogen to the combustion chamber through pores or of small holes. A technical problem arose as to how to make a pre-injector cavity and how to select an appropriate pressure differential such that, besides cooling, an additional damping of pressure oscillations in the combustion chamber could be obtained.

Let us assume that the injector face-plate thickness of the combustor is small compared to the length of the acoustic wave in hydrogen, such that the inertial properties of gases in pores can be ignored. By linearizing the velocity dependence of pressure differential $\delta P = \xi\rho U^2/2$, we obtain an equation for the specific impedance (i.e., impedance related to the bottom unit area), $R = 2\delta P/U$. In the combustion chamber and the pre-injector cavity upstream of the face plate, there occur pressure oscillations of longitudinal-transverse modes described by Eq. (7.3). The acoustic resistance W for such oscillations according to Eq. (7.6) depends on the frequency $W = \rho c\overline{\omega}_{mn}/\sqrt{\overline{\omega}_{mn}^2 - 1}$, where $\omega = \omega/\omega_{mn}$ is the relative frequency. The dimensionless active impedance becomes

$$\overline{R} = \frac{R}{W} = \frac{2\delta P}{\gamma PM} \frac{\sqrt{\overline{\omega}_{mn}^2 - 1}}{\overline{\omega}_{mn}}$$

The transposed impedance on the side of the intermediate bottom at the pre-injector cavity depth L is purely imaginary, $\overline{y} = i \cot(\sqrt{k^2 - v_{mn}^2}\, L)$. Because the cavity depth is limited, and the difference of the wave-number squared $(\kappa^2 - \nu^2)$ is small, the argument of the cotangent function is also small, leading to a large value of the cotangent function.

According to Eq. (7.8) for the reflection coefficient, it is advisable to reduce the imaginary part of the impedance to improve the absorption of pressure oscillations at the bottom (i.e., the combustor head end). Therefore it was suggested to divide the pre-injector cavity by partitions in such a manner that the waves with closely planar shapes will occur in the cavity. In this case, the imaginary part of the impedance at the porous bottom becomes $\overline{Y} = i\cot(kL)$, where kL is comparable with unity. It is also advisable to select a dimensionless active part of impedance close to unity. According to the conditions of $\overline{R} = 1$ and one-dimensional oscillations in the injector-head cavity $W = \rho c$, we obtain the optimal pressure differential $\delta P/P = \gamma M/2$. The Mach number in the pre-injector cavity is about 0.1 to give a dimensionless pressure differential of 0.07, that is, about 7% of the combustion chamber pressure.

Chapter 8

Determination of Stability of Oscillations from Natural Disturbances

I. Methods for Determination of Oscillation Decrement from Natural Disturbances (Noise) in a Chamber

In evaluating combustion stability, combustion-chamber noise can be used as a source of natural disturbances [38]. The assumption is made that these disturbances are small and that the pressure pulsations are described by linearized equations. These disturbances are introduced directly into the combustion-chamber volume and described by random functions whose spectral densities have large bandwidth compared to the frequency characteristics of the combustion chamber in the neighborhood of the natural frequencies being studied. Then the pressure pulsations in the general case can be described with a system of linear equations and presented by the block diagram shown in Fig. 8.1.

The block diagram of the combustion chamber as an oscillatory system with distributed parameters is represented by the sum of oscillatory systems of second order. A number of methods have been devised for determining the oscillation decrement from natural disturbances. The following methods will be considered: spectral, correlation, amplitude, and instantaneous period method.

The spectral method defines the oscillation decrement from the width of the "noise" power spectrum (see Fig. 8.2). In an actual case, the filter width Δ_f of spectrum analyzer should be taken into account for the combustion chamber process. For the case of a simple filter, we have

$$\delta T = \frac{\pi(\Delta - \Delta_f)}{f_o},$$

where δT is the oscillation decrement, Δ is the width of the power spectrum $S(\omega)$ at a level of half that of the maximum amplitude, and f_o is the resonant frequency. An algorithm was developed to take Δ_f into account for other types of filters as well. In an actual algorithm, more complicated relationships are used to take into account neighboring (close) peaks, background, and other interferences (three-point methods). An exhaustive research of point pairs lying to the

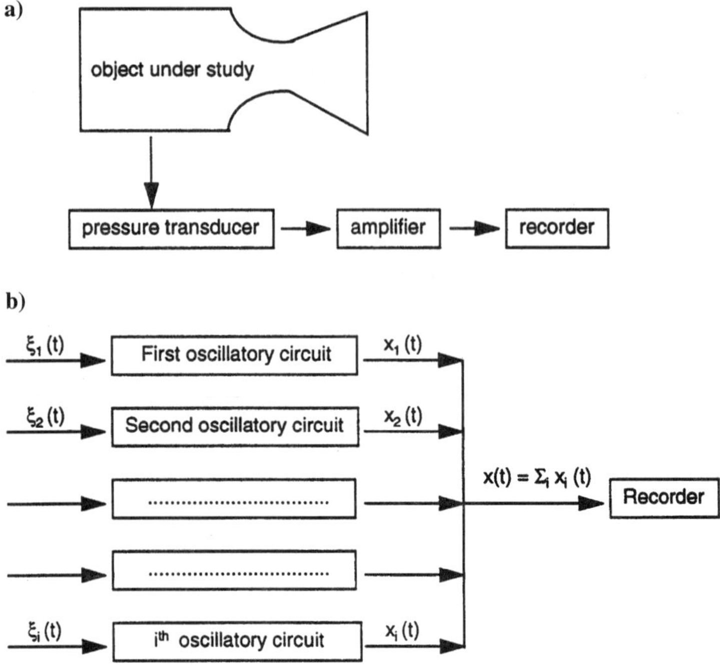

Fig. 8.1 Block-diagram of a) pressure pulsation recording of the object under study and b) equivalent electrical block diagram of signal recording.

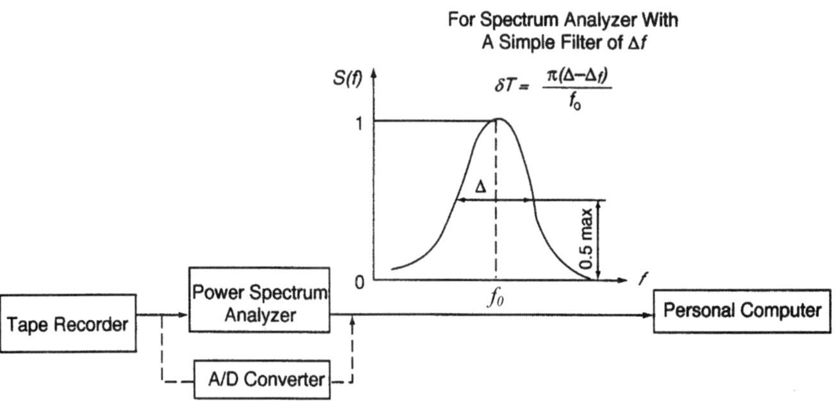

Fig. 8.2 Block diagram of the oscillation decrement δT determined by a spectral method.

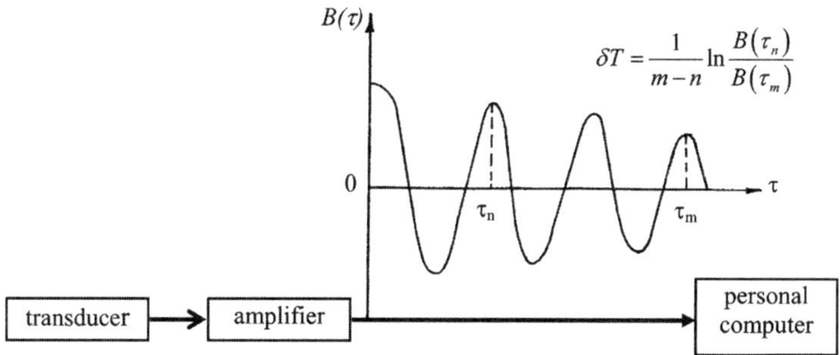

Fig. 8.3 Block diagram of the oscillation decrement δT determined by a correlation method.

left and to the right of the maximum is performed, and a pair is selected, which gives the minimum sum of the value of the decrement and its theoretical statistical spread.

The correlation method defines the oscillation decrement from the rate of decay of the autocorrelation and cross-correlation function $B(\tau)$ (see Fig. 8.3):

$$\delta T = \frac{1}{m-n} \ell n \frac{B(\tau_n)}{B(\tau_m)}$$

where τ_n and τ_m correspond to the maxima of the function $B(\tau)$ for the nth and mth periods of oscillations.

The amplitude method (see Fig. 8.4) defines the oscillation decrement as a function of the ratio between two amplitudes at the outlet of two different filters tuned to a resonance frequency of the oscillatory process being analyzed,

$$\delta T = F\left(\frac{\bar{A}_1}{\bar{A}_2}\right)$$

In a particular case, the decrement is defined from the ratio of amplitudes at the outlet and inlet of one of the filters (preferably a rejection one). Discrete values of the function F are determined by calibration on a simulator. The decrement is calculated from these values by linear interpolation. The simulator represents an oscillatory circuit with a rearranged decrement.

The instantaneous period method (see Fig. 8.5) defines the oscillation decrement from the mean module value (T') of instantaneous period fluctuations of the oscillations being studied,

$$\delta T = F(\bar{T}')$$

The instantaneous period of a random signal with the zero mean value is the sum of two adjacent intervals between successive zeros of the signal. As in the amplitude method, the discrete values of function F are determined by calibration on a simulator. The decrement is calculated from these values by linear interpolation.

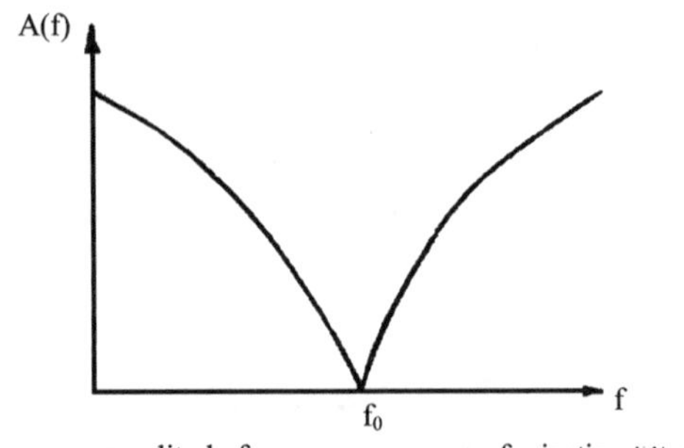

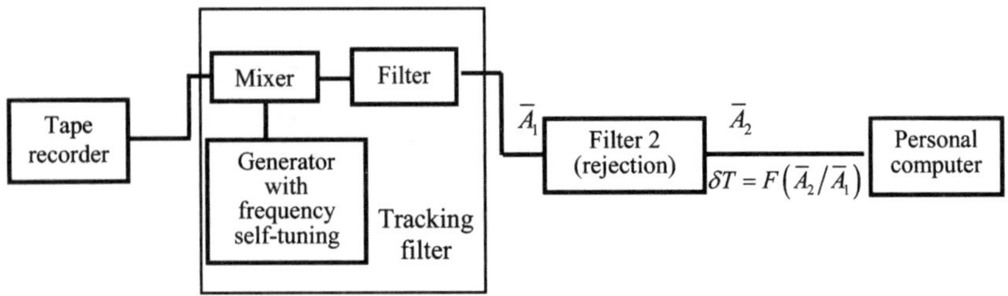

Fig. 8.4 Block diagram of the oscillation decrement δT determined by an amplitude method.

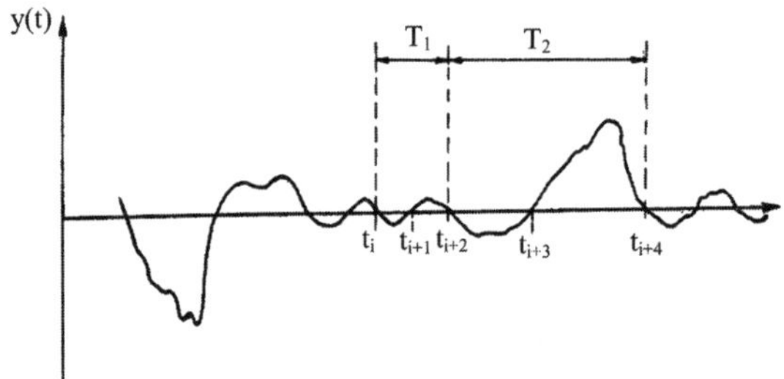

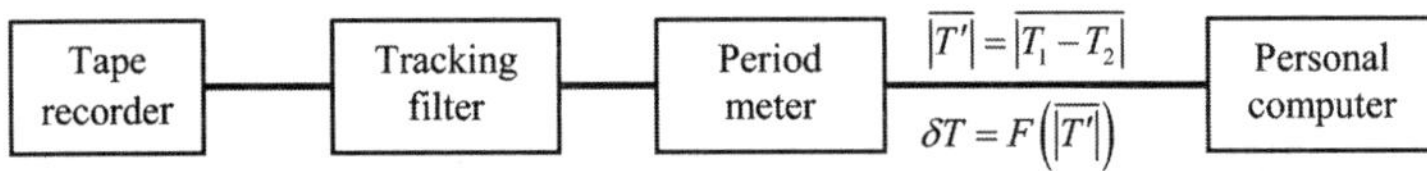

Fig. 8.5 Block diagram of the oscillation decrement determined by an instantaneous period method.

II. Accuracy of Determining Oscillation Decrement by Different Methods

There should be a clear idea of the decrement accuracy before using it for quantitative evaluation of stability margins. The error of the decrement determination consists of the statistical error caused by interfering factors, such as close frequencies, background noise, frequency instability, process unsteadiness, and perhaps others.

The theoretical estimate of statistical error requires tedious calculations and results in awkward final results. That is connected with the necessity to take into account the correlation of the spectrum points used in the decrement calculation and the correlation function. Therefore, the results of these theoretical studies are given in a most simplified form.

A. Error of Spectral Method

The value of the relative statistical error $\varepsilon_{\delta T}$ of the decrement estimate is equal to the statistical error ε_{S0} of the power spectrum estimate $S(\omega_0)$ (the square of the filtered amplitude) at the resonance frequency ω_0:

$$\varepsilon_{\delta T}^2 \approx \varepsilon_{S0}^2 \approx \frac{1}{(\Delta_{\text{ef}} \times T_{\text{aver}})}$$

where $\varepsilon_{\delta T}^2 = \overline{(\delta T - \delta I)^2} / (\delta T)^2$ is the square of the relative statistical error of the estimate of decrement δT; $\varepsilon_{S0}^2 = [\bar{S}(\omega_0) - S(\omega_0)]^2 / S^2(\omega_0)$ is the square of the relative statistical error of the estimate of the power spectrum at the resonance frequency; $\Delta'_{ef} = (\Delta'_f)^2 / \Delta''$; T_{aver} is the averaging time; Δ'_f is the effective width [AFC]^4 of the filter; and Δ'' is the effective width [AFC]^4 of the system (chamber + filter). The effective width of [AFC]^4 is the ratio of the area under the curve [AFC]^4 to the maximum of [AFC]^4.

Let us consider two extreme cases of a very wide $(\Delta'_f \gg \delta)$ and a very narrow $(\Delta'_f \gg \delta)$ filters. For the first case (a wide filter), we have minimum statistical error equal to

$$\varepsilon_{\delta T}^2 \approx \frac{1}{\delta T_{\text{aver}}}$$

but poor frequency resolution and large errors caused by the existence of closely spaced frequencies, background, etc.

For the second case (a narrow filter), we have good frequency resolution but large statistical error equal to

$$\varepsilon_{\delta T}^2 \approx \frac{1}{\delta T_{\text{aver}}}$$

It follows from these two extreme cases that there exists an optimal filter width that gives the least total error.

B. Error of Correlation Method

The formula for estimating the decrement error is given for the simplest case when the decrement is determined from two points $B(\tau_1)$ and $B(\tau_2)$ with $\tau_1 = 0$:

$$\varepsilon^2_{\delta T} = \frac{[1+f(\alpha)]}{\delta T_{\text{aver}}}$$

where $\alpha = \delta T \times n$, $n = \tau_2/T$ is the number of periods and $f(\alpha) = (e^{2\alpha} - 1 - 2\alpha - 2\alpha^2)/2\alpha$. The minimum statistical error of the decrement estimate is achieved with $n = 1$ (adjacent maxima). In this case

$$F(\alpha) \approx 0, \quad \varepsilon^2_{\delta T} \approx \frac{1}{\delta T_{\text{aver}}}$$

However, in processing an actual signal, using an initial part of the correlation function is unreasonable because of the background effect and filter existence. Using adjacent maxima is unreasonable because of the influence of instrument error and closely spaced frequencies. Thus, the initial part $B(\tau)$ is eliminated, and the remaining part $B(\tau)$ is used more completely by proper averaging of some successive values $B(\tau_n)/B(\tau_{n+1})$.

C. Error of Amplitude Method and Instantaneous Period Method

The relative statistical error of the amplitude method consists of relative estimates of the square of both filters' amplitudes. As the relative error for the rejection filter is an infinitesimal value, the relative statistical error is practically equal to the relative error of the squared amplitude at the filter outlet, $\varepsilon_{\delta T} \approx \varepsilon_{A_1^2}$. Thus, everything that was said for the spectral method holds true for the relative statistical error of the amplitude method.

As for the instantaneous period method, no theoretical study of the estimation statistical error was performed because of the complexity of the task. Only an experimental error estimation with the help of a simulator was carried out.

Actual statistical errors for all described methods amounted to about 10 and 30% with a realization time of 1 and 0.08 s, respectively. Thus, with a realization time of 1 s, statistical error is not a governing factor, and the total accuracy is mainly defined by the just-mentioned interfering factors. With an averaging time of 0.08 s, the statistical error is often the governing factor.

The amplitude method of the decrement determination from the combustion-chamber noise and the instantaneous period method are not practically used at the present time. However, some data obtained by these methods are shown in this book.

III. Peculiarities of Stability Evaluation Under Transient Conditions

The stability characteristics discussed in Sec. II of Chapter 6 require for their determination a definite stationary region of the engine operating conditions. To achieve uniformity in the stability estimate under transient and steady-state conditions, estimations during engine start and under transient conditions will be considered.

Because an LRE starting process is a transient regime, it differs from the stationary regime by some peculiarities, which might have a different effect on the stability of steady operation. The main peculiarities of the starting process are its short-time direction and a wide range of the operating parameters such as flow rates, mixture ratio of propellant components, and pressure and temperature of combustion products. Their values affect the dynamic nature of the whole stability estimation procedure. The total starting time $\Delta\tau_{max}$ can be estimated as the time interval from ignition in the combustion chamber to the moment when nominal conditions are achieved (at the top or at the bottom), that is, for example, $(100 \pm 10)\%$ of $P_{ch,nom}$. The highest starting speed is estimated by a value

$$\Delta\tau_{min} = P_{ch,nom} \Big/ (\partial P_{\kappa}/\partial\tau)_{max}$$

where $(\partial P_{\kappa}/\partial\tau)_{max}$ is the highest rate of the pressure rise in the chamber.

The starting time for modern engines varies from ≈0.02 to ≈4 s. Suppose that for steady-state conditions, the allowable time delay in determining the stability with respect to soft excitation (the oscillation decrement by an order of magnitude) is 1 s. Then for the startup process, this value is bound to be about 0.01s. Under this circumstance, researchers have suggested that it is impossible to obtain sufficiently accurate estimates of stability by passive methods (from noise) in such a short time interval.

Actually, according to the general theory of random processes, it is known that the error σ of estimating a characteristic such as the autocorrelation function $\kappa(\tau)$, from which the oscillation decrement can be determined, depends on several variables: the length of realization T_r of the part being processed, the averaging time, and the unknown decay coefficient δ. Those three variables are connected by the relation

$$\sigma\kappa(\tau) \approx \frac{1}{\sqrt{\delta T_p}}$$

Based on this relation, and on data for the principal error of determining the oscillation decrement by statistical methods under steady-state conditions, the researchers came to the conclusion that with a length of realization of 0.01 s, and with an averaging time of 1s, the error would be 100% and more.

In this case however, an essential difference was not made between the error $\sigma\kappa(\tau)$ in determining correlation function and the error $\sigma_{\delta T}$ of determining the oscillation decrement itself. In addition, the influence of a value of a shift in the correlation function on the value of the decrement error was not studied. On the other hand, some circumstances were known to be indicative of the fact that the error in the oscillation decrement should not be extremely high, if only for a quite short averaging time. The case in point arises when a wideband random signal is the input of the narrowband oscillatory system. Its spectral density $S_{in}(f)$ under the transient conditions remains constant $S_{in}(f) \approx$ const, and the decrement value is inversely proportional to the amplitude A_r of oscillations at the resonant frequency f_r. The amplitude method of decrement determination is based on this relation.

In turn, the accuracy of determining the oscillation amplitude does not become catastrophically worse with a shorter realization length, at least with $T_r > \tau_r$, where τ_r is the system relaxation time. And this allows looking forward to obtaining decrement estimates under transient conditions with acceptable accuracy under the preceding condition concerning $S_{in}(f)$.

Finally, in some cases there are relatively "slow" (about 1 s and more) transient conditions including the starting interval, for example, engine 11D56, for which the passive methods could be acceptable without any additional conditions. Therefore, direct theoretical evaluation was required rather than indirect evaluation of possible errors of the stability characteristic for short intervals of realization. On this basis, a conclusion could be made about the actual limits of applicability of passive methods for the transient conditions.

There is one more reason that induced us to study the possibilities of applying passive methods for transient conditions. Because of the relatively fleeting nature of the transient processes in general, and the starting conditions in particular, passage of the operation process through the possible instability regions can be very fast. Because of the final relaxation time of the system, the excited oscillations have no time to develop to such an amplitude level that they can influence the operational integrity of the structure or the time when the transition to vibrational combustion under steady-state conditions would take place.

According to available experimental observations, this time of the rise of oscillation amplitude to a "dangerous" level can come to several hundredths of the second. Therefore, practically there is no need to gain a small lag (for example, less than 0.01…0.02 s) of the passive methods.

The question of the permissible residence time of the system in the incremental region is discussed in Sec. III.B.

A. Selection of Optimal Duration of the Part to Be Processed in Startup

As a digital model, we take a system consisting of three resonance circuits connected in parallel, with "white" noise $A_n\xi(t + \rho_n)$ applied as input inlet (see Fig. 8.6).

The existence of the wideband background is simulated by applying white noise $\xi(t)$ to the system input, with an amplification factor B. The selection of three resonance circuits is dictated by the necessity to simulate two adjacent peaks in the signal spectrum to the left and to the right of the resonance peak being studied.

The coefficient A_n and phase shifts φ_n (see Fig. 8.6) characterize the conditions of excitation of the nth natural mode of the system. In estimations of the errors of

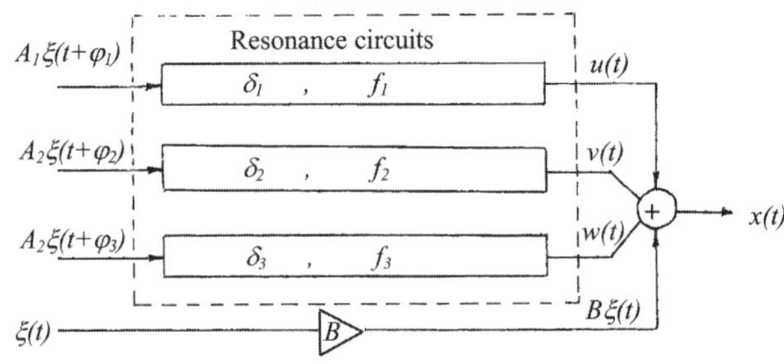

Fig. 8.6 Diagram of a digital model.

the spectrum, decay time, and determination of the oscillation decrement, with the averaging made for a long period compared to the oscillation period, an initial phase shift is not of great importance. The phases of the signals will change in a different way (at a different rate) because they have different resonant frequencies. Thus, hereafter, it will be assumed that $\varphi_1 = \varphi_2 = \varphi_3 = 0$.

For estimation of errors in the stability parameters, the height of the spectrum peak close to the frequency being studied plays a significant role. The height of the spectrum peak is dependent on the signal amplitude A_n, logarithmic decay coefficient δ, and resonant frequency f.

For convenience in further calculations, it will be assumed that $A_1 = A_2 = A_3 = 1$, and the spectrum peak height will be regulated by two parameters: δ_n and f_n. This means that the excitation conditions for all three oscillatory circuits are assumed to be same. In this case, the combustion-chamber model is simplified and has the form shown in Fig. 8.7. The model contains the desired signal and interfering background. For example, in determining the decrement and amplitude of oscillations for the nth spectrum peak, the desired signal is only the signal obtained by passage of white noise through one circuit with resonant frequency f_n. The signals obtained after passage of white noise through the other two circuits and additive noise ($B\xi_i$) form the interfering background.

In the preceding model it is assumed that the frequency f remains constant (or varies slightly) during realization time $t_r = 0.05$ s.

According to digital filtering theory, the following system of algebraic equations can be written:

$$
\begin{aligned}
U_i &= \alpha_{11} U_{i-1} + \alpha_{21} U_{i-2} + \xi_i \\
V_i &= \alpha_{12} V_{i-1} + \alpha_{22} V_{i-2} + \xi_i \\
\omega_i &= \alpha_{13} \omega_{i-1} + \alpha_{23} \omega_{i-2} + \xi_i \\
X_i &= U_i + V_i + \omega_i + B\,\xi_i
\end{aligned}
$$

where

U_i, V_i, ω_i = output signals of the resonance circuits
$l = 1,2,3\ldots$ are integer numbers ($N\theta$ the realization length)

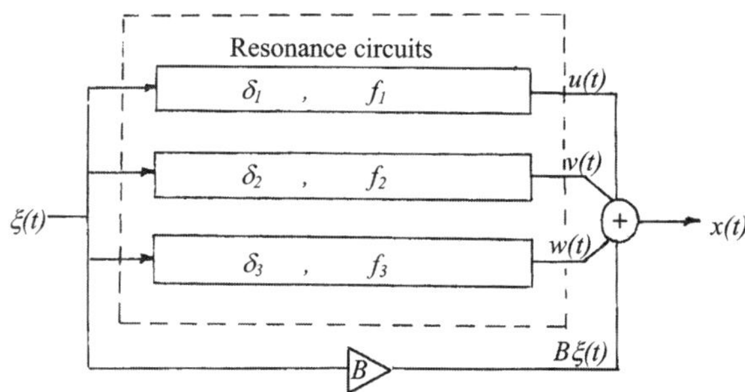

Fig. 8.7 Diagram of a simplified digital model.

ξ_i = the ith number obtained with the help of a random-number generator on a computer
$\alpha_{1n} = 4\cos(\omega_n\theta)/(1 + e^{2\delta_n\theta})$, $(n = 1, 2, 3)$ $\omega_n = 2\pi f_n$
σ_n = the logarithmic decay coefficient of a resonance circuit
$\alpha_2 n$ = decay coefficient of the nth resonance circuit
$\theta = 1/\Omega$ = sampling (step) time
$U_{-1} = U_0 = V_{-1} = V_0 = \omega_{-1} = \omega_0 = 0$

Such a model will be called three dimensional as it comprises three natural frequencies f_n.

Figure 8.8 shows mean values of δT obtained for the model and set (actual) values of (δT)s, depending on the processing time t_p, averaged from 40 realizations. It is seen from this figure that with small values the average value of δT increases and differs greatly from the set value. Such a difference of δT from the set value is accounted for by systematic errors, which should be taken into account, when using different methods of δT determination. For example, the preceding systematic errors can be almost completely compensated using calibration with the same realization time, resonance frequency, and oscillation decrement. Figure 8.9 shows the minimum theoretical error against random errors obtained for the model. It is seen from the figure that with $\beta \geq 7.5$ the minimum theoretical error and the error obtained with the model agree satisfactorily.

Figure 8.10 shows the decrement variation curve obtained with the model by the correlation method and the set (actual) curve. It is seen from this figure that when δT enters the negative region, the error increases significantly. The reason is that with the correlation method of determining δT it is impossible to determine its negative values. Furthermore, with high negative decrement values, the correlation method gives some positive values, and the lowest decrement value is obtained when its actual value is zero.

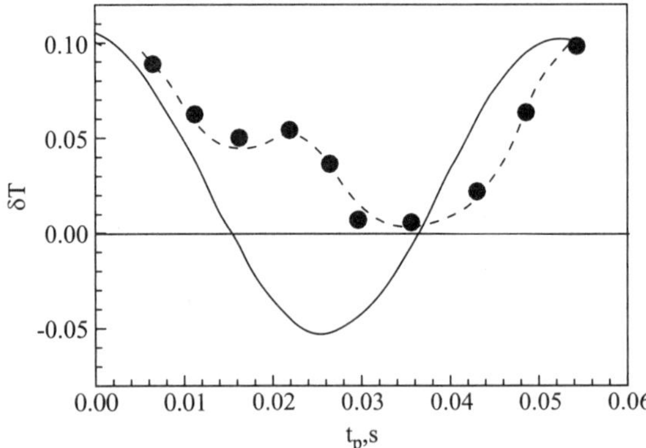

Fig. 8.8 Effect of processing time on frequency and phase distortions of the oscillation decrement obtained by a correlation method with a realization length of 0.05 s: specified__; calculated_ _ _.

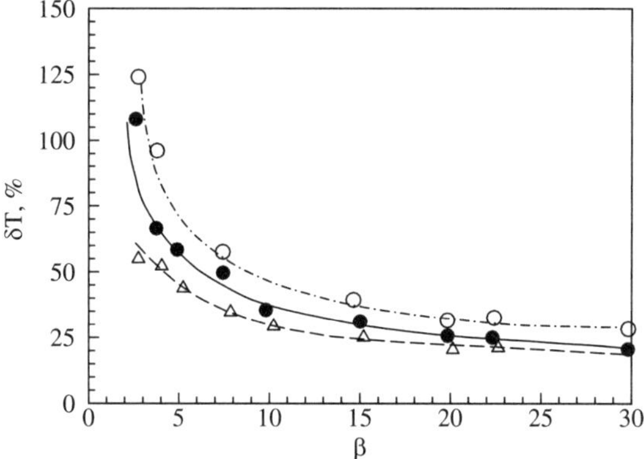

Fig. 8.9 Effect of $\beta = \delta t_r$ on the random error of oscillation decrement $\sigma_{\delta T}$ and squared amplitude $\sigma_A{}^2$ of oscillations determined by a correlation method; Estimates of $\sigma_{\delta T}$ on the model: ○—○ with $\tau = 8\ T_0$; ●—● with $\tau = 0.5\ T_0$; minimum theoretical - - - -; Δ—Δ, estimates of $\sigma_A{}^2$ on the model.

B. Estimation of Permissible Oscillatory System Residence Time in Region of Negative Values of Decrement

One of the problems solved with the model is estimation of the permissible oscillation decrement "residence" time in the negative region. The decrement behavior during starting, or any other transient regime, is unknown a priori. It is only known that high-frequency oscillations rather often occur during starting; sometimes they persist in the steady-state regime as well. This indicates that in the

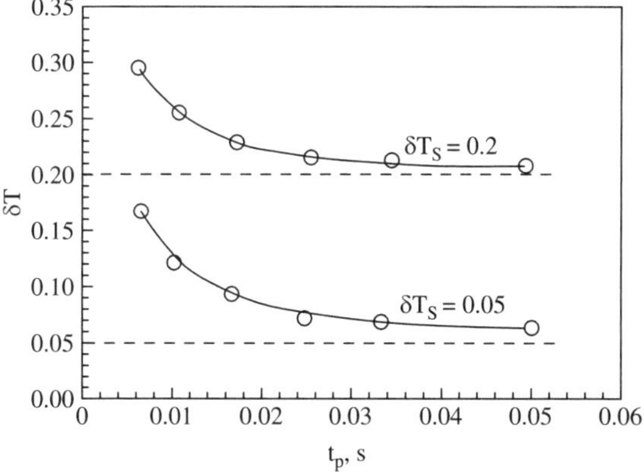

Fig. 8.10 Average decrement values δT obtained in the digital model with different realization time t_p; calculated and - - - specified δT_s decrement values.

starting period the decrement often can assume negative values; the testing result depends on what values the self-oscillation amplitude has attained in this case.

Two cases will be considered:

1) The oscillation decrement varies with time by the harmonic law $\delta T(t) = A \cos \omega t$. The δT variation curves are shown in Fig. 8.11.

2) The oscillation decrement varies by the step law, as shown in Fig. 8.12.

In each of these cases, the calculations were performed under two conditions: when the decrement varied from 0.1 to –0.05 and from 0.15 to –0.1 (see Figs. 8.11 and 8.12). These negative decrement values were chosen on the basis of experimental data.

Figure 8.13 shows the relative oscillation amplitude vs the residence time τ_0 of the decrement in the negative region when $\delta T(t) = A \cos \omega t + B < 0$. The values of τ_0 were changed by varying the frequency $f = \omega/2\pi$ from 10 to 120 Hz for both the harmonic and the step laws. The initial value of amplitude A_0 was determined for $\delta T = 0.1$ and 0.15, respectively. The amplitude rise was considered for the first oscillation period only. Every point in Figs. 8.13 and 8.14 has been obtained by averaging from 20 realizations. It is seen from Fig. 8.13 that the amplitude rise to dangerous values ($\approx 10\ A_0$) takes place with $\delta T = -0.05$ in the period $\tau = 24$ ms, and with $\delta T = -0.1$ in 10 ms.

Relative values of oscillation amplitude vs step changes of δT are shown in Fig. 8.14. It is seen from the figure that the amplitude rise by the step law occurs faster, but the amplitude rise has an exponential pattern. Thus, these plots have been reconstructed on the $\ell n(A/A_0)$ and $\eta = \delta\tau_0$ coordinates, shown in Fig. 8.15. The points calculated for the case when $\delta T(t) = A \cos \Omega t + B$ are clustered near one straight line, and when $\delta T(t)$ varies by the step law, they are clustered near the other straight line. Therefore, when δT varies by the harmonic law, the relative amplitude curve on the preceding coordinates can be approximated by a straight line with an accuracy of up to 10%

$$\eta = 1.4\ \ell n \left(\frac{A}{A_0}\right)$$

and with stepwise variation of $\delta T(t)$,

$$\eta = \ell n \left(\frac{A}{A_0}\right)$$

The relations just shown can be generalized by one formula:

$$\tau_0 \delta_{ef} = \ell n \left(\frac{A}{A_0}\right)$$

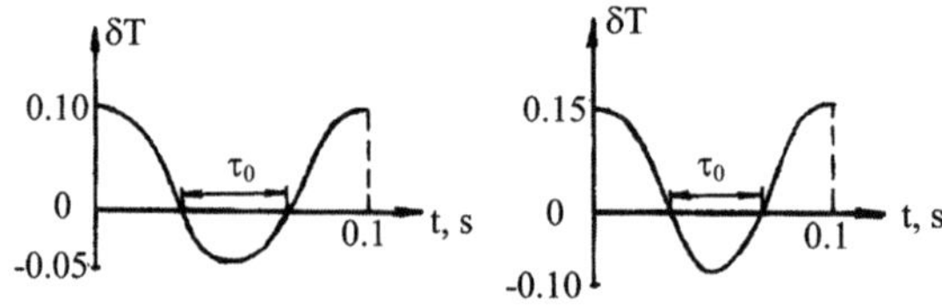

Fig. 8.11 Harmonic law of oscillation decrement variation.

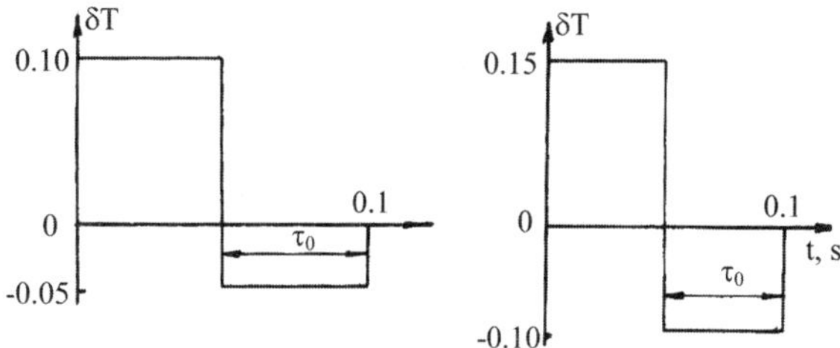

Fig. 8.12 Step law of oscillation decrement variation.

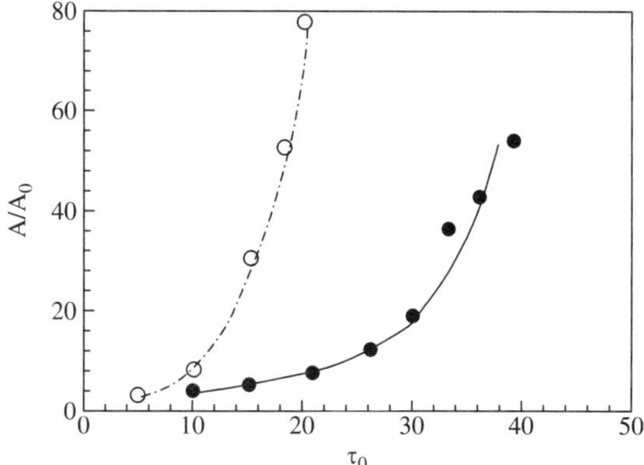

Fig. 8.13 Relative oscillation amplitude A/A_0 vs residence time τ_0 of decrement in the negative region by the harmonic law: ○—○ min δT = –0.10, max δT = 0.15; ●—● min δT = –0.05, max δT = 0.10.

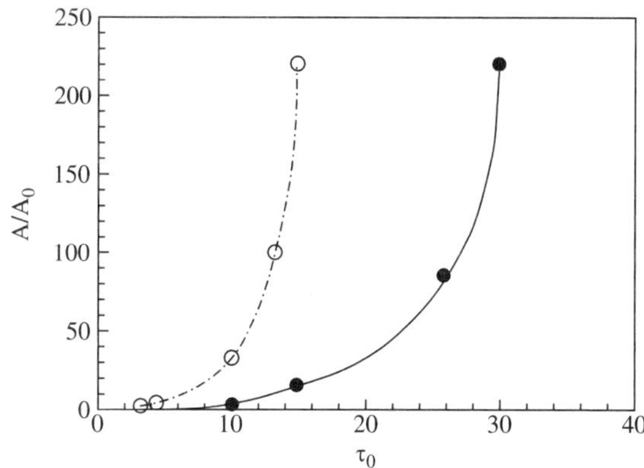

Fig. 8.14 Relative oscillation amplitude A/A_0 vs residence time τ_0 of decrement in the negative region by the step law: ○—○ min δT = –0.10, max δT = 0.15; ●—● min δT = –0.05, max δT = 0.10.

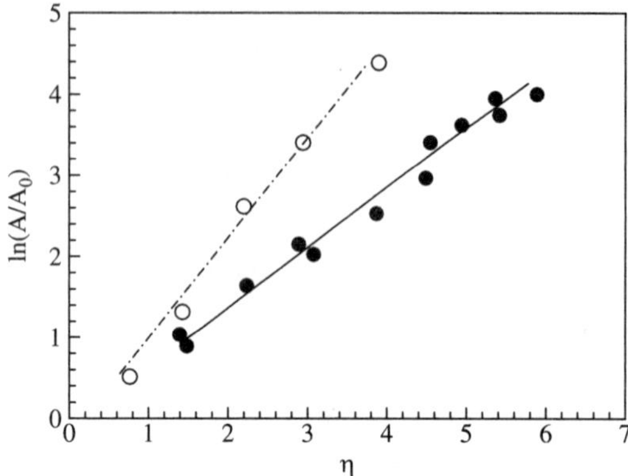

Fig. 8.15 Relative oscillation amplitude vs $\eta = \delta\tau_0$ (τ_0: residence time of δT in the negative region). Decrement variation law: ○—○, step law; ●—●, harmonic law.

where δ_{ef} is the effective value of $\delta T(t)$.

Dividing the relation $\eta = 1.4\ell n(A/A_0)$ by the coefficient $1.4 \approx \sqrt{2}$ in the left part, we obtain the effective value of $\delta T(t)$ in its harmonic variation.

Thus, the permissible residence time τ_p in the negative region can be calculated by the following formula:

$$\tau_p \approx \frac{\ell n\left(\frac{A_p}{A_0}\right)}{\delta_{ef}}$$

where A_p is the permissible value of the oscillation amplitude.

By way of example, τ_p is estimated for the case that can occur in practice: $f = 2000$ Hz, $\delta T = -0.1$, and $A_p/A_0 = 10$.

We obtain the following:

1) For the harmonic law,

$$\tau_p = \frac{(1.4 \cdot \ell n 10)}{(-\delta T/f)} = \frac{(1.4 \cdot 2.3)}{(0.1 \cdot 2000)} = 0.016 \text{ s} = 16 \text{ ms}$$

2) For the step law,

$$\tau_p = \frac{\ell n 10}{(0.1 \cdot 2000)} = 0.0115 \text{ s} \approx 12 \text{ ms}$$

IV. Methods of Determination of Stability from Natural Pressure Disturbances

The system of recording the RVP on a computer has led to the development of a number of up-to-date methods for obtaining stability characteristics. As an example, the basic principles of one of the methods for calculating stability characteristics with the averaging time from 0.08 s will be given.

As shown in Sec. III, and based on experience, this averaging time provides sufficient accuracy of the main stability characteristic, the oscillation decrement.

To process RVP, the method base on a simple digital filter (SDF) is used. The peculiarity of this method is the use of arbitrary values of sampling frequency f_{sampl}, the averaging time T_{aver} and SDF filter passband Δ_f. The accuracy of determining the stability parameters and primarily the main parameter, the oscillation decrement, depends on the values of the just-mentioned quantities. The statistical error depends to a great extent on the ratio between the passband of the oscillatory system being studied and that of the filter (see Sec. II), a fact that should be taken into account.

The processing method with the use of SDF is based on a fixed value of the filter decrement $(\delta T_f d_0)$, instead of the fixed value of its passband. This resulted in some specific features of the method: a varied step for SDF spectrum frequency, a more complicated method of determining its resonant rise and other features.

According to this program, the processing comprises the stages described next.

1) A file is selected for processing.

2) A parameter (RVP) to be processed is selected.

3) The scaling factor A_t, obtained in the processing of the respective calibration, is specified.

4) According to the specification, the averaging time T_{aver}, processing frequency ranges, the decrement d_0, the width of SDF filter passband Δ_{spec}, as well as the processing time interval, are specified.

5) The boundary frequency f_b, accounting for the specified value of filter passband Δ_{spec} and filter decrement d_0, $f_b = n \cdot \Delta_{spec}/d_0$, is determined. For frequencies below f_r, the SDF filter width Δ_f is assumed to be constant and equal to $\Delta_f = \Delta_{spec}$: The step for frequency h_f in calculation of the SDF spectrum is constant and equal to $h_f = \Delta_{spec}/2$ for these frequencies. For frequencies f_i higher or equal to frequency f_b, the width of SDF filter is assumed to be equal to $\Delta_f = d_0 f_i/\pi$, with the frequency step being variable and equal to $h_{fi} = f_{i-1} d_0/2\pi$. The filter passband width determined in this way is improved by applying the correction, which compensates the existence of the SDF filter acceleration part resulting from the initial zero conditions at the filter inlet.

6) The SDF spectrum is determined in the specified frequency range.

7) The spectrum maxima are determined, and four of them are selected for further processing.

8) Each of the four selected spectrum maxima is processed. The selected maximum and respective resonant frequency are refined with the use of parabolic interpolation. The respective mean-square amplitude is calculated and refined accounting for the existence of the acceleration part in the averaging range caused by the zero initial conditions in calculation of the SDF spectrum.

9) Three decrement values are determined for each selected spectrum maximum: the central one as well as to the left and to the right of the peak, with the algorithms of decrement determination being different for two frequency regions ($f < f_r$ and $f \geq f_r$).

10) In the averaging part with duration T_{aver}, the maximum and mean rectified value of the signal are determined.

To perform the processing, the following are specified.

1) The main frequency range of processing $f_0 - f_{in}$, is located within the range 0-$f_{samp}/3$. (Here f_{sampl} is the sampling frequency.) The signal spectrum in this range is derived later during processing (see Fig. 8.16).

2) The frequency range for processing four maximum spectrum values is selected in the main range. The range selection depends on the problem to be solved.

3) For the averaging time T_{aver}, the main values are $T_{aver} = 0.08$ and 0.25 s. Other values of T_{aver} can be specified. With the specified averaging time, the value of $L = T_{aver} f_{sampl}$ should not exceed 15,000. Otherwise, the value of T_{aver} is automatically selected, for which L = 15,000. If it is necessary to perform the averaging in large time intervals, additional processing programs should be used.

4) The SDF filter decrement d_0 is preset from the sequence 0.05; 0.1; 0.2. The main decrement value $d_0 = 0.1$.

5) The SDF filter passband width Δ_{spec} is specified from the sequence 12.5; 25; 50; 100; 200 Hz. The main value of Δ_{spec} = 50 Hz.

6) The processing time interval is specified.

Five columns are represented on the computer screen and on the graphic copies of the screen as a result of the processing (see Fig. 8.16). Each column corresponds to the specified averaging time. Then the next five columns are displayed and so on to the end of the specific processing time interval.

At the top of the screen, the spectrum of the parameter being processed is shown within the specified main processing range. The results given in the table are

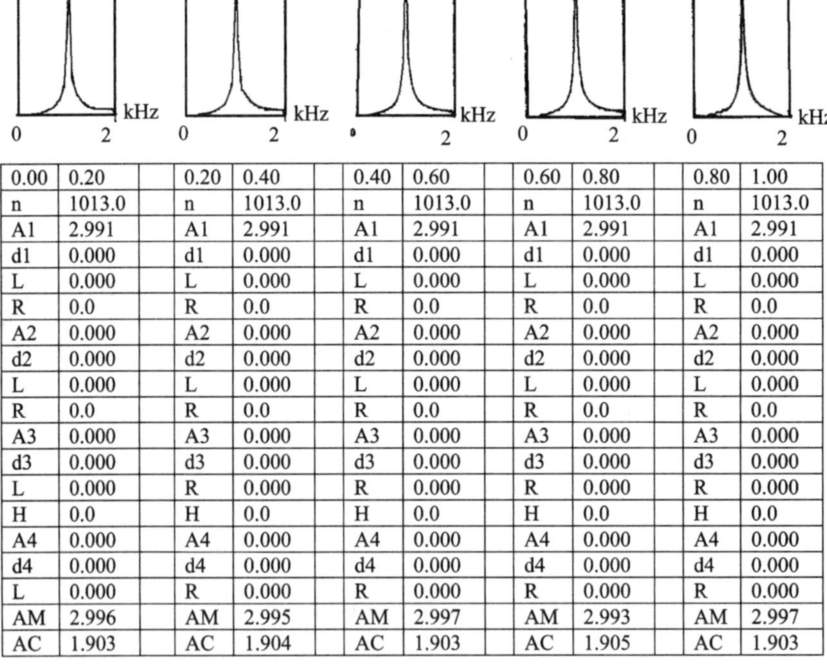

0.00	0.20		0.20	0.40		0.40	0.60		0.60	0.80		0.80	1.00	
n	1013.0		n	1013.0		n	1013.0		n	1013.0		n	1013.0	
A1	2.991		A1	2.991		A1	2.991		A1	2.991		A1	2.991	
d1	0.000		d1	0.000		d1	0.000		d1	0.000		d1	0.000	
L	0.000		L	0.000		L	0.000		L	0.000		L	0.000	
R	0.0		R	0.0		R	0.0		R	0.0		R	0.0	
A2	0.000		A2	0.000		A2	0.000		A2	0.000		A2	0.000	
d2	0.000		d2	0.000		d2	0.000		d2	0.000		d2	0.000	
L	0.000		L	0.000		L	0.000		L	0.000		L	0.000	
R	0.0		R	0.0		R	0.0		R	0.0		R	0.0	
A3	0.000		A3	0.000		A3	0.000		A3	0.000		A3	0.000	
d3	0.000		d3	0.000		d3	0.000		d3	0.000		d3	0.000	
L	0.000		R	0.000		R	0.000		R	0.000		R	0.000	
H	0.0		H	0.0		H	0.0		H	0.0		H	0.0	
A4	0.000		A4	0.000		A4	0.000		A4	0.000		A4	0.000	
d4	0.000		d4	0.000		d4	0.000		d4	0.000		d4	0.000	
L	0.000		R	0.000		R	0.000		R	0.000		R	0.000	
AM	2.996		AM	2.995		AM	2.997		AM	2.993		AM	2.997	
AC	1.903		AC	1.904		AC	1.903		AC	1.905		AC	1.903	

Fig. 8.16 Results of calibration signal processing.

shown in Fig. 8.16. It contains 1) the time interval of spectrum determination and 2) the results of processing each of the four (i = 1, 2, 3, 4) principal spectrum maxima within the specified frequency range.

For each of the four maxima, the following parameters are shown: f_i, frequency; A_i, mean-square amplitude; and d_i, the central value of the decrement. Below it there is the minimum decrement value from the left set of points (displayed with symbol L before it) or the right one (displayed with symbol R before it).

Equality of the central decrement value and the right or "left" one means the absence of interference during decrement determination, primarily the absence of closely spaced frequencies in the spectrum. Further, the values of the characteristic shown for the other three maximum values of the spectrum are represented in the decreasing order of amplitude A_i. At the end of the table, the following are represented:

1) AM ($\overline{A}_n$) is the maximum value of an initial signal in the processing area,

$$A_n = 1/m \sum_{i=1}^{m} A_{ni}$$

where m is the number of measurements in the time area T_{aver}.

2) $AC(A_{\text{mr}})$ is the mean rectified value of the signal in the specified area T_{aver},

$$A_{\text{mr}} = 1/n \sum_{i=1}^{n} |Y_i|$$

where n is the number of measurements in time area T_{aver} and $|Y_i|$ is the absolute value of a single measurement.

The results of processing a calibration signal with a frequency of 1000 Hz and amplitude of 3 V are shown in Fig. 8.16. The following values were specified for processing: the main frequency range of 0–2 kHz; the frequency range for determination of four maximum spectrum values 0–2 kHz; $T_{\text{aver}} = 0.2$ s; filter decrement of $d_0 = 0.1$; width of filter passband of $\Delta_{\text{spec}} = 50$ Hz; and time interval of processing of 0–1 s.

It is seen from the data shown in Fig. 8.16 that the results of determining the frequency, amplitude, and decrement practically coincide with the specified values (1000 Hz, 3 V, and $\delta T = 0$). The spectrum contains only one peak. This corresponds to the fact that the calibration signal having this shape represents a sinusoid without any remarkable distortions.

Chapter 9

Evaluation of LRE Process Stability by Use of Artificial Pressure Disturbances

I. General Requirements for Devices Generating Pressure Oscillations

DEVICES for generating pressure disturbances in main combustion chambers and gas generators must satisfy a number of requirements, which can be divided into the following groups: 1) characteristics of a pressure pulse introduced into a combustion chamber; 2) stability of the pulse; and 3) design reliability of disturbance-generating devices operating as part of an engine. The choice of pulse characteristics can be considered from two different perspectives: simulation of possible disturbances occurring in a combustion chamber and generation of disturbances for assessing the process dynamic stability (hard-excitation stability, see Chapter 6).

Under normal conditions, hard excitation of high-frequency oscillations can result from the effects of various disturbances on the combustion processes. The sources of such disturbances can be 1) pressure pulsation in the combustion chamber; 2) explosions of local volumes of propellant mixtures, which have been prepared but not yet reacted, in the mixing head cavity or in measurement-system connecting channels; 3) hydraulic shocks and pressure undershoots in propellant supply lines; 4) gas pockets in hydraulic lines; 5) pulsations in liquid propellant flowing through the chamber cooling lines at high thermal loads [18]; 6) pulsations in liquid propellant as a result of partial cavitation in propellant supply lines; 7) oscillatory processes in the hydraulic line resulting from regulating system operation; and 8) low-frequency oscillations, whose frequencies are less than the lowest natural frequency of oscillations in the combustion chamber.

Experimental experience has shown that reproduction of natural disturbances presents great difficulties and in most cases does not produce a desirable result. This approach does not allow one to compare the effectiveness of measures concerning several different natural disturbances. Moreover, this approach precludes the establishment of simple criteria for evaluating the process stability margin for all engines.

The introduction into the chamber of artificial disturbances with certain characteristics presents vast opportunities for obtaining uniform estimations of

Table 9.1 Disturbance-generating devices

DD type	Operating pressure 10^5 Pa	Disturbance pressure 10^5 Pa	Number of DO on engine	Period of application
Shock tube (gas) 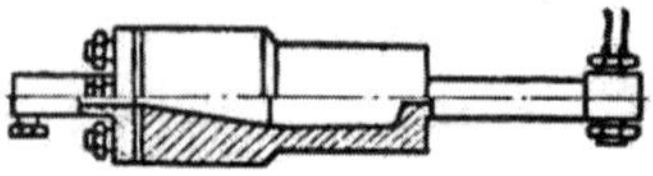	≤100	5–20	4	1966–1977
In-chamber (cond. explosive) 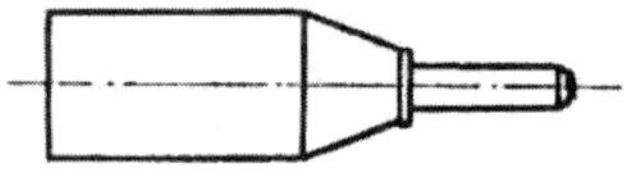	≤200	10–60	1	1971–1979
Four-charge device 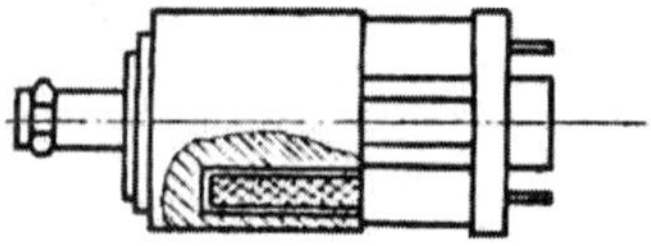	≤200	2–20	1	1974–1979
Two-three-charge device 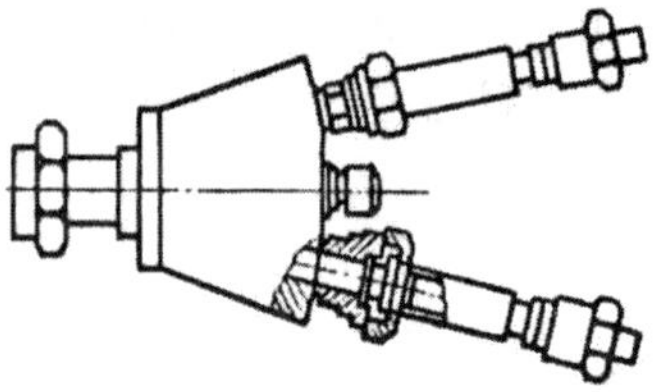	≤700	To 100	1	Since 1980
Five-charge device 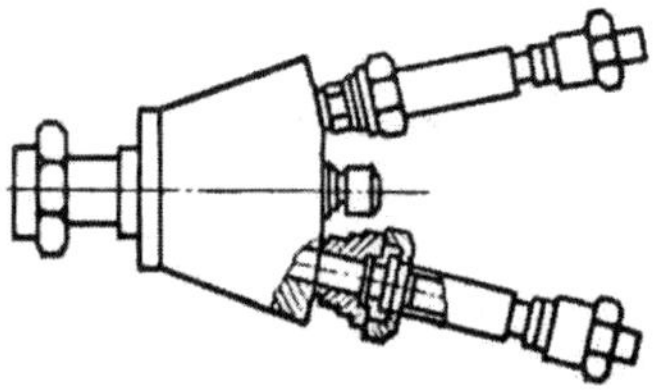	≤400	To 100	1	Since 1980

process stabilities. An artificial pulse, however, must be able to mimic a wide class of natural disturbances observed in a combustion chamber. The issue of simulating natural disturbances using artificial perturbations is discussed in Sec. III.

Various devices were developed to generate artificial pressure disturbances. A partial list is given in Table 9.1 [38], which does not cover all the devices used during laboratory tests and all modifications of these devices used for engine tests.

Each proposed device has design features that differ from the others in their specific effects on the chamber process. Strong pressure disturbances are generated with the help of in-chamber disturbance devices (IDD) and external pulsing devices. The disturbance generated by these devices is significant only at the moment of initiation. The time duration of disturbance action is very short. However, even a small disturbance induced by a spherical pressure wave from IDD has a significant effect on the chamber process. The in-chamber explosion usually produces undirected pressure disturbances. An external device generates disturbances propagating along the length of the transmitting channels, and thus the strongest disturbance is directed along the channel axis. As shown in Fig. 9.1. the disturbance intensity is significantly affected by 1) the distance between the disturbance introduction point and the chamber; 2) the orientation of the disturbance with respect to the chamber (i.e., tangential or radial); 3) physical states of propellants (i.e., liquid–liquid, gas–liquid, or gas–gas); 4) injector face-plate (mixing head in Russian literature) design (the presence of baffles and acoustic dampers); and 5) combustion-chamber dimensions (i.e., diameter and length). All of these factors must be taken into account when evaluating quantitatively the stability characteristics of a given combustion chamber by means of artificial disturbances.

The most important requirement imposed on introduced disturbances is pulse stability. Long-standing experience of the work with disturbance devices has shown that a shock wave degenerated from a detonation wave as a source of disturbance

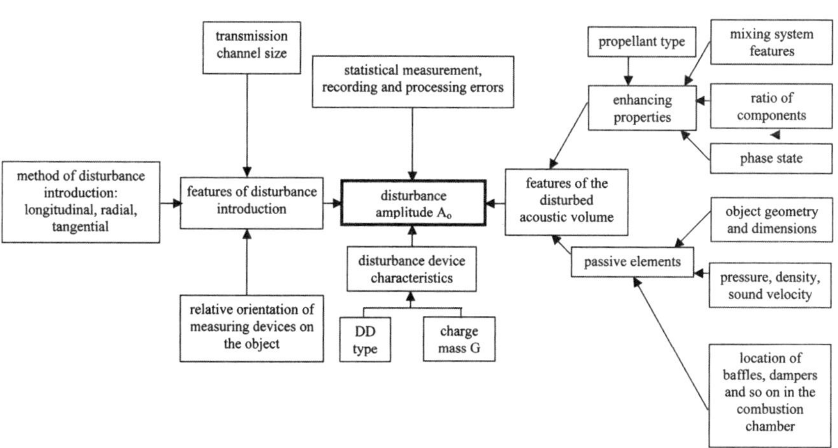

Fig. 9.1 Factors affecting the functional relationship between the disturbance amplitude in a combustion chamber and the charge mass in a disturbance-generating device.

gives the most reliable result as compared to other devices. The use of low-velocity detonation devices was especially successful.

The main requirements imposed on a pulse disturbance device are as follows:

1) An artificial pulse disturbance should be able to simulate (in terms of its effect on the combustion process) a wide class of natural pressure disturbances observed in combustion chambers.

2) A pulse disturbance should have spectral composition with such an energy distribution that can lead to the excitation of pressure oscillations of sufficient intensity in the required frequency range, even in combustion chambers with acoustic dampers.

3) A device should provide reliable and repeatable disturbance characteristics having disturbance amplitudes within required limits.

4) Possibility to supply several disturbances during one test at specified moments of time should exist.

5) Introduction of disturbances in combustion chambers and gas generators in the pressure range of 1–80 MPa when the gas generator operates at oxidizer-rich conditions should occur.

6) Minimum failures of combustion chambers and minimum permissible diameters of holes for introducing disturbances should be included.

7) A disturbance device should function reliably under LRE test firing conditions with elevated vibration, temperature, and so on.

8) A disturbance device should have minimum weight and overall dimensions.

9) Availability of production facilities should exist.

It follows from the aforementioned requirements that, owing to a wide variety of environmental factors encountered during engine operations and complexities of interactions between artificial disturbances and combustion-chamber processes, the selection of an optimal device, disturbance-introduction location, and optimal processing system, which provides the highest accuracy, can only be made on the basis of numerous experimental studies.

II. Propagation of Pressure Disturbances in Combustion Chamber

When a disturbance is introduced into a combustion chamber, the ensuing shock wave and the gas flow behind it interact with the following combustion processes in the chamber: 1) liquid discharge from injectors, 2) disintegration of jets and films, and formation of droplets, 3) evaporation of liquid propellant, 4) mixing of oxidizer and fuel vapor, and 5) chemical reactions.

As a consequence of pressure disturbances in the combustion chamber, low-frequency pressure pulsations can be initiated in the supply lines, along with variations of the propellant flow rate and mixture ratio through the injector.

The coupling effect depends on the disturbing pulse in terms of 1) its form (e.g., shock wave, sinusoidal wave, etc.), 2) duration, 3) amplitude of the initial pulse, and 4) characteristics of disturbance propagation in the passive medium.

Let us consider the characteristics of disturbances produced by a shock tube and an in-chamber disturbance device. The formation and release of a disturbance pulse from a shock tube were studied with the help of Tepler's photography of the gas volume into which the disturbance being studied was introduced. The high-speed photorecorder IAB-451 was used, and the setup diagram is shown in

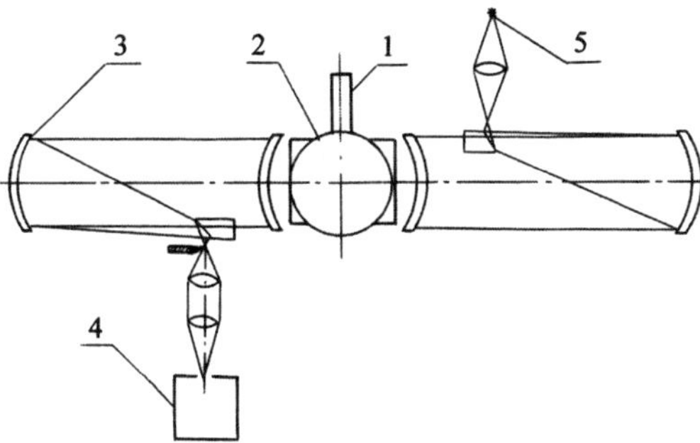

Fig. 9.2 Experimental setup for studying disturbances generated by a shock tube: 1, shock tube; 2, chamber with controlled pressure; 3, Tepler's device; 4, photorecorder SER-2M; and 5, light source.

Fig. 9.2. Decoding of high-speed photography results made it possible to reveal the properties of the disturbance released from the shock tube [39].

After the diaphragm bursts, a planar shock wave forms and travels to the open end of the channel. As in any shock tube with a burst diaphragm, at some distance from the shock-wave front, the contact surface moves, that is, an interface between the gas in the channel before the diaphragm burst and the combustion products in the test chamber. The planar shock wave exiting the channel starts propagating to the sides perpendicular to the direction of movement. As this takes place, a rarefaction circular wave forms and travels through the gas flowing behind the wave in the direction of the channel axis. Subsequently, the velocity of the wave front decreases at the edges, and the wave shape gradually changes over to a spherical configuration. The front velocity along the channel axis remains constant until the rarefaction wave reaches the axis. Further pressure drop behind the shock-wave front causes an increase in pressure differential between the shock tube and the open channel section and an increase in the gas velocity at the tube exit up to the sound velocity.

Because an underexpanded flow occurs in this case, the gas velocity becomes supersonic. Immediately after the shock wave leaves the open end, there starts the formation of a ring vortex at the tube edge moving together with the gas behind the front. The gas stream eddying and changing in discrete steps to a subsonic stream slows down near the ring vortex, transferring its energy to the wave. While the pressure differential between the channel open end and the shock tube decreases, the gas velocity in the stream drops. As the wave propagates through the combustion chamber, its intensity decreases rapidly. Under the conditions being studied, the shock wave degenerates to an acoustic wave over a length of 300 mm. With an increase in the chamber pressure, degeneration of the shock wave to an acoustic wave occurs at a shorter distance from the point of disturbance introduction.

The time history of the pressure disturbance generated by this device in the chamber was measured using a piezoelectric pressure transducer. It can be approximated well by the expression:

$$A = 0 \qquad t < 0$$
$$A - A_0 \exp^{\alpha t} \qquad t > 0$$

where A_0 is the maximum amplitude of the disturbance, t the time, and α the damping coefficient.

The disturbance characteristics (A_0 and α) are affected by the following parameters of the shock tube and combustion chamber: cavity volume V; filling pressure p_{st}; combustion chamber pressure p_{ch}; molecular weight of the gas in the chamber μ; and diameter of the channel whereby the disturbance is introduced to the chamber d_c. By changing these parameters, it is possible to change the disturbance parameters.

The duration of the disturbance pulse depends on the pressure ratio p_{st}/p_{ch} (p_{st} is the initial pressure of the mixture in the shock tube), the shock-tube volume V_{st}, and shock-tube outlet area. Figure 9.3 shows the effects of p_{st}/p_{ch} and V_{st} on the pulse duration τ, which increases as these values increase [39]. The solid lines in Fig. 9.3 denote the values of τ calculated by the formulas for a supersonic flow of detonation products from the shock tube:

$$P = \frac{P_0}{(1 + Bt)\, 2\chi/\chi - 1}, \quad B = \frac{\chi - 1}{2}, \quad \frac{\mu m S_{cr}}{V_{st}} R\sqrt{T_0}$$

where p is the instantaneous pressure of detonation products in the shock-tube cavity; p_0 the initial pressure of detonation products in the shock tube at the moment of diaphragm burst; χ the adiabatic exponent of expansion of detonation products of a methane-oxygen mixture; t the time; μ the flow coefficient; S_{cr} the critical cross-sectional area; V_c the conditional volume of the shock tube; R the basic constant; and T_0 the initial gas temperature in the shock tube:

$$m = \left(\frac{2}{\chi + 1}\right)^{\frac{\chi+1}{2(\chi-1)}} \sqrt{\frac{\chi g}{R}}$$

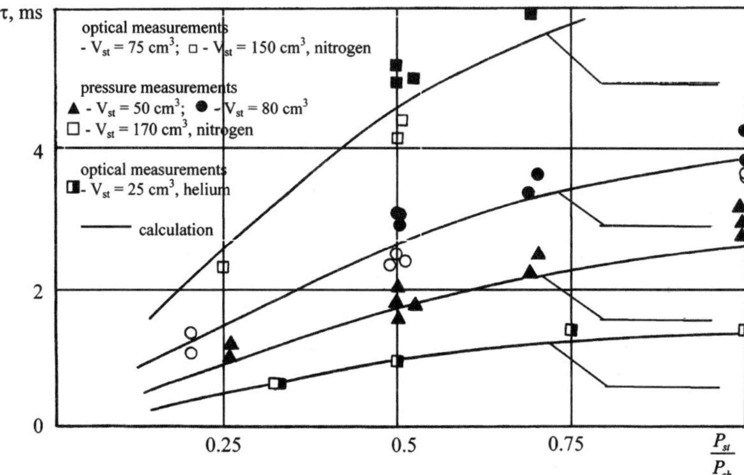

Fig. 9.3 Relationships between disturbance duration and diaphragm pressure differential p_{st}/p_{ch} as function of shock-tube volume V_{st} (d_t = 14 mm).

These relations were derived based on the assumptions that 1) the relation of the pressure between detonation products in the shock tube and the combustion chamber is supercritical (i.e., the pressure ratio is high enough to render the flow supersonic) and 2) there is no heat exchange between the gas and shock-tube wall.

The following quantities were employed in the calculations:

$$R = 3600 \frac{\text{kg cm}}{\text{kg degree}}; \quad T_0 = 3500\text{ K}; \quad \mu = 0.62; \quad p_0 = 11 p_{st}; \quad V_c = 1.6 V_{st}$$

in which the presence of the transmitting channel volume from the diaphragm to the shock-tube outlet is taken into account, and $\chi = 1.33$. The disturbances were introduced through a tube with a diameter of 14 mm.

The data determined experimentally from the disturbance duration measured by pressure transducers installed on the model chamber and from optical measurements with the use of device IAB-451 are shown in Fig. 9.3. The calculated curves for the time duration of a supersonic flow are in good agreement with experimental data on pulse duration. This provides a basis to use the aforementioned formula for determining disturbance pulse duration, with the chamber pressure p_{ch} taken as the instantaneous pressure p.

It has been obtained in experiments that the pressure pulse duration increases with decreasing outlet area. The duration τ remains unchanged when disturbances are introduced to a chamber filled with nitrogen, hydrogen, or helium. The result is also insensitive to the chamber diameter and the introduction direction.

The pressure disturbance generated by an IDD device has the pattern of a spherical shock wave. An ideal configuration of IDD is provided by an explosive charge in a spherical, thermal-insulated shell. Technically, however, it is difficult to manufacture such a charge. The waveform generated by a cylindrical IDD differs from that by a spherical one only in the close vicinity from the disturbance device. During explosion of a short cylindrical charge, a wave of large amplitude is generated with a rapid rise and drop of pressure. The explosion of a cylindrical charge with a length-diameter ratio L/D of less than six differs slightly from that of a spherical charge. The amplitude of an IDD-produced shock wave, to first approximation, is proportional to the explosion energy. Regretfully, existing shock-wave theories are unsuitable for calculating the amplitudes of disturbances generated by an in-chamber device.

Two assumptions made in shock-wave theories are not valid for IDD:

1) The location at which the wave intensity is measured is located far from an energy source, whose mass is negligible. For IDD, such a distance is comparable with a chamber diameter.

2) The entire explosion energy is transferred to the pressure wave. Energy losses caused by motion of the shell fragments are not taken into account.

The amplitude of the shock wave sharply decreases in a short time span during its propagation, for example, 50 or 60 μs during which the wave travels over a distance of 125–160 mm from the explosion epicenter. The wave structure and strength continue to change due to the interactions with the chamber operating process. Depending on the type of propellant and distribution of the atomized propellant near the injector head and IDD position, the wave either decays or grows more rapidly than that in an inert gas.

An essential drawback of IDD is a scatter of the shock-wave amplitude as high as 200%.

An uncertainty of disturbance intensity is caused by the variation of explosive charge density, heterogeneity of shell material, and nonuniformity of heat-insulating ablative material for the combustion chamber. Another drawback of IDD lies in the impossibility to provide a long residence time of the charge in the combustion chamber due to shell fracture.

Experimental studies on the effects of artificial disturbances have shown that at an initial moment of disturbance introduction the combustion-chamber process is significantly distorted. This phenomenon should be taken into account in processing and evaluating the chamber stability in response to the final disturbances.

A number of researchers made attempts to develop devices with minimum yield of gases following the shock wave, in order to reduce the disturbing effects on the chamber operating process. Such devices were developed on the basis of electrical discharges, various pulsators, and so on. But all of those devices could not be used for actual combustion chambers.

The amplitude of a disturbance recorded by a pressure transducer is determined by a number of factors listed in Fig. 9.1.

III. Choice and Substantiation of Pressure Pulses Generated in Combustion Chambers

Let us consider to what extent the effects of single disturbances in the form of shock wave or other types on the chamber operating process are equivalent. The degree of their discrepancy can be evaluated with an equivalence coefficient as follows.

A relation of maximum pulse intensities, provided those pulses of similar shape exceeding given pulses in intensity have the same effect on the operating process, is called a coefficient of pulse equivalence in their effect on the process. The equivalence coefficient of two pressure pulses with different shapes, P1 and P2, is defined as the ratio of the Weber numbers W1 and W2 at which fragmentation of the same droplets occur by pulses P1 and P2.

To calculate the coefficients of pulse equivalence, we will consider their effects on two specific operating processes as an oscillatory system. Such processes for liquid–liquid (and in some cases gas–liquid) combustion chambers include a liquid droplet breakup process defining the amplification of oscillations of rather high intensity and a process of disturbance propagation in the combustion chamber as in an acoustic resonator.

The process of droplet deformation and breakup for each type of disturbances shown in Table 9.2 (i.e., shock, sinusoidal, and triangular disturbances) can be described by their respective system of equations. One can then determine the critical values of the Strouhal numbers [19]

$$Sr = \frac{U_{\rm gm}T}{r_0}$$

where $U_{\rm gm}$ is the maximum value of the gas velocity in the disturbance pulse, T the time duration of pulse action (a disturbance period), and r_0 the initial radius of a droplet.

Table 9.2 Critical Strouhal numbers for disturbances of different types

Ref. no.	Disturbance	Analytical expression	Relative density	Critical Strouhal number ($We_0 = 70$)
1	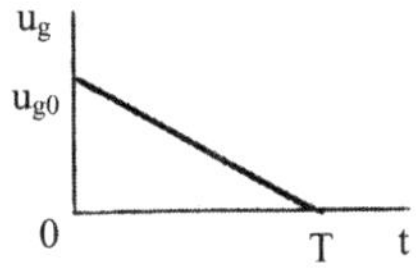	$U_g = \begin{cases} U_{g0}\left(1 - \frac{t}{T}\right) \text{ with } 0 \le t \le T \\ 0 \text{ with } t < 0 \text{ or } t > T \end{cases}$	$1 \cdot 10^{-2}$	124
			$1.5 \cdot 10^{-2}$	106
			$2 \cdot 10^{-2}$	94
2	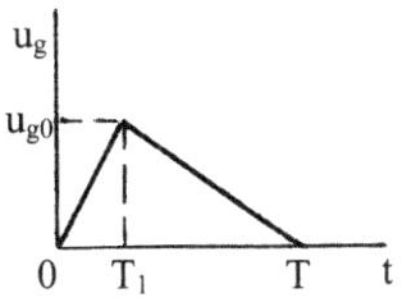 $\frac{T_1}{T} = 0.25$	$U_g = \begin{cases} \frac{U_{g0}t}{T_1} \text{ with } 0 \le t \le T_1 \\ U_{g0}\frac{(T-t)}{(T-T_1)} \text{ with } T \le t \le T_1 \\ 0 \text{ with } t < 0 \\ \quad t > T_1 \end{cases}$	$1 \cdot 10^{-2}$	93
			$1.5 \cdot 10^{-2}$	78
			$2 \cdot 10^{-2}$	70
3	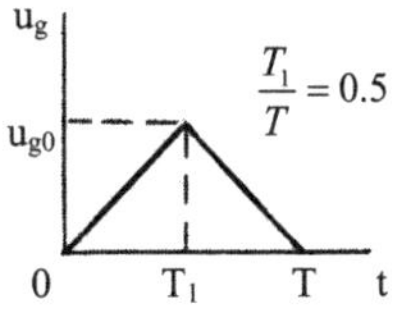 $\frac{T_1}{T} = 0.5$	$U_g = \begin{cases} \frac{U_{g0}t}{T_1} \text{ with } 0 \le t \le T_1 \\ \frac{U_{g0}(T-t)}{T-T_1} \text{ with } T \le t \le T_1 \\ 0 \text{ with } t < 0 \text{ or } t > T_1 \end{cases}$	$1 \cdot 10^{-2}$	77
			$1.5 \cdot 10^{-2}$	64
			$1 \cdot 10^{-2}$	57
4	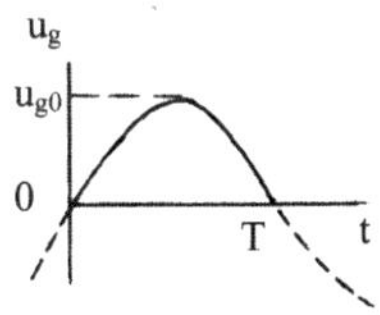	$U_g = \begin{cases} U_{g0} \sin\frac{\pi}{T} t \text{ with } 0 \le t \le T \\ 0 \text{ with } t < 0 \text{ or } t > T_1 \end{cases}$	$1 \cdot 10^{-2}$	63
			$1.5 \cdot 10^{-2}$	50
			$2 \cdot 10^{-2}$	47

These numbers define the critical intensity of disturbances that cause droplet breakup for given drop sizes and disturbance duration. A relation of critical Strouhal numbers for two disturbances is equal to the coefficient of their equivalence in droplet breakup.

The calculated critical Strouhal numbers are given in Table 9.2. The value of the coefficient of equivalence of two pulses in droplet breakup depends on pulse characteristics, droplet properties, and chamber conditions. In general, they can differ much from unity. The following notations are used in Table 9.2:

ρ_g = gas density near the combustion chamber injector head
ρ_l = liquid density
$We_o = \rho_g U_{gm} d_0^2/\sigma$ = initial value of the Weber number
d_o = initial drop diameter
σ = surface-tension coefficient

Disturbances with relatively high intensity, as a rule, have asymmetrical shapes. In Table 9.2 such disturbances are conventionally designated as disturbance no. 2. In some cases encountered in practice, the duration of pressure buildup in such disturbances is at least half the duration of its decay. The coefficients of equivalence for such disturbances with respect to the disturbances in the form of shock wave are generally less than 1.4 (Table 9.2). The preceding calculations are qualitative and show that the characteristics of disturbances, both natural and artificial ones, significantly affect the process. This fact should be taken into account in selecting a value for the final disturbance stability margin.

The coefficients of disturbance equivalence in their effects on acoustic resonators can be defined as a ratio of the maximum intensities of disturbances having index *i* or *k* in Table 9.3, provided that they have the same components in their frequency spectra at the frequency under consideration. Acoustic coefficients of disturbance equivalence depend on the frequencies at which the disturbances are compared.

The amplitude spectrum of disturbances is determined by decomposing them using a Fourier-series expansion, as shown for some typical disturbances in Table 9.3.

A sample calculation of acoustic coefficients of equivalence at different frequencies for disturbances of various forms presented in Table 9.3. is shown in Fig. 9.4. In general, there can be large deviations of acoustic coefficients of equivalence from unity. However, under actual combustion chamber conditions they do not deviate much from unity. For example, if artificial pressure pulses in a combustion chamber have a sinusoidal shape with a duration equal to the period of dominant self-oscillations, the acoustic coefficients of equivalence with artificial disturbances of form no. 3 (see Table 9.3) at the frequency of dominant self-oscillations are about 1.4.

It can be concluded from the preceding discussion that the difference of actual disturbances of relatively high intensity from a shock wave in their effect on a self-oscillatory system can be characterized by coefficients of equivalence, whose values ranges from 1 to 1.4.

Hence it follows that single disturbances in the form of a shock wave and other types, whose presence is most probable during the operation of an actual combustion chamber, are substantially equivalent to each other in their effects on some characteristics defining the operating process stability. It is desirable to compensate for possible differences of these disturbances in their effect on hard excitation of high-pressure oscillations, from the disturbances of different forms, by appropriate selection of a guaranteed stability margin. The guaranteed stability margin should allow for not only possible nonequivalence of pulses, but also various

Table 9.3 Characteristics of studied pulses: Fourier integral transformation of disturbances

Index i or k	Graphic presentation	Analytical expression	Fourier integral transformation
1	M, M_1, t_u, $T=2\pi/\omega$, t	$M(t) = \begin{cases} 0 \text{ with } t < 0,\ t > T_u = T/2 \\ M_1 \sin \frac{T_i}{T_u} \cdot \text{t with} \\ 0 \le t \le T/2 = T_u \end{cases}$	$S(w) = 2T_i T_u M_1 \cdot \left[\frac{1}{T_i^2 - T_u^2 w^2} \cdot \frac{\cos T_u W}{2}\right]$
2	M, M_2, T_t, T_u, t	$M(t) = \begin{cases} 0 \text{ with } t < 0,\ t > T_u \\ M_2/T_1 \cdot t \text{ with } 0 \le t \le T_1 \\ \frac{M}{T_u - T_1}(t_u - t) \text{ with } T_1 \\ T_1 \le t \le T_u \end{cases}$	$S(w) = \frac{M_2}{w^2}\left\{\left[-\frac{2}{T_1}\sin^2 T_1 \frac{w}{2} - \frac{1}{T_u - T_1}(\cos wT_u - \cos wT_1)\right] + \left[\frac{1}{T_1}\sin wT_1 - \frac{1}{T_u - T_1}(\sin wT_u - \sin wT_1)\right]\right\}^{\frac{1}{2}}$
3	M, M_3, t	$M(t) = \begin{cases} 0 \text{ with } t < 0 \\ M_3 e^{-\alpha t} \text{ with } t > 0 \\ \alpha > 0 \end{cases}$	$S(w) = \frac{M_3}{\sqrt{\alpha^2 + w^2}}$

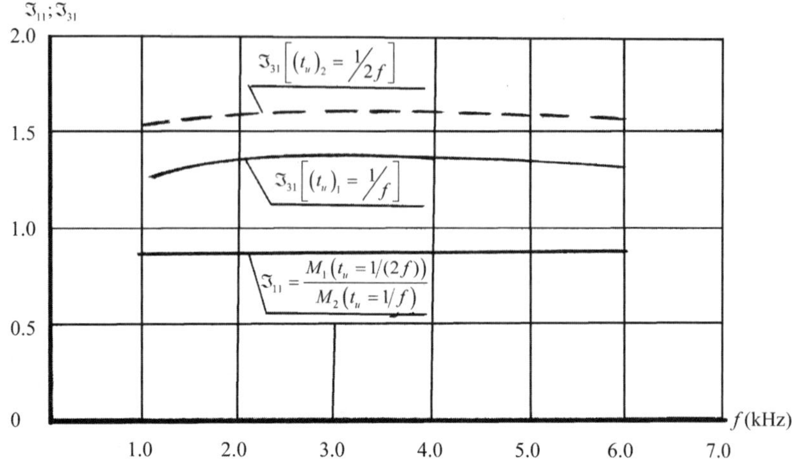

Fig. 9.4 Calculated values of pulse equivalence acoustic coefficients $\Im_{ik} = M_i/M_k$ (i and k are pulse numbers in Table 9.3).

measurement and data-processing errors, as well as the availability of actual stability margin. It is recommended to provide the margin in the form of a ratio between a minimum artificial disturbance intensity resulting in instability and a maximum natural disturbance intensity, which might be encountered during operation of the combustion chamber as part of the test procedure.

$$\frac{A_{\text{cr min}}}{A_{\text{noise max}}} = N$$

IV. Design Features of Disturbance Devices

This section provides a brief description of design features of disturbance-generating devices that have been used in Russia since 1966 for estimating the process stability margin with respect to hard excitation of high-frequency pressure oscillations. A partial list of these devices is given in Table 9.1.

The shock-tube design shown in Fig. 9.5 contains a cavity in the case (1) filled with a stoichiometric mixture of methane and oxygen and separated from the combustion chamber by a burst diaphragm (2). Upon initiating detonation in the cavity, the diaphragm bursts, and a disturbance is introduced into the combustion chamber. The cavity filled with the detonating mixture has an acceleration region (3) to initiate detonation. When the flammable mixture is ignited at the end of the acceleration region, the combustion is accelerated in this region and transits to detonation, which then initiates detonation of the remaining mixture in the cavity. Detonation of the mixture in this device results in a sharp rise of the pressure acting on the diaphragm at the moment when the detonation wave reaches it, that is, at the moment of actual burnout of the flammable mixture in the shock tube. The device enables good stability of disturbance characteristics over a wide range of pressure across the diaphragm. Such a requirement cannot be met in devices

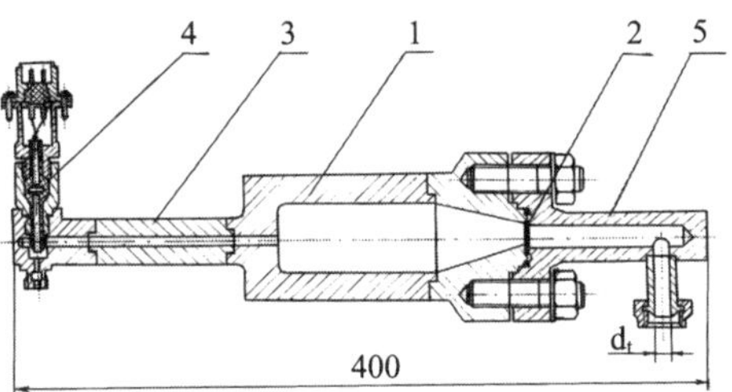

Fig. 9.5 Shock-tube design: 1, case; 2, diaphragm; 3, acceleration tube; 4, mixture ignition device; and 5, section connecting with combustion chamber.

using deflagration of gas mixtures and gun powder. In the present shock-tube design, the combustion products are practically inert. A loading system, which allows one to automatically maintain a stoichiometric composition of the loaded mixture, is used for charging the shock tube.

One of the disadvantages of the design is that the pulse amplitude is limited by the loading pressure in the gas volume. At a pressure over ~6.0 MPa from a spontaneous detonation, a bulky test stand is required for loading the shock tube by a stoichiometric mixture of methane and oxygen.

The IDD, as shown in Fig. 9.6, contains an explosive charge (hexagen, i.e., RDX/HMX) placed in a case made of a special thermal insulation material KTAF-10. The charge is ignited by a blasting cap of beam action (KD) TAT-1-1 with a power igniter. The blasting cap is initiated immediately after burnout of the protective shell, whose thickness regulates the time delay of the device initiation.

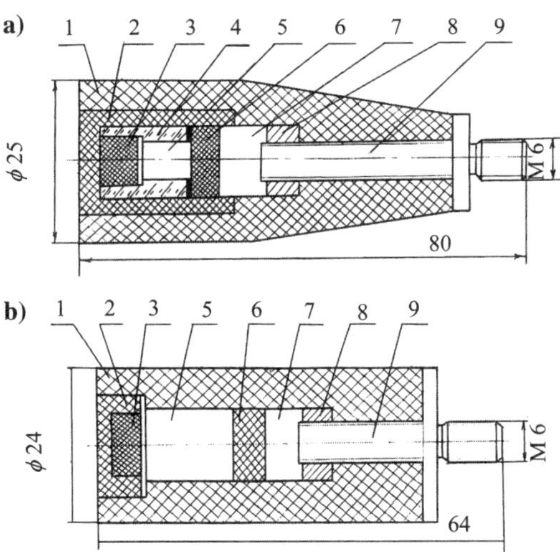

Fig. 9.6 Design of in-chamber disturbance-generating device a) for combustion chambers and b) for gas generators: 1, case; 2, sleeve; 3, gunpowder; 4, bush; 5, blasting cap; 6, explosive charge; 7, epoxy resin; 8, nut; and 9, screw.

The advantages of this device are 1) the possibility of placing it into a combustion chamber (or gas generator) without breaking the wall integrity and 2) its simplicity in that no support system is required for its operation. The disadvantages are 1) the lack of long residence in the combustion chamber, 2) the difficulty of introducing the disturbance at a desired time, 3) a large scatter of pressure disturbances, and 4) limitation of blasting cap initiation by a pressure of up to ~20.0 MPa.

An external multipulse disturbance device (MDD) with use of an explosive, hexagen, was developed to generate detonation waves for stability evaluation in 1974–1975. Because explosive charges were placed in the device cavity, the charges do not undertake such high thermal loads as do the charges for in-chamber disturbance devices.

The design of an MDD with four charges of an hexagen in the textolyte shell is shown in Fig. 9.7. Each charge consists of an igniter, a blasting cap, and an explosive. The pressure pulse is introduced into the combustion chamber through the connecting channel at a prespecified time. The charge is secured on a threaded sleeve.

A double sealing is used to prevent gas leakage from the explosion chamber. The first sealing prevents slow gas flow from the explosion chamber to the cavity underneath the cap after charge explosion. Tightness is provided by pressing the cap through the gasket to the device case by means of a pressure disk. With such design of assembly, the sealing part is not subject to the action of a powerful shock wave. Such a design, however, complicates the supply of an electrical pulse to the igniter.

The charge explosion results in a very high pressure in the explosion chamber (Fig. 9.7) that necessitates the use of an MDD with thicker walls and hence of larger weight.

Figure 9.8 shows the pressure disturbances generated by different excitation devices as a function of charge mass. Details of combustion chambers U107 and S5.23 are given in Chapter 11.

For charge mass from 1 to 4 g in the combustion chamber U107, the MDD and shock tubes provide approximately the same pressure disturbances. The amplitude of disturbances generated by IDD in the combustion chamber S5.23 is substantially higher than that by shock tubes by a factor of 3–4. The pulse intensity generated by MDD depends on the relationship between charge mass and explosion-chamber volume.

In addition to the heavy weight of the device, the disadvantages, as well as disadvantages, of IDD are 1) the application is limited by the pressure in the combustion chamber because the detonator is under pressure because of a diaphragm, 2) reliable initiation occurs up to a pressure of ~20.0 MPa, and 3) there are high energy losses for detonation of the charge case.

The most successful development is an external disturbance device based on low-velocity detonation, which was studied in detail quite recently [40 and 41]. In general, three different combustion scenarios can occur in the condensated explosive: convective combustion, low-velocity detonation (LVD), and normal detonation (ND). Convective combustion is unstable and under some conditions changes over to LVD. Past studies have shown that at a particular charge diameter and length, along with certain types of shell material and shell thickness, the LVD behavior becomes quasi-stationary.

The velocity and pressure at the wave front during combustion of moderated hexagen are shown in Table 9.4. Because of considerable pressure reduction in the

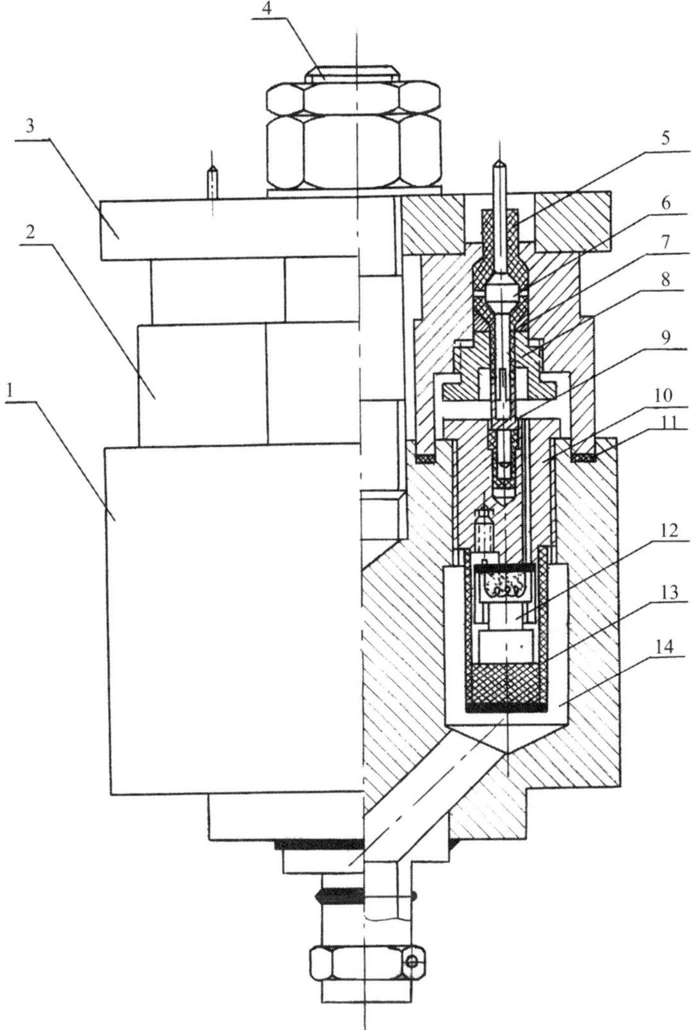

Fig. 9.7 Multipulse disturbance-generating device: 1, case; 2, sleeve; 3, flange; 4, pin; 5, insulator; 6, contact conductor; 7, bush; 8, bush; 9, wiping gland; 10, threaded bushing; 11, gasket; 12, charge (igniter, detonator, explosive); 13, charge case; and 14, explosion chamber.

detonation wave and decrease in its propagation velocity, LVD occurs in enclosed volumes (shells) at 100% charging density without mechanical damage of the metal shell with a thickness of 2–5 mm.

For certain types of explosives, values of charge length and diameter and shell thickness were selected such that the transition of LVD to normal detonation is completely ruled out. Initially LVD was used with a bulk charge of moderated

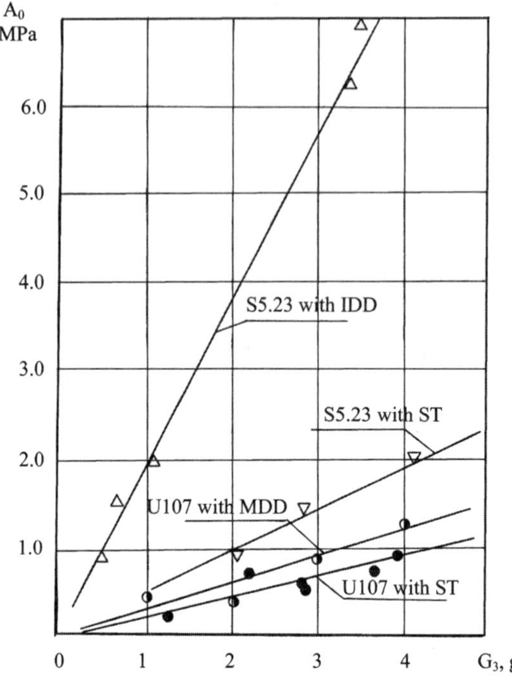

Fig. 9.8 Pressure disturbance generated by different excitation devices against explosive charge mass.

hexagen. In spite of some advantages of this DD as compared to MDD, the studies on its improvement were stopped because of major deficiencies of the bulk charge. Storage and transportation of charges, and DD vibrations during engine tests result in uncontrollable change of explosive charge density. Further initiation of such charges in DD results in considerable change of the disturbance pulse and even in the transition to normal detonation.

A device with a pressed porous charge has been developed to eliminate the drawbacks of a bulk charge (see Fig. 9.9). An explosive charge consisting of 1–4 pressed blasting tablets made of moderated hexagen is placed into the case (3) with a diameter of 6 or 8 mm. To increase the efficiency of the main charge

Table 9.4 Velocity and pressure at wave front during combustion of moderated hexagen

Substance	Combustion behavior	Propagation velocity, m/s	Pressure at wave front, kgf/cm^2
Moderated hexagen	Convective	0.2–5.0	1…10
	Low-velocity detonation	800–3000	(3.5–15)10^3
	Normal detonation	8000–8500	(150–200)10^3

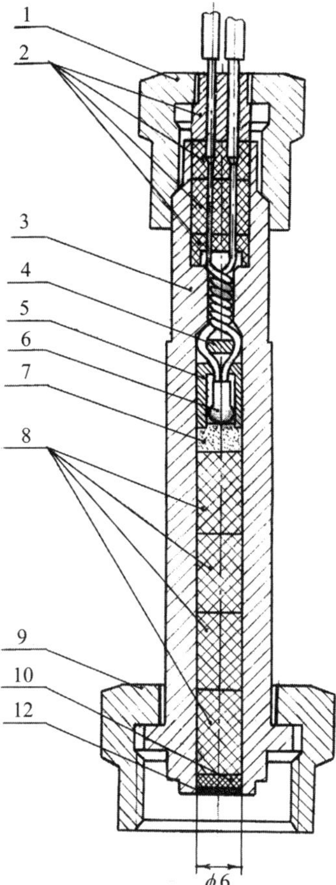

Fig. 9.9 Explosive charge made of four pressed porous moderated hexagen with a diameter of 6 mm: 1, cap nut; 2, seals; 3, case; 4, labyrinth; 5, bush; 6, electric initiator MB-2N; 7, granulated hexagen; 8, pressed hexagen; 9, cap nut; 10, tamping plut; and 11, hermetic.

initiation, 0.2 g of hexagen powder is placed between the electric igniter MV-2H located in this part of the case and pressed explosive charge. Because of its low bulk density ($\rho = 0.83–1.0$ g/cm^3), the hexagen powder affords a "soft" mode of the main charge ignition. The sealing elements (2) provide tightness of conductors brought out of the case.

The explosive tablets placed in the case are covered with a wad and encapsulated in the epoxide compound. Structurally, a single-charge device is made such that it becomes an element of a multicharge device.

A five-charge device based on LVD is shown in Fig. 9.10. The case (1) of the device is a collector connected to the device outlet channel (8). A nipple (6) located at the center of the case and a combustion chamber nipple are employed to purge residuals from the device and to prevent accumulation of residuals.

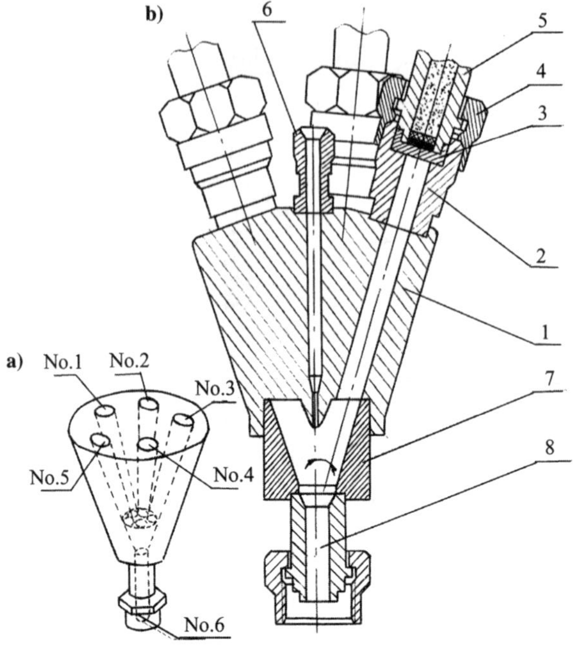

Fig. 9.10 Five-charge device based on LVD: a) five-charge (chamber) disturbance device and b) scheme of relative positions of channels in the device, where 1, case; 2, nipple; 3, diaphragm; 4, cap nut; 5, charging chamber; 6, purging nipple; 7, reducing tube; and 8, outlet channel.

Five-charge DDs are available in two different sizes using charges of 6 or 8 mm. The corresponding charge mass changes 0.6 to 3.6 g, allowing a wide-range variation of the amplitude of the disturbance introduced into the combustion chamber.

In the assembly connecting the charge with the DD case, a duralumin diaphragm (3) is mounted. It protects the charge from the combustion chamber and serves as a soft sealing element in the connection. The membrane thickness depends on the pressure loads arising from the initiation of adjacent charges. Trapping of diaphragm fragments to prevent them from entering the turbine is required, when using LVD DD in LRE gas generators.

Figure 9.11 shows the design of a two-charge DD with a trap channel for diaphragm fragments. Two charging chambers (4) with pressed porous charges are mounted on the device case (1) at an angle of 17.5 deg to the central axis. The chambers are separated from the channels in the case by a burst diaphragms (5). From the turn-off point of the inclined channels, the shock-wave disturbance propagates into the object under investigation through the channel (8) turning around 90 deg, and the diaphragm fragments move to the trap through the channel (7).

A comparison of pulse magnitudes of MDD and LVD DD is shown in Fig. 9.12. MDD produces a considerably weak pulse as compared to LVD DD and has a wide scatter of an initial pulse. LVD DD devices met most requirements and led

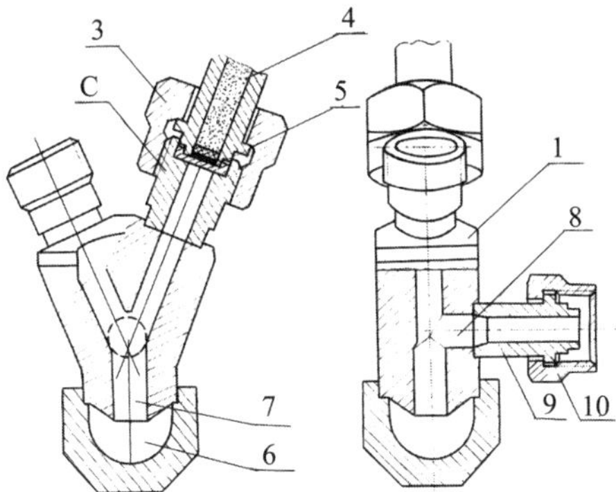

Fig. 9.11 Two-charge disturbance device for gas generator: 1, case; 2, nipple; 3, cap nut; 4, charging chamber; 5, diaphragm; 6, trap; 7, trap channel; and 8, outlet channel.

to successful engine tests with a pressure in the combustion chambers up to 27.0 MPa and in gas generators to 70.0 MPa.

V. Dependence of Pressure Disturbance Characteristics on Physical and Design Parameters

The characteristics of pressure disturbances were studied using the experimental setup shown in Figs. 9.13–9.15. The facility shown in Fig. 9.13 includes a cylindrical combustion chamber with a diameter of either 280 or 400 mm, and is equipped with transducers of the LH-601-type (PT_0–PT_4) filled with inert gas. Disturbances were introduced into the model combustor with shock tubes located at different distances from the head and in different directions. The facility

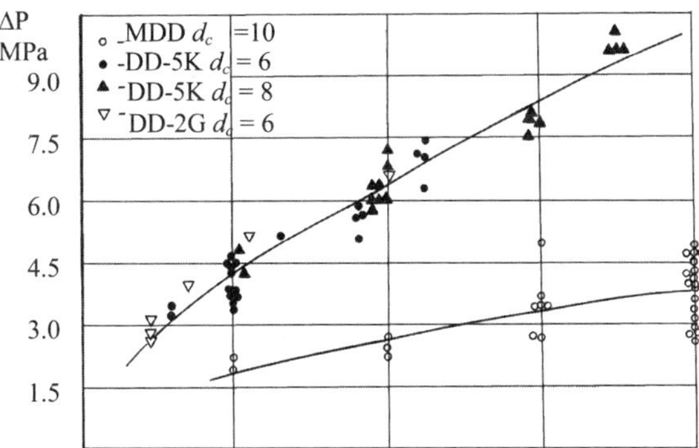

Fig. 9.12 Variation of differential pressure ΔP in the shock wave as a function of charge mass (G) in the devices: d_c = 10 mm for MDD; d_c = 6 and 8 mm for DD LVD; and d_c = 6 mm for generator DD LVD (medium-air: backpressure P_{ch} = 5.0 MPa; d_c = channel diameter in DD case).

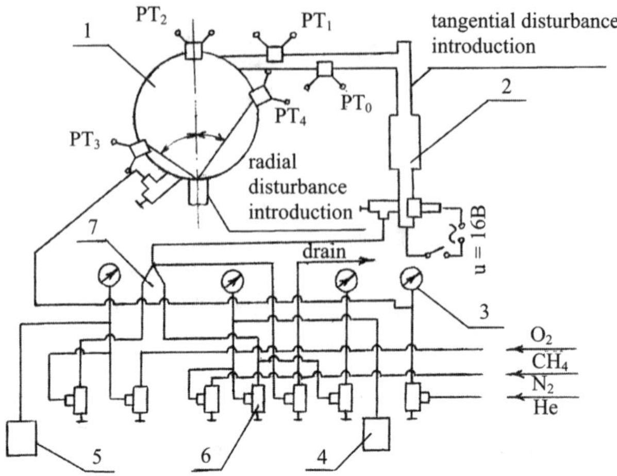

Fig. 9.13 Experimental setup for studying pressure disturbances in a model combustion chamber: 1, model chamber; 2, shock tube; 3, gas filling system; 4,5, measuring tanks; 6, filling value; and 7, mixer.

represented in Fig. 9.14 allows one to determine with high accuracy characteristics of pulses generated by the disturbance devices and hence to observe the influence of some relatively weak factors on them. It consists of a cylindrical pipe with a length of 210 cm and a diameter of 56 mm connected to a shock tube or other disturbance devices. The disturbance is introduced into the pipe in the axial direction from one end. The pipe is filled with an inert gas. Disturbances are recorded by the LX-601-type of pressure transducer. The disturbance characteristics, such as the relations between shock-tube parameters, obtained in the both setups, agree qualitatively well.

The dependence of the relative intensity of pressure disturbance recorded by the transducer PT_2 (see Fig. 9.13) on the initial pressure in the shock tube is shown

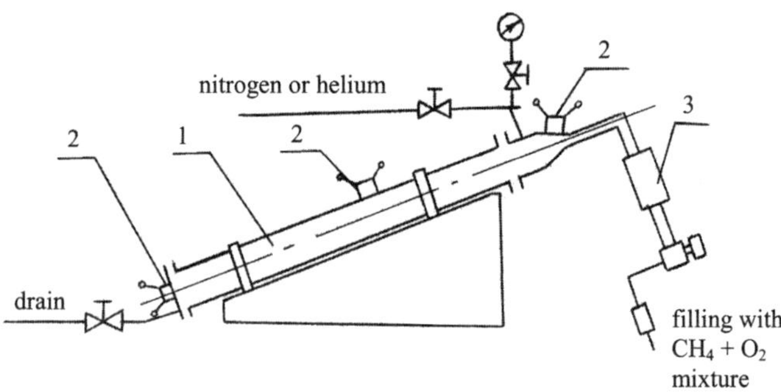

Fig. 9.14 Experimental setup for studying the effects of various parameters on pulses generated by a shock tube: 1, cylindrical pipe; 2, pressure transducers; and 3, replaceable shock tubes.

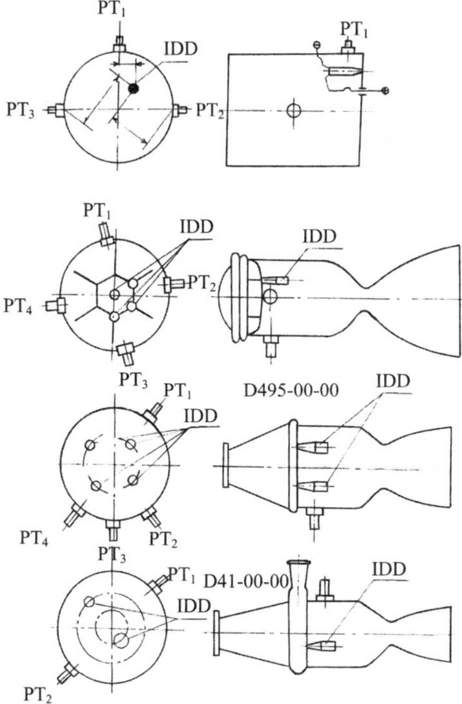

Fig. 9.15 Experimental setups for various in-chamber distance devices (IDD); PT is pressure transducer.

in Fig. 9.16. Disturbance was introduced into the chamber tangentially through an orifice with a diameter $d_t = 14$ mm. The chamber pressure was $P_{\text{ch}} = 2.0$ MPa. Shock tubes with volumes of 75 and 150 cm^3 were used.

It can be seen that the relative intensity of the disturbance increases linearly with increasing charging pressure (i.e., the mass of the flammable mixture in the shock tube). This result takes into account the increase in detonation wave intensity as the initial pressure in the tube increases and hence the rise of pressure disturbance produced by it. To illustrate this, the disturbance intensity in the transmitting channel as a function of the gas-mixture charging pressure in the shock tube is given in Fig. 9.17. The shock-wave intensity in the transmitting channel increases linearly with the increase in the initial pressure of the gas mixture in the shock tube.

The effect of shock-tube volume on the disturbance intensity is shown in Fig. 9.16. The disturbance intensity slightly increases with increasing volume of the shock tube. The phenomenon can be attributed to the replenishment of the shock wave propagating through the chamber with the energy of the gas stream flowing out of the shock tube behind the disturbance front. The replenishing effect manifests itself in the chamber volume at some distance from the shock-tube exit and has no effect, for example, in the transmitting channel itself of the disturbance device.

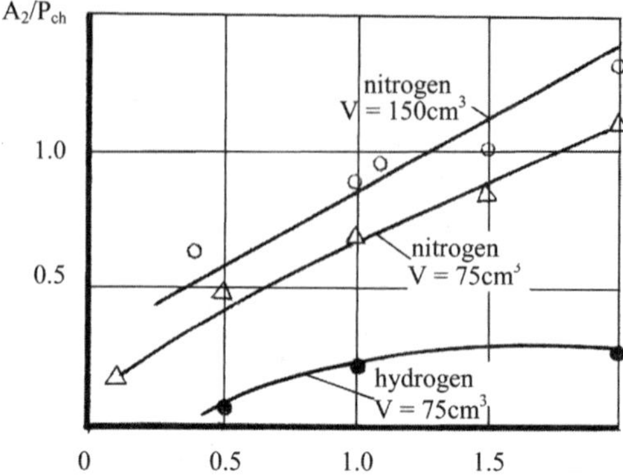

Fig. 9.16 Relative pressure amplitude measured by transducer PT2 against initial pressure in the shock tube (average values for a series of experiments are shown).

The latter is seen in Fig. 9.17, where the points corresponding to the pressures in the transmitting channel with different volumes of the shock tubes coincide.

The replenishment of the shock wave in the chamber with the energy of the outgoing gas stream is influenced by a number of factors. These include, for example, the presence and the shape of the wall along which the disturbance propagates, the angle at which the transmitting channel is directed to the combustion chamber, etc. These effects were observed by means of the apparatus shown in Fig. 9.13 and the high-speed photograph of the process of the disturbance entering the volume using Tepler's device IAB-451. In particular, when the disturbance is introduced radially into the combustion chamber, the disturbance recorded by all of the pressure transducers is much weaker than that in the case with a tangential introduction.

The molecular weight of products filling the combustion chamber has a quite strong effect on the intensity of generated disturbances. Thus, for disturbances

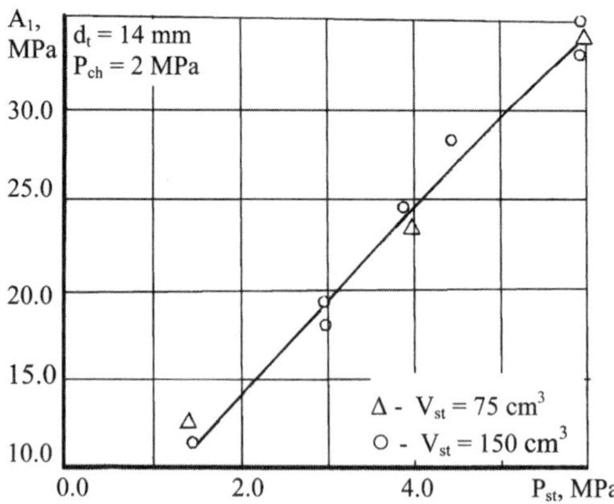

Fig. 9.17 Disturbance amplitude in the transmission channel against gas mixture pressure in the shock tube.

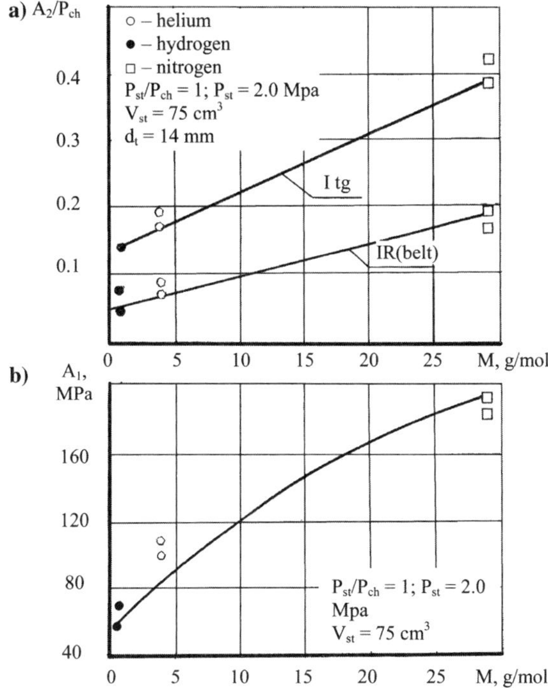

Fig. 9.18 Disturbance value against a type of gas filling the model chamber: a) relative value of disturbance in the chamber and b) value of disturbance in the transmitting channel.

propagating in the tangential direction in the model combustion chamber filled with hydrogen and helium, their intensity has a value of $A_2/P_{ch} = 0.15$ and 0.2, respectively, as opposed to $A_2/P_{ch} = 0.4$ in nitrogen under the experimental conditions marked in Fig. 9.18. (A_2 is the pressure in the shock-wave front, and P_{ch} is the chamber pressure). Simultaneously, the pressure-disturbance intensity in the transmitting channel assumes a value of 17.0–18.0 MPa in nitrogen, 10.0–11.0 MPa in helium, and 6.0–7.0 MPa in hydrogen.

An essential change of the magnitude of pressure disturbance in the volume can be obtained by varying the transmitting channel area at the disturbance entrance into the combustion chamber. Figure 9.19 shows the relative disturbance intensity against the transmitting channel diameter d_t. The disturbance amplitude A_2 decreases sharply with decreasing diameter d_t. With d_t decreasing to 4 mm, a "shockless" gas flow behavior was observed with constant parameters during some period of time.

Figure 9.20a shows some general results obtained from the experiments with tangential introduction of disturbance into the chamber. The plots indicate the weakening of the pressure wave as it moves along the chamber wall around the circumference $l_{circ.}$, whose starting point is on the edge of the tangential introduction

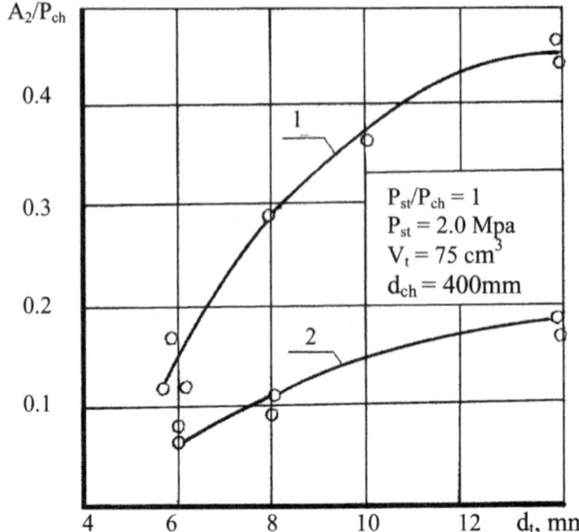

Fig. 9.19 Value of the highest pressure disturbance against transmission channel diameter.

hole. The distances of 0.1, 0.6, and 1.15 m correspond to the locations of pressure transducers PT_2, PT_3, and PT_4. It is also observed from the relationships between the amplitude of the pressure wave directed tangentially and the outlet diameter, as well as that kind of gas filling the chamber, that the relative degree of wave weakening is almost independent of the pressure differential across the shock-tube diaphragm (i.e., the wave intensity). In addition to the distance from the disturbance introduction point, the direction of the disturbance entering the chamber has a considerable effect on the measured value of disturbance.

The experiments made it possible to ascertain that with introduction of a disturbance into the combustion chamber in the tangential direction (or in the direction tangential to the circumference with a radius constituting a part of the chamber radius) the weakening of shock wave is less than that with radial introduction. For example, for the experimental conditions shown in Fig. 9.19, the value of pressure disturbance recorded by transducer PT_3 with radial introduction is about 60% of its counterpart with tangential introduction. This ratio is valid in spite of the fact that the shortest distance from transducer PT_3 to the radial introduction point is less than that of tangential introduction.

The experiments with the use of burst diaphragms in a shock tube, which have different thickness in the range of 0.1–0.9 mm, have shown that the disturbance generated by a shock tube practically does not depend on the diaphragm thickness over a wide range of the diaphragm thickness. Thus, the use of detonation in a shock tube does not impose any special requirements on the diaphragm quality in order to provide stable disturbance characteristics. Because there might be some deviations of a gas mixture from its stoichiometric composition during experiments, the decrease in pressure disturbance intensity with the change of mixture composition within its detonation limits was evaluated experimentally. Results have shown that within the detonation limits of the mixture deviations of disturbance intensity with respect to the methane content in the mixture are observed.

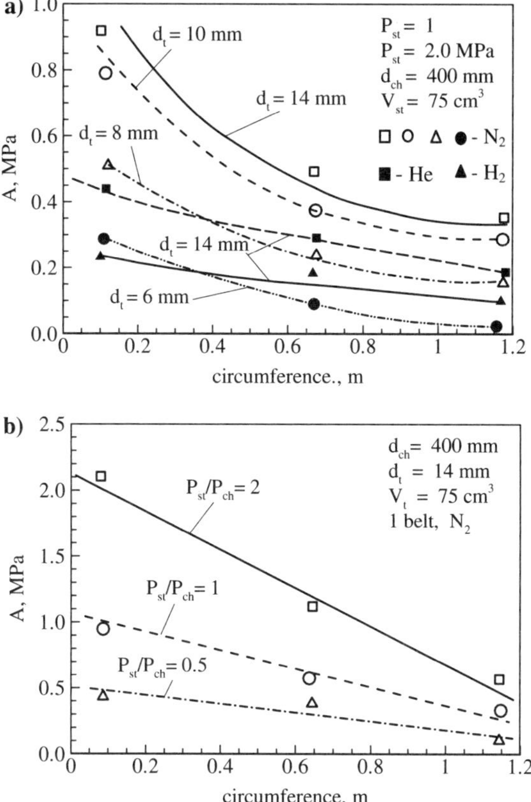

Fig. 9.20 Relationships between values of pressure disturbances and distance covered by the shock wave along the chamber wall when the disturbance is introduced tangentially. (The average values for a series of experiments are shown.)

The deviations are insignificant near the stoichiometric ratio, but become more significant near the detonation limits. The decrease in disturbance intensity is less for rich mixtures than that for lean mixtures.

The characteristics of disturbances produced by an IDD were studied using the experimental setups shown in Fig. 9.15a. Each setup consisted of a cylindrical vessel filled with either nitrogen or hydrogen at a pressure of 0.1–10.0 MPa. The speed of sound at a normal temperature is $C_{N_2} = 340$ m/s and $C_{H_2} = 1290$ m/s in nitrogen in hydrogen, respectively. To study the operability of the blasting cap and explosive charge system under a pressure of up to 5.0 MPa, a cylindrical vessel with a volume of 5 liters was used with nitrogen filled to the required pressure.

The in-chamber device was installed according to the design specifications for combustion chambers (i.e., it was attached to the sealing cover of the vessel at 35 mm from the cylinder axis). In this case, the center of the explosive charge was at 500 mm from the vessel bottom. IDD was initiated by heating up a wire by an electric current at a voltage of 10 V. The layout of pressure transducers (Fig. 9.15a)

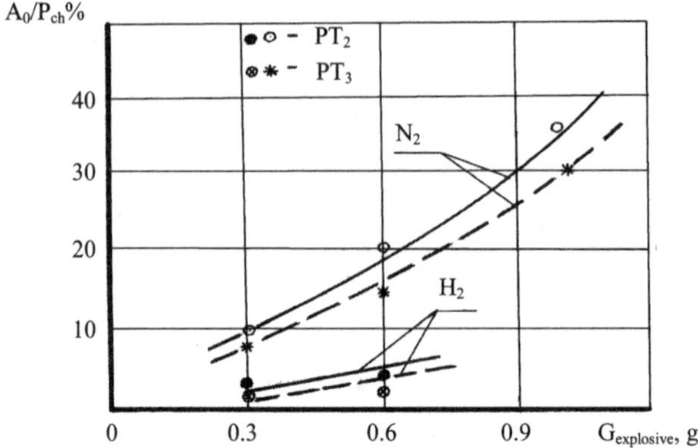

Fig. 9.21 Relative disturbance amplitude as function of IDD charge mass.

in the experimental setup and their radial distance from the center of the chamber are similar to those used in the combustion chambers of engines 15D117, 15D79, S5.23, and 4D75 (see Chapters 11 and 13). Figure 9.21 shows the effect of explosive weight on the disturbance intensity under a gas pressure of $P_{ch} = 5.0$ MPa in the chamber filled with nitrogen or hydrogen. It is clear that the disturbance intensity increases linearly with the explosive weight, increasing from $A_0 = 0.5$ MPa for $G = 0.3$ g to $A_0 = 1.7$ MPa for $G = 1.7$ g. With the speed of sound in the gas increasing from $C = 340$ to 1290 m/s, a reduction in disturbance intensity by a factor of five is observed.

The results obtained indicate a weak effect of the distance from the explosive to the pressure transducer on the disturbance intensity. (The average values of disturbance amplitude for two transducers are shown on the plot.) Thus, for the accepted layout of transducers, a distance increase of 1.18 times results in an amplitude decrease of 1.1 times, which is within the limits of measurement error and should be considered as a qualitative indication of the disturbance weakening with the distance from IDD to the transducer.

To determine the value of artificial disturbance responsible for combustion instability, the stability of the disturbance produced in a combustion chamber (or in a gas generator) is an important factor. For more accurate determination of the disturbance amplitude, it is necessary to know the values of its scatter in a number of successive blasts of IDD. In the experiments with the explosive weight from 1.0 to 9.9 g in an inert medium at a pressure of 1.0–4.0 MPa, the scatter of disturbance amplitudes did not exceed 40% with the distance between the measuring locations of transducers and the explosive being 200 mm.

Disturbances produced from an explosive charge mass of 0.3–1 g are required to estimate the process stability in combustion chambers with high operating pressures (e.g., $P = 15.0$ MPa). When IDD is made with explosive of this weight, more stringent requirements are imposed on the manufacturing accuracy for all of the operating components of an in-chamber disturbance device. The explosion initiator, blasting cap TAT-1-1, is an off-the-shelf commercial product. According to the nameplate data, the deviation of explosive weight in the detonation chamber does not exceed 1–2%. The charge of hexagen is weighted with an accuracy of

2–3%. The reliability of detonation transmission from the detonation chamber to the explosive charge is guaranteed by the fact that the explosive charge mass is always greater than its critical value. Therefore, possible scatter of the disturbance amplitude mainly results from the process of manufacturing a standard IDD.

Initiation of numerous IDD with explosives of 0.3, 0.6, and 1.0 g in nitrogen and hydrogen under a pressure of $P_{ch} = 5.0$ MPa was made in order to determine the magnitude of pressure disturbances. An analysis of the results has shown that the widest scatter of pressure disturbance amplitudes does not exceed 40%.

Within measurement error limits, the influence of the thermal-insulation material thickness in the range of 5–7 mm, and the mechanical properties variation caused by the action of combustion products in a LRE chamber in the period of 5–8 s, on the disturbance characteristics was not observed. With an explosive mass increase beyond 1.4–1.5 g, the effect of the case material becomes vanishingly small compared with the case thickness in the range of 0–7 mm.

The experimental study of IDD operating characteristics for liquid–liquid systems was carried out in the combustion chamber S5.23 operating on NTO and UDMH propellants. Bipropellant centrifugal injectors were installed on the combustor bottom around the concentric circles. The total flow rate of the propellants was $G_{\Sigma} = 30$–52 kg/s. The combustion chamber pressure P_{ch} ranged from 4.0 to 10.0 MPa, and the oxidizer-to-fuel ratio K is 2.4–2.9. There were separating baffles with a height of 30 and 60 mm in some of the combustion chambers. A diagram showing the IDD charges is given in Fig. 9.15b.

Developmental testing of IDD for gas–liquid chambers was carried out on the experimental setup D495-00-00, which was arranged according to a closed circuit (i.e., staged-combustion cycle) with afterburning of oxidizing mixtures from the gas generator, as shown in Fig. 9.15c. The total flow rate of propellants was $G_{\Sigma} = 34$–46 kg/s. The chamber pressure P_{ch} ranged from 6.5 to 14.0 MPa; the oxidizer-to-fuel mass ratio K is 2.1–3.2; the relative flow rating q is 1.08–4.4; and the gas velocity at the injector outlet is 25–240 m/s.

Experimental studies on the device performance for gas–gas engines were carried out on the setup D41-00-00 (components are TG-02 and NA-27I). The layout is shown in Fig. 9.15d.

The overall results of disturbance amplitudes vs explosive charges are summarized in Fig. 9.22. The data for the combustion chambers of engines 4D75, 15D79, 15D117 are included. With the help of the IDD design, artificial disturbances can be produced in combustion chambers in the entire intensity range required for stability estimation, according to different operating procedures.

From the analysis of the preceding results and the comparison with the data obtained from the experiments in the model system (Fig. 9.15a), it follows that the disturbance intensity increases as the explosive charge mass increases. With an equal mass of the explosive charge, the intensities of disturbances recorded by the pressure transducers are different for different combustion chambers. The parameter $K = A_0/A_M$, where A_0 is the pressure amplitude in the combustion chamber being tested and A_M the pressure amplitude in the model chamber, characterizes intensification of the process.

The thickness of the end part of the thermal-insulation case made of textolite was used for adjusting the device initiation time. The experimental studies on engines 15D117, 15D79, 4D75 have shown that with the end wall thickness of 2–4 mm the IDD initiation time delay is 1.5–3 s from the moment of process

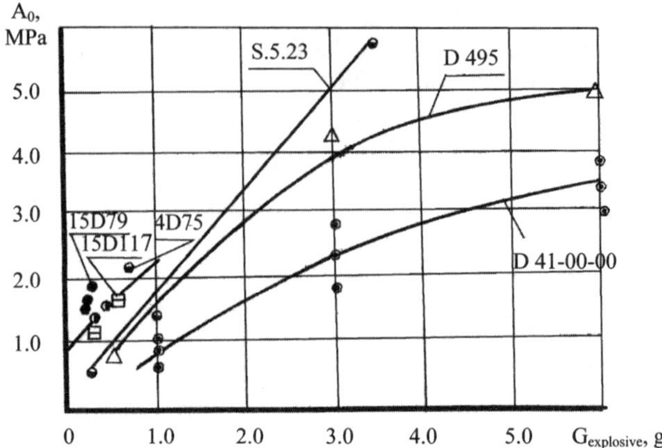

Fig. 9.22 Disturbance intensity as function of IDD charge mass.

"setting up." Such a time period is sufficient for combustion chambers whose warm-up time τ is less than 1.5 s. When IDD are used in gas generators, the initiation time can be increased by thickening the end wall made of the material KTAF-10.

By analyzing the relative intensities of disturbances produced in a combustion chamber by a disturbance device, and comparing the results with the disturbances produced by a shock tube with a gas-mixture weight equal to the explosive charge weight, it was found that disturbances generated by IDD are three to four times higher than those generated by a shock tube under the same condition. This is a natural consequence of the fact that during explosion of condensed explosives more energy passes into the shock wave than during gas explosion. In addition, introduction of disturbances into the combustion chamber from a shock tube or other sources located outside the chamber involves the use of a transmitting channel, in which significant explosion energy losses always occur.

From the data set forth in this chapter, it follows that a device with the use of LVD charge as a source of a pressure pulse fully satisfies all of the requirements for generating artificial disturbances in combustion chambers and gas generators. This is further confirmed by the long-term applications of the devices for evaluating the stability margin with respect to hard excitation of actual and experimental chambers and gas generators.

The characteristics of disturbance propagation inside of the hexagen charge with LVD and its ensuing effect on the chamber process are similar to those provided by a shock-tube device. In a shock tube, the shock wave propagates. In the charge with LVD, the shock wave propagates through the porous charge of hexagen. The influence on the process is also similar to a shock tube.

Devices with LVD, however, have the following features: the requirement of introducing disturbances into the chamber through a small-diameter channel, very high pressures in the combustion chamber of 25–30 MPa, the presence of vibration baffles in the combustion chamber, the interaction of charges in five-pulse device, and so on. Additional studies are needed. Experimental results for disturbance devices with LVD were obtained in laboratory setups including

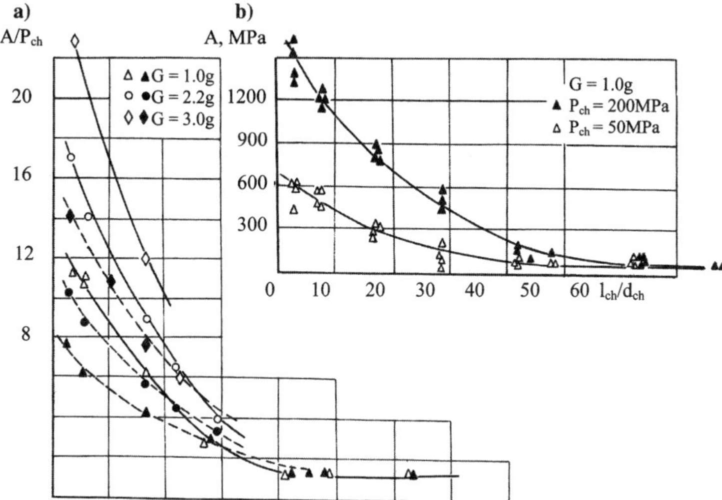

Fig. 9.23 Damping of disturbance amplitude in the transition channel between DD and the chamber when it is filled with the air under a pressure of $P_{ch} = 5.0$ and 20.0 MPa: a) variation of disturbance intensity and b) variation of disturbance amplitude.

cylindrical chambers with a diameter of 56, 90, 130, 155, 255, and 400 mm filled with air, nitrogen, or helium up to 15.0–35.0 MPa. Five-charge and two-charge devices with explosive mass of $G = 0.52$–3.6 g generate shock-wave disturbances in the studies.

Figure 9.23 shows the experimental relationship between the disturbance amplitude and the dimension of the straight channel through which the disturbance propagates. The experiments were carried out with channels filled with air under a pressure of 15.0 and 20.0 MPa. The disturbance was produced by a charges of 1.0–3.0 g.

The relative length was defined as the ratio of the length of the channel segment, over which the disturbance passed, to its diameter, that is, $\bar{l} = l_c / d_c$. Results indicate that the disturbance decays rapidly (approximately by 80–90%) over a relative length $\bar{l} = 32$–35. In an initial section of the channel (~15–30 mm from the diaphragm) under an air backpressure of $P_{ch} = 5.0$ MPa, the disturbance pressure was 56–90 MPa from charges with $G = 1.0$–2.0 g, and 120 MPa from a charge with $G = 3.0$ g. A similar result was obtained when channels were filled with nitrogen.

The studies carried out under laboratory conditions to determine the relationship between the disturbance amplitude A_0 and the initial backpressure in the chamber P_{ch} made it possible to obtain the relation for air, nitrogen, helium, and combustion products of nitrocellulose powder after processing the experimental data: $A_0 = \alpha P_{ch}^{\mu} G^{\nu}$. This relation is commonly known as the Rankine–Hugoniot equation. After straightforward manipulations and introduction of variables, we obtain

$$A_0 = \left[1 - \frac{\kappa - 1}{\kappa + 1} \frac{\rho_{ch}}{\rho} \Big/ \frac{\rho_{ch}}{\rho} \frac{\kappa - 1}{\kappa + 1} - 1\right] P_{ch}$$

where ρ_{ch} and ρ are the densities of the medium in the chamber and that behind the disturbance front, respectively. The preceding equation shows that in the pressure range of concern (i.e., $P_{ch} = 1$–80 MPa), the ratio of specific heat capacities $(\kappa - 1)/(\kappa + 1)$ decreases from 0.2 to 0.0816, and the ratio of ρ_{ch}/ρ in the range of disturbances for normally used DD ($M = 1.8$–3.0) slightly increases by 10–15%. As a result, the ratio in the square brackets slightly decreases by 5–7%, which, in turn, results in a practically linear increase in the disturbance amplitude A_0 with the chamber pressure growth.

The disturbance increase with increasing chamber pressure of $P_{ch} = 2.5$–60.0 MPa has been found in all cases during experiments for the same energy source (i.e., $G = \text{const}$), as shown in Fig. 9.24.

The propagation of a shock-wave pulse over a chamber with vibration baffles was experimentally studied in a chamber with a diameter of 130 mm. Even for such a small chamber for which the linear distance from the disturbance introduction point to the transducers installed around the circumference and along the axis is still small ($l = 100$–170 mm), the change of the bottom configuration as a result of the presence of baffles results in a considerable change in the amplitude of the disturbance (see Fig. 9.25). A comparison of the average amplitudes of pressure pulses produced by charges $G = 0.6$–1.4 g has shown that they are lower than the amplitudes recorded in the chamber under similar conditions but without baffles at the bottom. Therefore, baffles used for suppressing high-frequency pressure

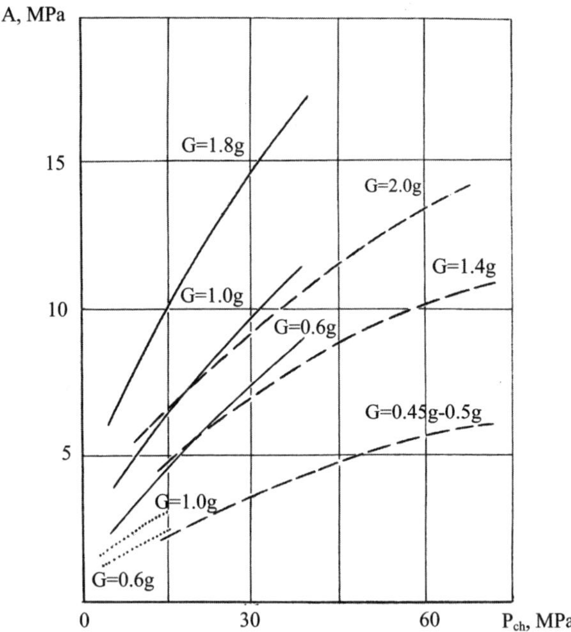

Fig. 9.24 Dependence of disturbance amplitude on the pressure in chamber filled with helium (...............), pyroxylin combustion products (– – – –), and air (———).

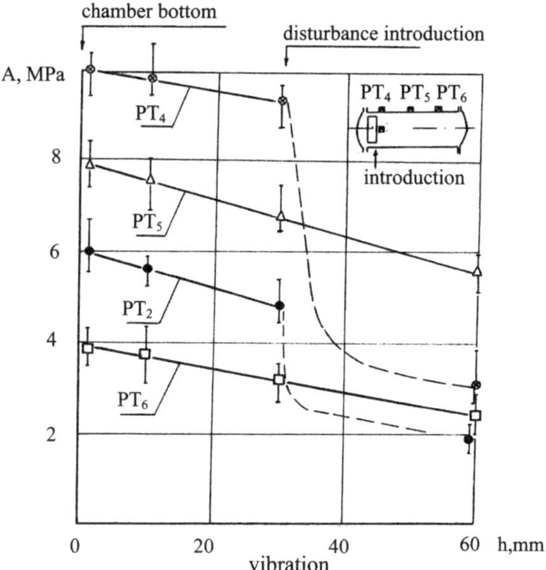

Fig. 9.25 Effect of length of vibration baffles (with the number of ribs m = 6) on disturbance amplitude distribution in the chamber.

oscillations affect the value of primary disturbance amplitude. This effect must be taken into account for generation of disturbances with sufficient amplitudes in the combustion chamber. The vibration baffles made from plates and tubes (imitating injectors protruding into the combustion chamber) were used in such experiments.

Attenuation of disturbances with increasing number of baffles was verified over a wide range of the initial pressure in the chamber $P_{ch} = 15.0–20.0$ MPa. The comparison of the disturbance amplitudes in chambers with and without baffles can be quantitatively evaluated using the ratio $A_{baffle}/A_{without\ baffle}$. For baffles with a length of $h = 30$ mm and with an increase in the number of blades, this ratio monotonically decreases in the range of 0.6–0.91. The disturbance introduction channel and transducers were located at a distance of 30 mm from the chamber bottom. Under these conditions, for a given type of baffle (tubes or plates) and fixed chamber pressure P_{ch}, a larger charge weight systematically causes a slight (2.5–3.0%) decrease of the ratio $A_{baffle}/A_{without\ baffles}$. The change of the chamber pressure P_{ch} with a fixed value of the charge practically does not have any effect. In addition, with an increase in the number of baffles, the ratio $A_{baffle}/A_{without\ baffles}$ decreases to 30–40%. The present case has a scatter of experimental results as high as 15–20%. The effect of P_{ch} with respect to the effect of the number of baffles can be neglected. In the case of clear orientation of baffles (the number of blades is clearly defined, that is, plates and tubes are located along the radii of the chamber), the relationship for tubes-injectors and plates has an identical pattern, for example, with the number of baffles of three or six. When the number of tubes-injectors at the chamber bottom increases, clear orientation is blurred, and one can talk about the similarity of baffles with a multiblade structure. The comparison of disturbance amplitudes between baffles with the number of blades of 9–12 and baffles consisting of 28 and 31 tubes-injectors uniformly arranged on the chamber bottom has shown that they have practically the same effect within the measurement error. Thus, an increase of baffle blades over 12 has a slight effect on the attenuation of artificial disturbance.

With an increase in baffle length, the distortion of the shock wave becomes more severe. This can be clearly seen in the case of the longest baffle ($h = 60$ mm). Under these conditions (see Fig. 9.25), the disturbance amplitude recorded by transducers located within the baffles sharply drops by a factor of two to three. A much lower extent (15–20%) was noted for the transducers located at 30 and 90 mm from the edge of the baffle blades. The type of baffles (plates or tubes-injectors) has a minor effect on disturbance damping with respect to the change of baffle length. The result is confirmed by the same behavior of the disturbance amplitude and by their close values.

VI. Selection of Minimum Artificial Pulse for Estimation of Stability Margin to Hard Excitation

It has been shown in Chapter 6 that the system response to calibrated disturbances should be used to estimate the nonlinear characteristics of the combustion-chamber process. The desired properties of devices for generating pulse disturbances in combustion chambers and gas generators, as well as a brief description of their design, have been given in the preceding sections. Justification of the selection of optimal criteria for evaluating of the process stability to pulse disturbances and the attainment of proper accuracy of their determination is discussed in this section and in Sec. VII.

Because the pulse characteristics were examined earlier, it remains to determine the magnitude of the initial pressure pulse suitable for estimating the chamber stability behavior and the decay rate of pressure oscillation induced by the pulse. The selection of the initial amplitude of the disturbance can be made in the following manners: 1) from estimation of the actual stability margin for combustion chambers and gas generators, which have 100% of reliability, and 2) on the basis of physical concepts of the effects of final disturbances on stability.

An optimal accuracy of the pulse characteristics obtained must be attained in all cases. Both considerations just listed were used to solve this problem. During engine operation, the distribution of natural pressure oscillation amplitude in the combustion chamber has a probabilistic pattern. The probability of the presence of a certain amplitude is associated with a specific time interval of engine operation and process characteristics. Certainly, the value of the externally introduced disturbance must exceed this level of natural pressure oscillation. The upper limit of the introduced disturbance is connected with limitations caused by the distortion of the process resulting from the externally impressed oscillations. In addition, the optimal value of the introduced disturbance is defined by conditions of attaining the optimal accuracy.

On the basis of extensive studies and long-standing experience, it is shown that the best results of quantitative estimation of the stability margin with respect to final disturbances can be obtained when calibrated disturbances are introduced into the chamber,

$$N_1 \leq n = \frac{A_0}{A_n} \leq N_2$$

and when the averaged oscillation coefficient δT_{ef} or the relaxation time τ_r (i.e., the time during which the pressure oscillation amplitude decreases by a factor of e after disturbance introduction) is determined as a process stability characteristic. Let us consider these statements in detail.

The problem of self-excited oscillations in a combustion chamber under the influence of noise is of dual interest. On the one hand, the initial pressure pulse introduced into the combustion chamber should exceed A_{noise}. On the other hand, self-excitation of process oscillations under the influence of turbulent combustion in a combustion chamber is of separate interest.

The experimental data on probabilistic excitation of process instability as a result of noise are presented in Sec. III of Chapter 3. The problem of unstable oscillations in a system can be formulated as follows.

At an initial moment of time t_o, process $A(t)$ has a certain value $A(t) = A_o$, where A_o is less than A_{cr}. It is required to find the probability of system excitation during time $t' \leq t$, that is, the probability of exceeding the current amplitude A_{cr} during time $t' \leq t$. Such a transition occurs because of the action of random noise $z(t)$. The probability is approximately equal to

$$P(t) = 1 - e^{1/t_0} \tag{9.1}$$

where $t' = 2<t>$ and $<t>$ is the average of the time during which the process $A(t)$ attains A_{cr} for the first time corresponding to the level of an unstable limiting cycle or a specified amplitude. The coefficient 2 is used because, after attaining the specified amplitude, the process can make a transition to an unstable region or return to a stable region.

The average time of attaining A_{cr} by the Markovsky process $A(t)$ is calculated by the following expression:

$$t'\ (t, A_0, A_{\kappa p}) = 2\int_{A_0}^{A_{\kappa p}} e^{-\psi(t)} \times \int_0^x \frac{e^{\psi(y)}}{by}\,\mathrm{d}y\,\mathrm{d}x \tag{9.2}$$

where

$$\psi(t) = 2\int \frac{a(A)}{b(A)}\,\mathrm{d}A$$

The coefficients of drift a and b in describing the Markovsky process are defined next:

$$a(A) = -\delta(A) + \omega_0^2 \frac{N}{8A}$$

$$b(A) = \omega_0^2 \frac{N_0}{4}$$

Denoting $x = A/A_{rms}$, $\delta(x) = \delta_0 \Phi(x)$, we obtain

$$t'\delta_0 = \int_0^x \frac{\exp \int \Phi(x)\,x\,\mathrm{d}x}{x} \int_0^x \frac{y}{\exp \int \Phi(y)\,y\,\mathrm{d}y}\,\mathrm{d}y\,\mathrm{d}x$$

Table 9.5 Hypothetical relationships between damping coefficient and amplitude

Function	$t' \cdot \delta_0$	$P \simeq \dfrac{t_r}{t'}$
$\delta = \delta_0$	0.1429×10^{30}	1.75×10^{-25}
$\delta = \delta_0 (1 - x^2/12^2)$	0.4658×10^{15}	0.536×10^{-10}
$\delta = \delta_0 (1 - x/12)$	0.4650×10^{10}	0.537×10^{-6}
$\delta = \delta_0 (1 - \sqrt{x/12})$	0.6465×10^{6}	0.387×10^{-1}

The instability of a system is fundamentally stochastic in the sense that it is in principle impossible to determine an exact time of transition from a noise to an oscillatory condition. Such a transition can only be described in a probabilistic sense.

The calculated data on the average period during which the process $A(t)$ first attains the level of $12A_{\rm rms}$ (rms amplitude of noise) for a number of possible relationships between the damping coefficient and amplitude are given in Table 9.5 (Fig. 9.26). The probability of occurrence of self-excited oscillations in an initially nonexcited system P was estimated for an initial damping coefficient $\delta_0 = 100$ and an engine operation time $\tau_r = 500$ s. The value of $12A_{\rm rms}$ was selected on the basis of experiments with combustion chambers, in which instabilities were very rare (on average one case of high-frequency oscillations per 300 startups). Table 9.5 and Fig. 9.26 indicate that the probability of exceeding a specified amplitude depends on the mode of characteristics.

Both passive and active methods were used to estimate the amplitude dependence of the damping coefficient. A shortcoming of the passive methods lies in the impossibility to estimate the relation $\delta(A)$ by the value $A = 4$–$5\ A_{\rm rms}$. If the amplitude $A_{\rm cr}$ exceeds essentially the noise level in the combustion chamber, the accuracy of predicting a limiting-cycle amplitude is quite low.

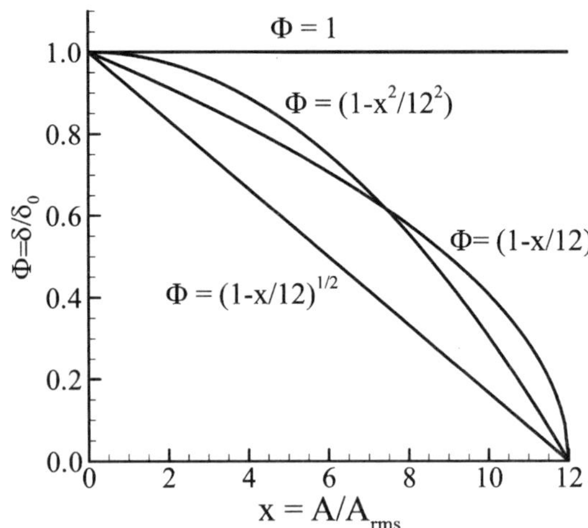

Fig. 9.26 Hypothetical relationships between damping coefficient and amplitude.

The experimentally obtained relationship $\delta = \delta(A)$ in the region of $A \leq 5\ A_{\text{rms}}$ could be approximated by the function $\delta = \delta_0[1-(A/A_{\text{cr}})^2]$. According to the estimated probability of hard excitation presented in Table 9.5 and Fig. 9.26 and the proposed relation $\delta = \delta_0(A)$ on the assumption that the function $\delta = \delta_0(A)$ does not undergo chance variations during engine operation for an actual time of $\tau_p = 500$ s with $\delta_0 = 100$, the probability of occurrence of self-oscillation is $P \approx 10^{-10}$, that is, this event is improbable.

These conclusions served as one of the justifications of the sufficient stability margin for introducing disturbances exceeding $N_1 = 12 \leq n = A_0/A_{\text{rms n}}$ or $N_2 \approx 15 \leq n = A_0/A_{\text{rms n}}$. The disturbance damps rapidly.

VII. Accuracy of Determination of Stability Margin to Hard Excitation

On the basis of the studies described in the preceding sections, a conclusion was reached that to estimate the stability margin with respect to hard excitation it is necessary to introduce a calibrated disturbance of $N_1 \leq n = A_0/2A_{nm} \leq N_2$, and the decay time of pressure oscillation damping of e times τ_p should not exceed 15 ms.

The main sources of error in determining the relative disturbance value n are as follows: 1) error caused by fluctuation of explosive characteristics, 2) error in measuring artificially initiated oscillations, and 3) error in measuring the natural noise of the operating process in the chamber.

Let us consider the possibility of reducing the error of estimating pulse characteristics. The error associated with charge characteristics for a low-velocity detonation is defined by the specifications of charges. The stability of a disturbance produced by charges in a passive medium has been attained by the strict manufacturing technology of charges. The specification requirements are given in Table 9.6. The maximum permissible deviation should not exceed ±20%.

If A_0 and A_n are considered to be statistically independent, the random relative error of index n measurement is equal to

$$\varepsilon = \sqrt{\varepsilon_{A_0}^2 + \varepsilon_{A_{nm}}^2}$$

where ε_{A_0} and $\varepsilon_{A_{\text{nm}}}$ are the random relative measurement errors of A_0 and A_{nm}, respectively. A random rms error of measuring pressure variations under steady-state conditions is calculated by the expression:

$$\varepsilon = \sqrt{\Sigma\varepsilon_i^2 + \varepsilon_0^2 + \varepsilon_{\text{met}}^2}$$

Table 9.6 Specifications of explosive charges

Mass of charge, g	Average value of disturbance, MPa	Maximum permissible deviations, MPa
0.6	3.3	±0.65
1.0	4.4	±0.9
1.4	5.4	±1.1
1.6	6.4	±1.3
2.2	7.3	±1.6

where $\Sigma\varepsilon_i$ is the error introduced by transducers and the electrical part of the measurement and recording system, ε_0 the processing system error, and ε_{met} the principal systematic error.

When stationary pressure pulsations are measured, the error $\Sigma\varepsilon_i$ is mainly defined by the metrological characteristics of the measurement and recording system, and for standard systems it accounts for about 15% of the total error. When measuring the shock-wave type of pulses in unsteady processes, this error can increase. The error ε_0 is mainly defined by the number of digitizing levels and sampling frequency. For operating systems this error is no more than 2%. The error ε_{Mem} is associated with the statistical nature of a pressure pulsation process.

Let us consider the random component of the systematic error in determining the characteristics of pressure pulsation amplitudes. It is important to establish practical methods for data reduction and to acquire the relations between the mean rectified, rms, and maximum values of pressure pulsations. Only the random component of the systematic error caused by the statistical nature of pressure pulsations will be considered next.

To first approximate, the circuit for pressure transducers and signal processing can be represented by the following linear dynamic model (see Fig. 9.27). In this model the combustion chamber, which represents a frequency-filtering acoustic system with an infinite number of degrees of freedom, is replaced by n integral resonance circuits (one circuit for each mode of free oscillation with frequency ω_i). Here $x(t)$ is an input signal conventionally represented in the form of white noise with a finite frequency range. Its spectral density $S_x(\omega)$ is constant within the band $\Delta\Omega = \Omega_2 - \Omega_1$, with $\Omega_1 < \omega_1$ and $\Omega_2 > \omega_n$. Here ω_1 and ω_n are the lower and upper bounds of the frequency band excited in the combustion chamber, respectively. Here, $x_\xi(t)$ is part of wideband noise, which has not passed through the resonance circuits and is the consequence of the spatial distribution of the input signal sources. And $\psi(t)$ is the normal-mode interference, represented by an arbitrary spectrum (i.e., signals generated by various engine subsystems, in particular, narrowband vibration interference resulting from TPU at fixed frequencies multiple of the rotor speed, cooling system noise of pressure transducers, self-oscillations in the connecting cavities of transducers, and test hardware).

The spectrum of in-chamber pulsations processed by this method is as a rule represented by a wideband background, a number of narrowband components in the vicinity of the chamber natural frequencies, and the frequencies of externally impressed oscillations, if they exist (see Fig. 9.27c). For steady-state engine operation, the random pressure pulsation process formed according to the diagram shown in Fig. 9.27 can be considered approximately stationary.

Using the approximate formula of one-dimensional probability density $W(h)$ for a random number $h = A_M/\sigma$, which characterizes the distribution of the absolute maxima of the normal narrowband noise envelope in different realizations $Y(t)$ of the specified duration t_r, we obtain

$$W(h) \approx (\delta t_r + 1) h e^{\frac{h^2}{2}} \left[1 - e^{\frac{h^2}{2}}\right]^{\delta t_p} \tag{9.3}$$

Here σ is the rms value of a random process, δ is the damping coefficient, and t_r is the time of realization.

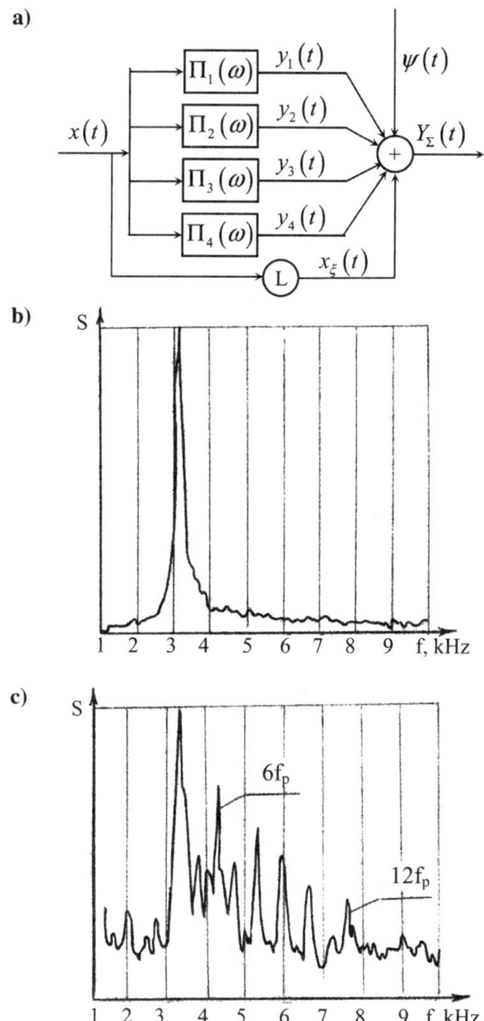

Fig. 9.27 Model of pressure-pulsation transducer signal shaping: a) amplitude spectra of signals, b) y_i, and c)Y_Σ.

Figure 9.28 shows the probability density of the absolute maximum $Y(t)$ of a narrowband process envelope for different $\beta = \delta t_r$. It can be seen from this figure that the probability density of the absolute maximum is not stationary, that is, it depends on the realization length t_r. With all other conditions being fixed, the probability density converges and reaches a higher value as t_r increases. These properties of the absolute maximum of a narrowband noise envelope are seen more distinctly in Fig. 9.29, where the relations of the mathematical expectation of the absolute maximum $\overline{N}$, the maximum dispersion $D(N)$, the normalized rms deviation of the absolute maximum ε_N, and the complex $\beta = \delta t_r$ are shown.

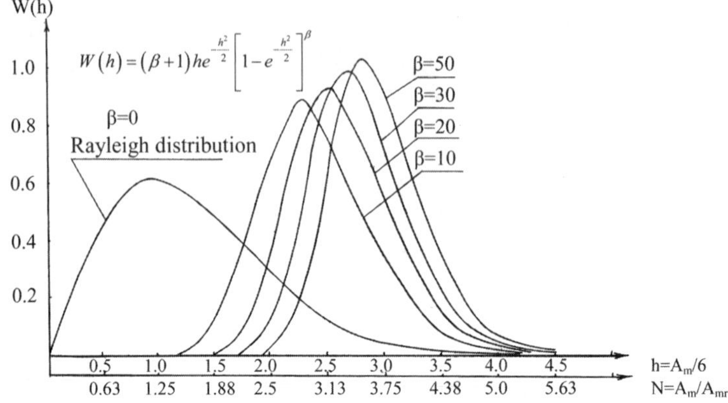

Fig. 9.28 Probability density of absolute maximum of narrowband noise envelope.

The statistical characteristics of the absolute maximum of a narrowband noise envelope are calculated using the following formulas:

$$\overline{N} = \int_0^\infty W(N) \cdot N \cdot \mathrm{d}N = f(\beta) \tag{9.4}$$

$$D_N = \int W(N) \cdot (N - \overline{N})^2 \, \mathrm{d}N = \varphi(\beta) \tag{9.5}$$

$$\varepsilon_{\mathrm{N}} = \frac{\sqrt{D_N}}{\overline{N}} = \frac{\sigma_N}{\overline{N}} = \psi(\beta) \tag{9.6}$$

$$N = \frac{A_m}{A_{\mathrm{mr}}}$$

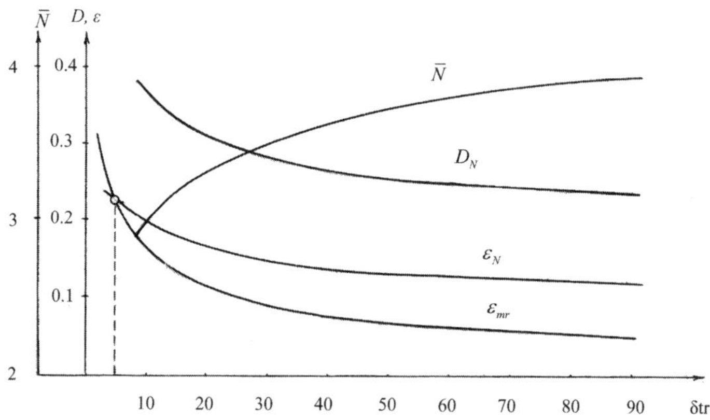

Fig. 9.29 Theoretical relationships between the average value of $\overline{N}$, dispersion D_N, relative error of envelope of absolute maximum, relative error of mean rectified value of narrowband noise, and parameter δ_{tr}.

The normalized rms deviation of the absolute maximum $\varepsilon_N = \sigma_N/\bar{N}$ decreases as the realization length t_r increases, mainly because of the decreased dispersion and its increased mathematical expectation. The dependence of the realization time on the absolute maximum mathematical expectation is an undesirable property when it is used as a measure of the chamber response to artificial disturbance. The value of $n = A_0/2A_{nm}$ decreases with the growth of the realization time t_r. If t_r is not fixed, n becomes ambiguous. In this connection, it is advisable to pass from the characteristic A_{noise} to the characteristic A_{rms} (or A_{mr}), with $n' = A_0/A_{\text{mr}\,n}$ and $n'' = A_0/A_{\text{rms}}$, where A_{mr} and A_{rms} are the mean rectified and rms values of pulsations, respectively.

It is known that the mathematical expectation of estimation of A_{rms} (or A_{mr}) does not depend on the realization time t_r. In the particular case of a normal narrowband process with the correlation coefficient of the form $\rho(\tau) = e^{-\delta|\tau|}\cos\omega_0\tau$, the normalized rms error of the rms value estimation becomes

$$\varepsilon_{\text{rms}} = \frac{\sigma_{\text{rmsA}}}{A_{\text{rms}}} = \frac{1}{t_p}\sqrt{\int_0^{t_p}(t_p-\tau)e^{-2\delta|\tau|}\cos^2\omega_0\tau\,d\tau} \qquad \omega_0 > \delta \quad (9.7)$$

With $t_r \gg \tau_{\text{corr.}}$, we get from Eq. (9.7)

$$\varepsilon_{\text{rms}} \approx \frac{1}{2\sqrt{\delta t_p}} \qquad (9.8)$$

Equation (9.8) gives the upper estimation of the normalized rms error. Figure 9.30 shows the calculated relationship between the normalized rms error and the complex $\beta = \delta t_r$. The calculation was made by the exact formula (9.31) and by the approximate formula (9.32), respectively. It can be seen from Fig. 9.30 that with $\delta t_r > 20$ the discrepancy of estimations by formulas (9.31) and (9.32) is less than 1%.

In practice, instead of the characteristic A_{rms}, A_{mr} (i.e., the mean rectified, or mean-modulus, value of a random process) is used because of the relative simplicity in determinating this value. The normalized rms errors of A_{rms} and A_{mr} are the

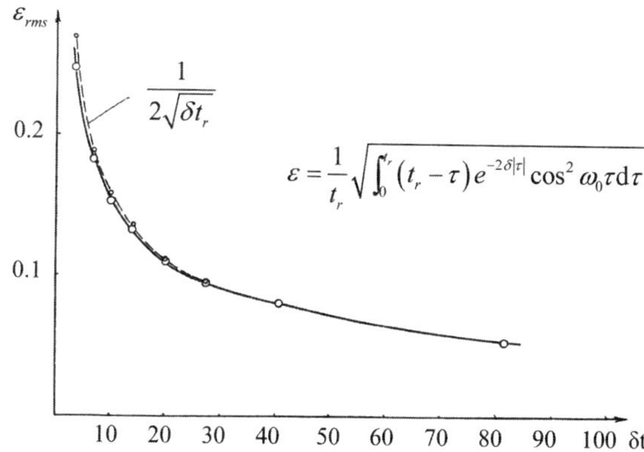

Fig. 9.30 Theoretical relationship between the normalized rms error of rms value of linear narrowband noise.

same. The ratio $\mathcal{K} = A_{\rm rms}/A_{\rm mr}$ depends on the distribution of the instantaneous process values. For a normal distribution, $\mathcal{K} = 1.25$.

For multicomponent wideband signal processing, according to the diagram in Fig. 9.27, the normalized rms error of the rms (mean rectified) value estimation is calculated using the following formula:

$$\varepsilon = \frac{1}{2\sqrt{\Delta f_{\rm eq} t_p}} \tag{9.9}$$

where $\Delta f_{\rm eg}$ is the equivalent statistical width of the spectrum and is defined as

$$\Delta f_{\rm eq} = \frac{1}{2\pi}\left[\int_0^\infty s\,(\omega)\,{\rm d}\omega\right]^2 \Bigg/ \left[\int_0^\infty s^2\,(\omega)\,{\rm d}\omega\right]$$

where $S(\omega)$ is the spectral intensity of the process $Y(t)$. In the particular case of a narrowband process with the correlation function $\rho(t) = e^{-\delta|\tau|}\cos\omega_0\tau$, we get

$$\Delta f_{\rm eq} = 1\Big/2\pi\cdot\frac{1}{\pi^2}\int_0^\infty\left[\frac{\delta}{\delta^2 + (\omega+\omega_0)^2} + \frac{\delta}{\delta^2 + (\omega-\omega_0)^2}\right]{\rm d}\omega$$

$$= 1\Big/\frac{2\delta^2}{\delta}\left[\frac{1}{2\delta^2} + \frac{1}{2(\delta^2+\omega_0^2)}\right] \tag{9.10}$$

with $\omega_0 \gg \delta$. As is seen, for a narrowband process, Eq. (9.9) coincides with Eq. (9.8).

The equivalent statistical width of the spectrum $\Delta f_{\rm equiv}$ in Eq. (9.9) represents the signal spectrum width at the output of a hypothetic rectangular filter under the effect of white noise at its inlet. The filter gives the same value of mean-squared statistical error as that of an actual dynamic system. Thus, Eq. (9.9) for the normalized rms error of the mean rectified (rms) value of a normal random process $Y(t)$ with an arbitrary spectrum is rather general. It follows from the same equation that the relative random error $\varepsilon_{\rm rms}$ decays with the increase of the equivalent width of spectrum $\Delta f_{\rm equiv}$ and the realization time t_r. Generally speaking, the error of the absolute maximum of a narrowband also has this property. As shown in Fig. 9.29, the distribution $\varepsilon_{\rm AM} = \psi(t_r)$ is more flat than that of $\varepsilon_{\rm rms} = F(t_r)$. At a higher δt_r, the error of the absolute maximum $\varepsilon_{\rm AM}$ exceeds the error of the rms value $\varepsilon_{\rm AM}$. For a smaller $\beta = \delta t_r$, the trend can reverse, and $\varepsilon_{\rm AM} < \varepsilon_{\rm rms}$.

The problem now becomes one of calculating an optimum average time t_r, which gives an acceptable error in practice. Shown next are the results of experimental determination of statistical characteristics of signals generated by an integral resonance circuit (i.e., a chamber simulator), as well as pressure pulsations in chambers and prechamber cavities of some actual engines. The statistical characteristics of pressure pulsations were determined for steady-state engine operation. A region of random process realization was subject to high-pass filtering to eliminate low-frequency components and dc shift. The cutoff frequency $f_{\rm cutoff}$ for the filter is 590 Hz. Then for this region, the values of $A_{\rm rms}$, $A_{\rm mr}$, $N = A_{\rm M}/A_{\rm mr}$, and $\mathcal{K} = A_{\rm rms}/A_{\rm mr}$ were calculated.

These characteristics were determined successively for realization regions with durations $t_r = 0.01$, 0.02, 0.04, and 0.08 s. The statistical characteristics of $\overline{A}_{rms}$, $\overline{A}_{mr}$, $\overline{A}_M$, $\overline{N}$, $\overline{K}$, σ_{rms}, σ_{mr}, ε_M, and ε_{mr} were calculated from the ensemble of 20 ($m = 20$) successive nonoverlapping realizations with a duration t_{ri}. An ensemble from 100 ($m = 100$) uncorrelated realizations was used for calculation of statistical characteristics.

A normal narrowband centered random process (quasi-harmonic noise) generated by an integral resonance circuit (R-L-C), into which a wideband signal from a noise generator was fed as the input, was used as a model signal. The natural frequency of the circuit f_0 is 3165 Hz. The circuits with decrements $\delta T = 0.036$, 0.057, 0.11, and 0.32 were used.

Figures 9.31 and 9.32 show the relationships between $\overline{N}$, ε_M, ε_{mr}, and the parameter $\beta = \delta t_r$. The variations of $\overline{N}$ and ε_M qualitatively correspond to their theoretical relationships. The relative value of the absolute maximum $\overline{N} = \overline{(A_M/A_{mr})}$ increases

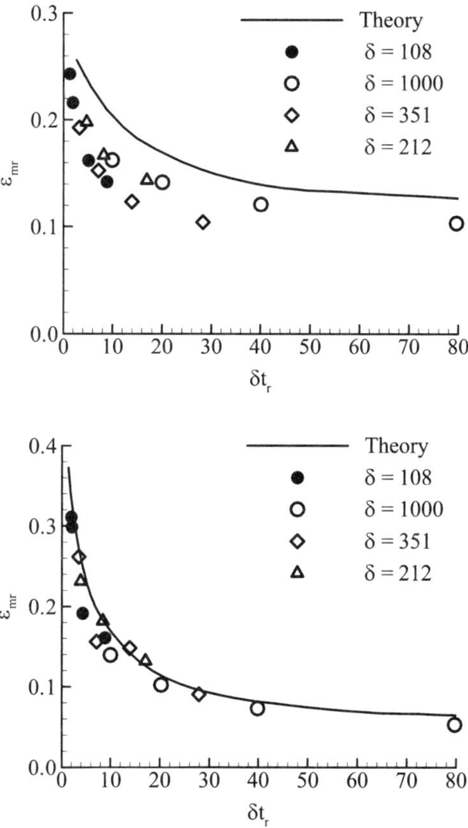

Fig. 9.31 Experimental relationships between the relative error of absolute maximum ξ_m as well as mean the rectified value ξ_{mr} of quasi-harmonic noise and complex δt_r (averaging from 1000 realizations).

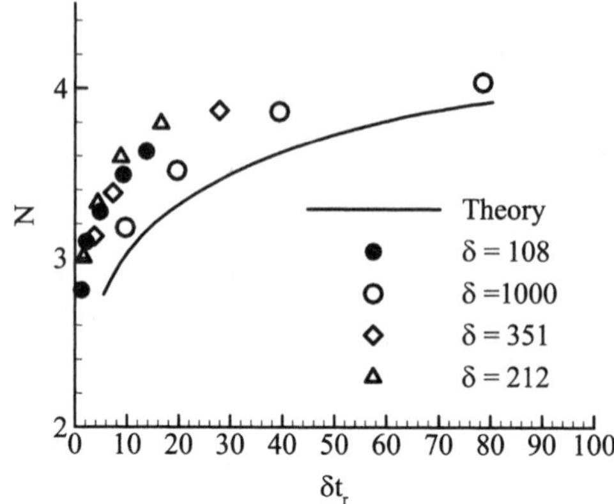

Fig. 9.32 Experimental relationship between the average value of absolute maximum of quasi-harmonic noise and complex δt_r (averaging from 100 realizations).

with the increase of the realization length t_r, but the relative error $\varepsilon_M = \varepsilon_M/\overline{A}_M$ exhibits an opposite trend. At a fixed length of realization, with the growth of circuit decrement the relative value of the absolute maximum $\overline{N}$ increases, and the relative rms deviation σ_M decreases. A relative rms error of the mean rectified value decreases as the length of realization t_r increases practically in accordance with the theoretical relationship, as shown in Fig. 9.32.

Figures 9.31 and 9.32 show that with $\delta t_r < 10$ the relative rms error of the mean rectified value exceeds the corresponding error of the absolute maximum. The possibility of such event realization was predicted by the theoretical relationships shown in Fig. 9.29. As can be seen from Fig. 9.32, experimental estimations of the absolute maximum $\overline{N}$, are on the average systematically overestimated by 15% with respect to theoretical estimations.

Actual oscillatory processes in LRE combustion chambers and gas generators very rarely have single-frequency, narrowband, quasi-harmonic noise (of the type of simulator signals discussed above). As noted earlier, the spectrum of actual signals contains multiple harmonics with the presence of a rather distinct wideband background. The dispersion of amplitude characteristics of such signals is a function of $\beta = \Delta f_{\text{equiv}}\, t_r$, that is, the product of the equivalent statistical width of spectrum and the realization length, where Δf_{equiv} is using Eq. (9.8).

The results of processing actual signals for engines (RD-0120, S5.92, 15D264, and 15D300) are given in Figs. 9.33–9.35.

Engine RD-0120 [12] is a single-chamber liquid oxygen-hydrogen rocket engine. It is made on a closed circuit (i.e., staged combustion cycle) with afterburning of the reducing gas from the turbine in the combustion chamber. It is designed for use as the main engine of the second stage of the space launch vehicle Energya-Buran. The primary engine characteristics are $P_N = 2.0$ MN (203.8 tf); $P_{\text{ch}} = 21.8$ MPa; $J_N = 4462$ m/s; $K_m = 6.0$; and $t = 500$ s.

Engine S5.92 [12] is made on a open circuit and uses nitrogen tetroxide and unsymmetrical dimethylhydrazine as propellants. It is a two-thrust chamber engine. For high-thrust operation, $P_v = 19.6$ kN (2.0 tf); $P_{\text{ch}} = 9.8$ MPa; $J_n = 3270$ m/s;

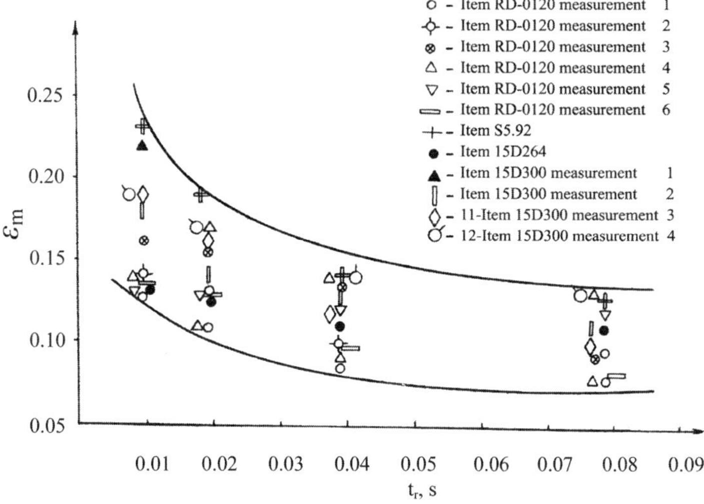

Fig. 9.33 Cumulative relationships (based on a set of engines studied and tests) between the relative error of absolute maximum and length of realization.

$K_m = 1.95-2.05$; and $t = 2000$ s. For low-thrust operation, $P_v = 13.72$ kN (2.0 tf); $P_{ch} = 6.85$ MPa; $J_V = 3160$ m/s; $K_m = 2.0–2.1$; and $t = 2000$ s.

Engine S5.92 is designed as part of the existing and advanced carrier rockets. It can be used as the final stage for insertion of a payload into various orbits of artificial Earth satellites and for planetary missions.

Engine 15D264 [12] is made on an open circuit and consists of a single-chamber high-thrust LRE and 16 low-thrust LREs. Propellant components are nitrogen

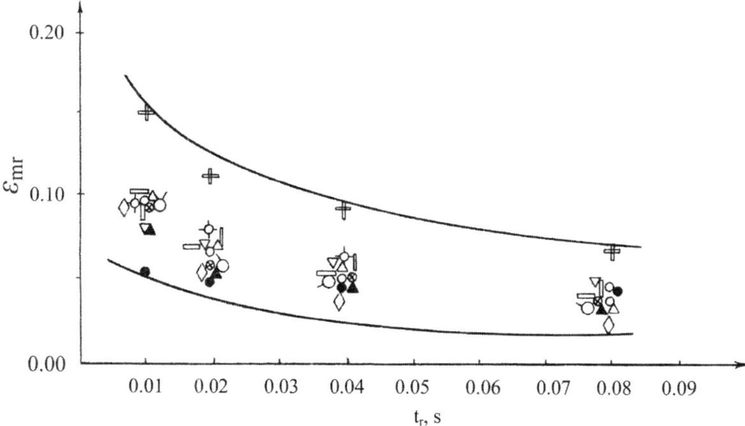

Fig. 9.34 Cumulative relationships (based on a set of engines studied and tests) between the relative error of mean rectified value and length of realization.

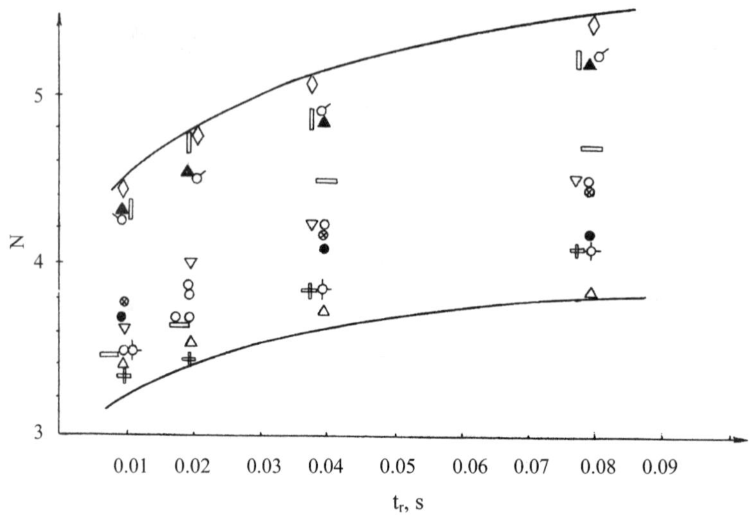

Fig. 9.35 Cumulative relationships (based on a set of engines studied and tests) between the average value of absolute maximum and length of realization.

tetroxide and unsymmetrical dimethylhydrazine. The primary engine characteristics are $P_v = 0.93 - 5.04$ kN (94.4−513.5 kgf); $P_{ch} = 4.07$ MPa; $J_v = 3055$ m/s; $K_m = 2.03$; $T_{tot} = 330$ s; and $n_{start} = 14$. The engine is to be installed in the front rocket compartments.

Engine 15D300 [12] is a four-chamber, dual-mode, single-start engine designed for flight control of the postboost rocket stage 15A18M (R-36M2). Propellant components are nitrogen tetroxide and unsymmetrical dimethylhydrazine. The main operation parameters are $P_v = 20.47$ kN (2087 kgf); $P_{ch} = 4.1$ MPa; $J_v = 3130$ m/s; $K_m = 1.8$; and $t = 700$ s.

From Figs. 9.33–9.35 it follows that with an increasing length of realization t_r, the statistical characteristics of pressure pulsations vary as follows:

1) The mean value of the absolute maximum $\overline{A}_m$ increases as the length of realization t_r increases.

2) The average of the mean rectified value $\overline{A}_{mr}$ (estimation of expectation) practically does not change with the change of t_r.

3) The rms deviation of the mean rectified values σ_{mr} decreases as t_r grows.

4) The mean rectified deviation of the absolute maximum σ_M decreases as t_r grows.

5) The relative rms errors of the absolute maximum and the mean rectified value of ε_{mr} decrease as t_r increases. In addition, ε_{mr} decreases approximately following the law of $\varepsilon_{mr}(t) \approx c/\sqrt{t_r}$.

6) The relative rms error of the mean rectified value of ε_{mr} decreases with the increase of the realization time t_r as a result of the decrease in dispersion of σ_{mr}^2. Furthermore, the relative error of the absolute maximum decreases as a result of the growth of its mean value and the decrease of its dispersion.

7) The relation $K = A_{\text{rms}}/A_{\text{mr}} \approx 1.25$ is satisfied between the rms and mean rectified values of pulsations, serving as an indicator of the normal law of distribution of pressure pulsation momentary values.

Thus, the experimental relationships between the statistical characteristics of the absolute maximum and mean rectified values and the length of realization t_r agree qualitatively with theoretical and experimental results for simulator signals.

Let us study numerical values of errors:

$$\varepsilon_e = \sqrt{\Sigma \varepsilon_i^2} = \sqrt{\varepsilon_t^2 + \varepsilon_r^2 + \varepsilon_a^2 + \varepsilon_{\text{rep}}^2}$$

where ε_t is the error introduced by the transducer LH (≈12%); ε_r the error by the recorder (≈5%); ε_a the error by the amplifier (≈5%); and ε_{rep} the error by the reproduction system (≈5%). Therefore $\varepsilon_e = \sqrt{12^2 + 5^2 + 5^2 + 5^2} \approx 15\%$.

Figures 9.33 and 9.34 indicate that for the set of engines concerned and the basic length of realization $t_r = 0.01$ s, the relative errors of A_M and A_{mr} (point estimations) lie in the ranges of $\varepsilon_M \approx 13$–22% and $\varepsilon_{\text{mr}} \approx 5$–$10\%$, respectively. That is, the random calculation error of A_M for a series of tests, on the average, exceeds twice the calculation error of A_{mr}. As noted earlier, the estimation of errors ε_M and ε_{mr} was made from an ensemble of 20 realizations with a duration of t_r, taken from the records of the steady-state operation during a particular test with a sufficient duration. The operating conditions of the measurement and recording system during a particular test, as well as the reproduction system during record processing, can be considered unchanged. Therefore the determination of errors ε_M and ε_{mr} practically can be considered random systematic, based on the statistical nature of the signals concerned. With a basic length of realization $t_r = 0.01$ s, the systematic random error exceeds the instrumental error of measurement (ε_M up to 22%, $\varepsilon_e \approx 15\%$, $t_r = 0.01$ s). In this connection, it becomes evident that for steady-state engine operations it is desirable to pass from the estimation of the absolute maximum A_M to the estimation of the mean rectified (rms) value, which has a lower random systematic error and a higher interference protection.

Thus, when passing from estimations with the use of A_{nm} to estimations of A_{mr}, to save the available statistical data, it is advisable to select a conversion factor $N = A_{nm}/A_{\text{mr}}$ with the same length of realization as earlier, that is, $t_r = 0.01$ s. The ratio $\overline{N}$ with $t_r = 0.01$ s is within $4.5 > \overline{N} > 3$. The average value of $\overline{N}$ is 3.8. On the basis of these considerations, the formula for the minimum required disturbance value assumes the form:

$$N = \frac{A_0}{2A_{nm}} \approx \frac{A_0}{8A_{\text{mr}}} \geq [n], \qquad \text{or} \qquad n' = \frac{A_0}{A_{\text{mr}}} \geq 8[n]$$

provided that $A_{nm} \approx 4\,A_{\text{mr}}$ with $t_r = 0.01$ s.

When using the rms (effective) signal value as a criterion of artificial disturbance estimation, the formula assumes the form: $n'' = A_0/A_{\text{rms}} \geq 6.4[n]$ provided that $A_{\text{rms}} \approx R \times A_{\text{mr}}$, where $K \approx 1.26$ for all statistical test data. Thus, to reduce the random error of the relative amplitude of an artificial disturbance (n), it is desirable to use the estimation of the mean rectified (or rms) value of natural noise of the process prior to the disturbance introduction as a measure of disturbance.

For engine steady-state operations, the length of realization t_r that provides an acceptable random systematic error of a mean rectified value ($\varepsilon_{\text{mr}} \leq 7\%$) is

$t_r = 0.15$ s. In this case, the total random rms measurement error of the mean rectified (rms) values of signals ε_Σ is about 17%.

If the time average t_r is not limited by the capability of RCP (rapidly changing parameters) processing technique, the random systematic determination error of the mean rectified signal values can be reduced by the corresponding increase of the realization length t_r when needed.

VIII. Procedure of Analyzing LRE Process Stability by Use of Artificial Pressure Pulses

On the basis of long-standing experience with estimation of the process stability with respect to final disturbances, and the results of studies given in the preceding sections of this book, an optimal technique for estimating process stability to hard excitation by means of artificial disturbances has been developed.

A. Selection of Optimum Value of Artificial Pressure Disturbance in Combustion Chamber

It was stated in Chapter 6 that the stability margin with respect to hard excitation is sufficient, if the pressure oscillations are damped rapidly after the introduction of pulse disturbances, and if a certain relation between the deviation of the initial peak pressure from the average value and the total high-frequency signal prior to the introduction of pulse disturbances is met. The selection of optimal values of these parameters, however, called for long-term studies and processing of a large volume of statistical data for numerous engines.

In determination of the minimum pulse from an artificial disturbance, the following should be taken into account: 1) a pressure pulse must exceed a critical value, below which the probabilistic excitation of process instability can occur (Sec. III of Chapter 3); and 2) a pulse must exceed such a level of noise oscillations to allow for the calculation of parameters of stability to hard excitation with a sufficient accuracy.

In the early studies in the 1960s, it was accepted on the basis of experimental data that the minimum pulse must exceed at least twice the maximum value of noise amplitude, that is, $N > A_0/2A_{nm}$. However, the dependence of realization time (i.e., engine operation time) on the theoretical absolute maximum value of noise pressure oscillation is an undesirable property of A_{nm} when it is used as a measure of the chamber response to an artificial disturbance. In this connection, it is advisable to pass from the characteristic A_0/A_{nm} to A_0/A_{mr}.

According to the analysis for different engines (see Fig. 9.35), the ratio A_{nm}/A_{mr} can be as high as ≈ 5 with the realization time of over 0.08 s. The ratio corresponding to the minimum pulse $N > A_0/A_{\mathrm{mr}}$ should be taken to be greater than 10. The selection of a minimum pulse from the standpoint of probabilistic excitation of pressure oscillations in the chamber $A_0/A_{\mathrm{mr}} \geq 15$ is substantiated in Sec. VI. This ratio was accepted as a minimum value in estimation of the stability to hard excitation, with the results of the studies shown in Sec. III taken into account.

The maximum value of an initial pressure pulse was limited by value of $A_0/A_{mr} = 25$, as it was stated earlier, because of the necessity to minimize the disturbance of the operating process when the artificial disturbance is introduced into a combustion chamber or a gas generator.

A relaxation time t_{relax} was determined from the statistical data for a great number of experimental and actual engines with high reliability. According to the available statistical data, for most liquid–liquid and gas–liquid combustion chambers and gas generators t_{relax} (τ_r) was less than 10 ms. Only in special cases did it reach 15 ms. This value was accepted as a criterion of estimation of stability margin to hard excitation of pressure oscillations.

Because the first peak of pressure is not necessarily maximum after the introduction of an artificial disturbance into the chamber, and in the analysis the amplitude after termination of pulse disturbance is taken as the maximum amplitude for calculation of τ_r, the ratio A_0/A_{mr} is replaced by A_M/A_{mr}, with all other values taken earlier. Thus, $15 < A_M/A_{mr} < 25$ remains unchanged. To estimate the process stability to final disturbances, it is necessary 1) to introduce a pressure pulse within $15\ A_{mr} < A_M < 25\ A_{mr}$ and 2) to determine a relaxation time.

A combustion chamber is considered to be stable with respect to final disturbances if $t_{relax} < 15$ ms and $t_{relax} = \tau_1 + \tau_e$, where τ_e is the time for decreasing the pressure oscillation amplitude by a factor of e and τ_1 is the time delay for disturbances to exert influence on the process.

It is shown in Sec. V that devices using low-velocity detonation of explosives (e.g., hexagen) meet the requirements for producing pulse disturbances. The specifications for standard charges with a nominal diameter of 6 mm are shown in Table 9.6. In addition, standard charges for cartridges with a diameter of 8 mm were developed. Slightly reinforced cases of disturbance devices with special diaphragms were developed for charges with 8-mm cartridges.

Statistical data on pulses were obtained from measurements in a unit similar to that shown in Fig. 9.14. The five-charge device was connected to a 50-mm-diam pipe with a conical adapter. Piezoelectrical transducers were installed at a distance of 100 mm from the conical part with a spacing of 120 mm between them. The pressure at the shock-wave front was determined from the time for the wave propagation between two transducers over a distance of 120 mm. The standard test conditions were maintained: pressure in the pipe is 5.0 MPa, boost air charging, and temperature is 20°C. The charges comply with all requirements for devices with explosives. They must survive in the temperature range between –40 and 50°C after being exposed to all transportation conditions and are reliable in operation.

In some cases, however, the design of the pulse device had to be modernized during tests without changing the design of explosives. Thus for combustion chambers with a diameter less than 200 mm, a grid with holes (i.e., perforated plate) was mounted between the explosives and outlet nipples to attenuate the pulse and trap the burst diaphragm fragments. It is practically impossible to make a charge with an explosive weight of less than 0.6 g, which would reliably guarantee low-velocity detonation.

The amplitude of a pressure pulse can be regulated by the ratio of the number of holes in the grid to its diameter. As an example, a disposable pulse device for regulating (decreasing) the pressure disturbance amplitude is shown in Fig. 9.36.

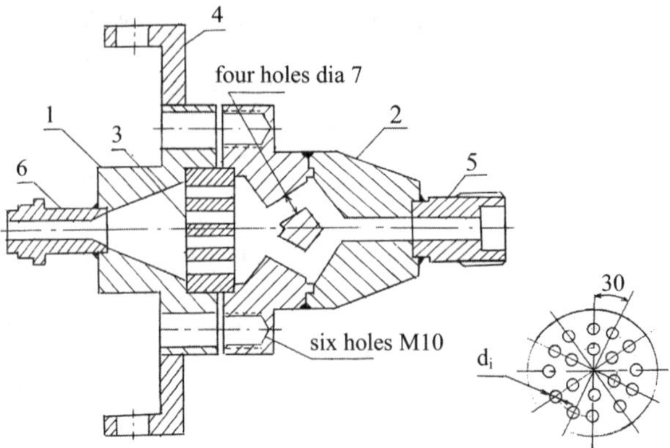

Fig. 9.36 Nonresable pulse disturbance device with a controllable disturbance amplitude: 1, 2, case; 3, detachable grids, where $d_1 = 4$ mm, $d_2 = 4.5$ mm, and $d_3 = 5$ mm; 4, bracket; 5, nipple; and 6, nosepiece for attachment to the chamber.

B. Data Processing of Artificial Disturbance Records

An optimal procedure for processing artificial disturbance data has been established based on the results presented in the preceding sections and experimental experience.

The method of artificial disturbances defines the following process parameters:

1) A_{mr} is the mean rectified value of signal in the region prior to the introduction of the disturbance (with a length of ~0.15 s for steady-state conditions and ~0.01–0.02 s for transient conditions).

2) A_0 is the initial peak of pressure caused by pulse disturbance.

3) A_M is the maximum peak of pressure caused by pulse disturbance.

4) A_M/A_{mr} is the relative peak of pressure caused by pulse disturbance.

5) Here t_e is the relaxation time, that is, the time elapsed for artificially initiated oscillations to decay by a factor of e, where e is the base of natural logarithm.

6) The energy spectrum for the time span corresponds to the relaxation time for the damped oscillations caused by pulse disturbance.

7) Here t_1 is the time span of direct influence of pulse disturbance on the process.

8) The $t_{relax} = t_1 + t_e$ is the total time interval of process relaxation.

9) The average decrement $\overline{\delta T}$ of oscillations is caused by artificial disturbance.

10) The values of three frequencies and amplitudes correspond to three main maxima of energy spectrum in the specified frequency range.

The recorded RCP processing program for determination of stability parameters from pulse disturbances proceeds in the following steps.

A calibration signal is processed to calculate a reference quantity using the following formula:

$$Mc = \frac{A_c}{B_c} \cdot \frac{2}{\pi}$$

where A_c is the value of the calibration signal in the unit of the parameter being measured and B_c is the average of the calibration signal module obtained as a result of the analog-digital conversion. To verify the time of pulse disturbance initiation, the pressure oscillation signal containing the pulse is plotted in the time span of 50 ms. Preliminary high-pass filtering of signal is then performed to eliminate the low-frequency components, with spikes caused by pulse disturbances being retained using the formula

$$Y_{f,i} = Y_i - \frac{1}{n+1} \sum_{j=i-n}^{i} Y_j$$

where $Y_{f,i}$ are the instantaneous values of the filtered signal, Y_j is the same as the initial signal, and n is a prime number. The cutoff frequency of the high-frequency filter is calculated by the formula

$$f_{\mathrm{hf}} = \frac{\Omega}{2(n+1)}$$

where Ω is the sampling rate of the analog-digital converter. The value of f_{hf} is selected on the condition that $f_{\mathrm{hf}} \leq 0.7 f_{\mathrm{r \cdot min}}$, where $f_{\mathrm{r \cdot min}}$ is the lowest frequency being studied.

To avoid random high-frequency oscillations precluded from the calculation of amplitudes and peak-to-peak amplitudes, low-frequency filtering of the signal is performed, with the values of peaks caused by pulse disturbances retained. For this purpose, discrete values of the signal are divided into groups with m points in each group. The points with maximum signal module are found in each group. These points are put to the last places in the groups with respective signs, and intermediate values of the signal are determined by linear interpolation.

The cutoff frequency of the low-frequency filter is determined by the formula

$$F_{\mathrm{lf}} = \frac{\Omega}{2m}$$

where f_{lf} is selected from the condition that $f_{\mathrm{lf}} \gg f_{\mathrm{r.max}}$, with $f_{\mathrm{r.max}}$ being the frequency having the fastest rate of increase in the amplitude spectrum within a range containing the frequencies being studied. To determine f_{lf}, the amplitude spectrum is approximately calculated for a time span of 2 ms containing pulse disturbance by the bandwidth (prime) digital-filter method.

Thereafter A_{mr} is calculated in the time span $t_r = 0.15$ s for the steady-state operation and $t_r = 0.01$–0.02 s for transient conditions immediately before artificially initiated oscillations by the formula

$$A_{\mathrm{mr}} = \frac{1}{n} \Sigma |Y_i| \cdot M_c$$

where $n = \Delta t_r \Omega$ is the number of readings of signal values Y_i.

The amplitude characteristics of pulse disturbance are calculated. A_0 is the initial peak of pressure caused by pulse disturbance, and A_m is the maximum peak of pressure caused by pulse disturbance. The relative peak of pressure caused by pulse disturbance is determined by the formula $n = A_m / A_{\mathrm{mr}}$, and the relaxation time t_{relax} is determined.

The value of t_{relax} is determined in the region of free damping of oscillations. Thus, the region with a duration t_1 from the initial peak of pressure A_0 is excluded from its calculation. This exclusion is made on the basis of the experimental data obtained during the time duration in which the direct effect of pulse disturbance upon the process takes place. In the remaining region, the maximum peak of pressure A_{rm} and the corresponding moment of time t_1 are determined. Then (with $N \geq 15$) in the region of the initiated oscillations, the best-defined peak with a value lower than that of A_{rm} by no more than e times and the corresponding moment of time t_2 are determined. Both times t_1 and t_2 are counted from the initial peak. The difference between t_2 and t_1 is equal to t_r. With $N < 15$, the time for damping the amplitude by a factor of e is determined as the time during which A_M decreases 1.5 times multiplied by 2.5 (assuming that damping proceeds following the exponential law). Time t_r is determined automatically in the case, if no more than one amplitude ($\gamma < 2$) exceeding the values of A_{rm}/e (with $N \geq 15$) and $A_{rm}/1.5$ (with $N < 15$) is encountered in natural noise of the process within a duration of 0.003–0.010 s immediately following the artificially initiated oscillations.

The energy spectrum is determined for the time span corresponding to the relaxation time with damped oscillations and in the specified range with undamped oscillations from pulse disturbance. The method with the use of a prime bandwidth digital filter is used, with its frequency characteristic similar to the frequency characteristic of a single resonance circuit.

Such a filter is described by the equation

$$Z_{ji} = \alpha_{ij} Z_{j,i-1} + \alpha_{2j} Z_{j,i-2} + Y_{fi}$$

where Y_{fi} and Z_{ji} are input and output signals, respectively; $i = 1, 2, \ldots, N$ is the number of processed points; and $j = 4, 5, \ldots, K$ is the number of energy spectrum components;

$$Z_{ji} = Z_{j,0} = 0$$

$$\alpha_{1j} = \frac{4\cos(2\pi f_j \Theta)}{1 - \exp(2\delta_f \Theta)}$$

$$\alpha_{2j} = -\exp(2\delta_f \Theta)$$

where

$f_j = \Delta f \cdot j$ th e filter tuning frequency
$\Delta f =$ 50 Hz the spectrum scan increment
$\Theta = 1/\Omega$ the sampling interval
$\delta_f =$ the logarithmic filter damping coefficient

To obtain each component of the energy spectrum S_j, the region of filtered signal Y_f containing artificially initiated oscillations is converted with the use of the just-described digital filter equation tuned to frequency f_j. The conversion results are squared, summed up, and multiplied by a weighting function, as follows:

$$\mathrm{S_j} = \frac{1}{N} \sum_{j=1}^{N} Z_{ji}^2 \beta_i$$

$$\beta_j = \begin{cases} \left[\frac{f_j}{1000}\right]^2, & f \leq 5000 \text{ Hz} \\ 0{,}006 f_j, & f \geq 5000 \text{ Hz} \end{cases}$$

where β_j is the weighting function used for spectrum equalization.

From t_r and $f_{r.M}$, the average decrement of oscillations caused by artificial disturbance is calculated by the formula

$$(\delta T)_0 = \frac{1}{t_r f_{\text{rm}}}$$

where f_{rm} is the frequency corresponding to the primary maximum of the energy spectrum.

As an example, the results of a pulse disturbance obtained in an experimental combustion chamber are shown in Fig. 9.36. The actual record of disturbance is shown in the upper plot of Fig. 9.37. Also shown are the record with filtering of low-frequency components and the spectrum of oscillations in the chamber after the introduction of pulse disturbance. Major results are summarized next: 1) the value $n = 40$ exceeds the optimal value of 25, and the pulse caused undesirable low-frequency oscillations; and 2) the relaxation time $t_{\text{relax}} = 5.8$ ms was less than the permissible value of 15 ms, and the system provided a sufficient stability margin to hard excitation. The subscript $q = 0$ indicated that high-frequency oscillations were damped to a noise level.

During tests of actual engines, the decrements and spectra of oscillations are estimated before introduction of artifical disturbances and after damping of oscillations. The decrements of pressure oscillations and spectra should not differ within the accuracy of measurements. A considerable difference means process instability.

The dependence of the relaxation time $\tau_r(t_r)$ on the operating mode for engine 11D56 is shown in Fig. 9.38. Engine KVD-1 (11D56) was developed in Russia (Isayev's DB) in 1960–1977 [12]. It was designed for upper stages (unit "R" RN-N1) and involved afterburning of reducing gases from the gas generator. The KVD-1 engine was retrofitted, and its design was sold to India [12]. Propellant

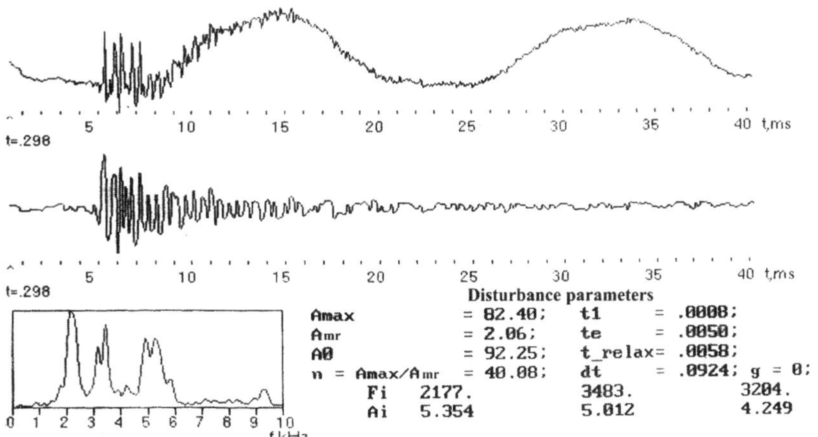

Fig. 9.37 Example of computer processing of a disturbance pulse obtained from an experimental study on a model combustion chamber.

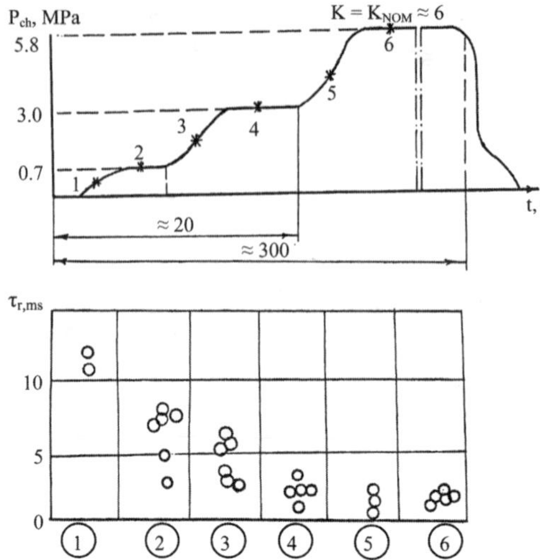

Fig. 9.38 Disturbance relaxation time τ_r with the change of hydrogen–oxygen engine operating conditions (11D56): cyclogram of engine operation (i = 1–6, serial numbers of disturbance introduction); values of τ_r are shown for different i.

components are liquid oxygen and liquid hydrogen. The engine operation conditions and performance parameters are given here: P_v = 69.6 kN (7.1tf), J_v = 462 s, t = 800 s, P_{ch} = 5.7 MPa, κ_m = 6, and D_{ch} = 200 mm. The number of startups is 3. It follows from Fig. 9.38 that the combustion chamber of engine 11D56 has the lowest stability margin to hard excitation during the startup. As the pressure in the chamber increases and the engine reaches its nominal operation, the relaxation time decreases and becomes less than 5 ms.

Chapter 10

Model Firing Tests for Selection of Injector Head Elements

AS DISCUSSED in Sec. II of Chapter 5, one of the LRE development stages involves laboratory tests of hydraulic, acoustic, and combustion models for combustion chambers and gas generators. Those tests are carried out to provide understanding of the stability. The results of studies of the combustion-chamber acoustics as applied to the problem of operating process stability are described in Chapter 7.

A method of firing model tests became particularly effective for closed-circuit (gas–liquid) engines with high pressure in the combustion chamber [42 and 43]. This method is based on the following conditions: a low pressure level practically equal to atmospheric pressure; a small weight flow rate of propellant (two and more orders of magnitude lower than actual ones); and replacement of a liquid propellant component by a gaseous one to provide actual volumetric flow rates at small weight flow rates. The simplicity and mobility of model units allows their use as convenient tools for checking the effectiveness of structural alterations in injectors and mixing heads on stability. Experience has shown that an acute necessity for solving such problems arises at all times and especially in early stages of engine development.

The method of firing tests of single bipropellant injector specimens installed on the mixing heads under study is most effective for obtaining information on stability. The use of a single injector (instead of a mixing head) and its laboratory tests cut the duration as well as material costs by more than an order of magnitude as compared to model tests of a full-size mixing head. This procedure affords prompt testing of a great number of injector versions to select a better version in respect to stability at any stage of engine development. If a design of injectors does not allow their independent testing, the use of full-size mixing heads in firing tests is justified, for example, in the case of mixing heads with monopropellant injectors and heads that cannot be separated into individual injectors.

Numerous special checks of qualitative and quantitative compatibility of the results of model and actual tests on stability have been made. These estimations have confirmed the validity of the methods chosen for approximate simulation of high-frequency oscillation excitation conditions in LRE. One advantage of using model units over firing actual engines consists of incomparably less strained conditions of operation. So additional special studies on the physics of combustion instabilities in LRE were carried out. The results of such studies are important for deliberate action on stability. Thus, the mechanism of probabilistic excitation disturbances and their stability in the combustion chamber was demonstrated with the help of a model unit as described in Sec. III of Chapter 3.

I. Basic Principles and Rules for Approximate Simulations

The design and operating parameters of models for study of high-frequency combustion instability were selected on the basis of general principles and rules used in the development of any simplified physical models.

Actual complex processes composed of a wide variety of actual phenomena can be represented with an approximate scheme. Then, after analyzing the particular conditions of the operating process, a phenomenon that has a dominant influence on the final result is selected. It is assumed that other phenomena play secondary roles; therefore, observance of all similarity conditions is not obligatory. With such an approach, in each particular case a simulation error will be defined by how correctly the researcher manages to reveal a governing phenomenon. Subsequently, it is a matter of reproducing under model conditions the main parameters defining this phenomenon.

A physical model must not only reflect correctly an actual process under study, but at the same time it must be essentially less complicated than the actual phenomenon. Correlation between the results for the model and actual tests is the most complete check of the assumptions made. It is most convenient to compare the results of the model and actual tests if the scale of the main parameters defining the process being studied is taken equal to unity.

It follows from the principles just listed that an informal approach should be used for successful approximate simulation of combustion instability. Its informality is concerned with the necessity to use physical concepts of a complex process under study such as high-frequency combustion instability in LRE. In the general case the combustion instability results from interactions of acoustic oscillations in the chamber with such phenomena as spraying, evaporation, mixing, and chemical kinetics. The accuracy of an approximate simulation depends on the accuracy of evaluation of the impact of each of these phenomena on the combustion instability and on the accuracy of reproduction under model conditions of a selected governing phenomenon.

The studies carried out by the approximate simulation method involve the following stages: 1) processing and analyzing the available information on tests of actual units or the units close to the actual ones in design and operating conditions; 2) choosing a general outline of a model; 3) developing a special working model designed for the specific studies; 4) experimenting on this model with improvement of its outline and simulation data; and 5) extending the data from model conditions to actual conditions and carrying out, when possible, control checks of qualitative and quantitative agreement between model test data and actual data.

At the first stage, the similarity conditions and all criteria characterizing the process are determined by analysis of the process under study. Furthermore, the similarity criteria and those parameters that have the most essential effect upon the process are determined.

In this way we identify the governing parameters and criteria forming the basis for solution of the second-stage problem and selection of a model type. In our case setting the general outline of a model primarily depends on answering the following questions:

1) Is it necessary to use full-scale mixing heads for a model chamber as a test object or is it sufficient to perform tests of single mixing elements? The latter is possible if the mixing heads being studied are equipped with bipropellant injectors of the same type operating practically independently.

2) Does the accepted method, based on combustion of gaseous components in the model chamber, allow identification of the governing similarity criteria already established at the first stage from analysis of processes in the actual chamber?

Thereafter the development of a special working model (the third stage) with consideration for its practical engineering realization is started. Providing the maximum similarity of a model and a simulated object to obtain the minimum simulation error is a natural measure of the efficiency of theoretical and technical solutions used. During experimentation on the working model, it is necessary to check the actual significance of each governing parameter identified in the analysis. Further, it is necessary to make a preliminary estimation of the expected error if for some reason there are slight deviations of the governing parameters made (the fourth stage). At the final stage, correlations of model and actual test data are made.

Following the principles and rules just discussed must result in devising a simplified physical mode of the chamber. The organization of the operating process is such that the reliability of predicting combustion stability is provided with minimum expense for development of this predictive model. The final appearance of such a model should be formed on the basis of a compromise between the expected accuracy of simulation results sufficient for making a decision on the selection of the mixing head versions providing the best stability on the one hand and on simplicity, portability, and cheapness of model tests, on the other hand. Such a compromise results in the procedure of firing tests of models of the actual mixing heads. Separate actual injectors were tested practically at atmospheric pressure in the chamber with the use of gaseous components to determine under model conditions the regions of high-frequency combustion instability.

II. Concept of Simulating Combustion Instability at Low Pressures

The method of simulating the conditions for exciting high-frequency oscillations in LRE at low pressures, is based on the following provisions:

1) The selection of a model is made according to the foregoing principles and the rules for approximating simulation of a complex process.

2) Full-scale mixing heads or single bipropellant injectors of LREs are the test objects. The application of natural geometry mixing devices sets the simulation scale equal to unity. It is assumed that bipropellant injectors operate to a great extent independently. The mutual influence of adjacent injector flames on the operating process in the initial flame section most sensitive to disturbances can be ignored.

3) To simplify the model design and the technique for determining stability boundaries, approximate atmospheric pressure must be maintained in the model

chamber. At low pressures the unit can operate for a long time under unstable combustion conditions without danger of chamber or mixing head breakdown. The absolute values of oscillation amplitudes $\overline{P'}$ are proportional to the low average pressure P. In addition, under low pressures, special devices (for instance, optical devices) for measuring pulsating and average characteristics of the combustion processes can be used. A low pressure should be provided by using a low mass flow rate of propellants $\dot{m}$. However, a level of a volumetric flow Q (and, accordingly, of an exit velocity U) of the oxidizer and fuel must be at the actual levels. These two requirements can be combined only through the use of oxidizer and fuel with low densities. Under low (atmospheric) pressures this results in the necessity to use gaseous components in the model chamber. At the same time, the use of gaseous components represents satisfactorily the combustion process conditions of actual LRE. Most of the properties of the liquid components in the injectors are close to dense-gas properties as a result of critical pressure and temperature conditions or conditions close to the critical state.

4) In the general complexity of physical and chemical processes proceeding in LRE chambers and gas generators, the mixing processes, which can be simulated with use of gaseous components, have a determining influence on the combustion process.

5) To bring the model of the mixing process as near to actual conditions, the method of diluting active components (oxidizer and fuel) with neutral gases (nitrogen N_2 and helium He) is used. At a constant value of the coefficient of excess oxidizer, $\alpha = (m_o/m_f)/L_o = \text{const}$, the velocities and densities of the interacting components can be changed in this way within certain limits.

6) The frequencies of the natural acoustic oscillations in the model chamber are kept close to the actual values. Tests of full-scale mixing heads are carried out in a full-scale model chamber. In tests of single-injector model chambers, the condition of $f = \text{idem}$ for transverse oscillations is met by an appropriate selection of the model chamber diameter D_m with consideration for an effective sound velocity during combustion: $D_m = D_n(c_m/c_n)$.

7) Stability in the model chamber is controlled by changing the volumetric flow of oxidizer and fuel. Information on the change of stability to soft and hard excitations for different versions of mixing heads and injectors follows from the analysis of the locations of stability boundaries and hysteresis.

8) In model tests the similarity of oscillatory processes in the injector channels is approximately provided by selection of an actual value of wavelength $\lambda = c_i/f$ of disturbances propagating along the channels. The value of c_i, the sound velocity in the channel, can be controlled by changing the temperature of the active gas, or by dilution with a neutral gas.

9) The selection of boundary conditions and governing similarity criteria is made by analysis of the physical concepts of the process being studied.

III. Methods of Evaluating Agreement Between Model- and Full-Scale Test Results

A final check of the accuracy of decisions made during selection of design and operating parameters of the model unit is carried out by comparing the results of model and actual tests in the course of improving the stability of a particular LRE. The agreement of the data obtained for a sample (reference) comparison of two or three versions of injectors is considered grounds for judging their agreement in other cases as well.

The comparison can be qualitative or quantitative. Qualitative comparison can be made by different criteria; quantitative comparison is made by the same parametric criteria. The purpose of comparison consists in checking the reliability of predicting under model conditions the trend of stability in an actual LRE. Thus, for instance, suppose that the operating parameters or design of the mixing head are changed. The tendency for change of stability at a level of the natural pressure disturbances (noise) in an actual chamber can be tracked from the variation of oscillation decrement δT and relative oscillation amplitude A. These can be determined from the spectral characteristics of chamber noise for the most "dangerous" frequency of free acoustic oscillations. In most cases the most probable dangerous frequencies are the frequencies of the first tangential mode f_{1T} or the first longitudinal mode f_{1L}, depending on combustion-chamber geometry.

The most dangerous is generally the frequency of free acoustic oscillations for which the oscillation amplitude in the chamber noise spectrum is the highest. A change of stability in an actual chamber caused by the action of finite-amplitude disturbances can be determined from change of relaxation time τ_r of damped oscillations after applying calibrated pulse disturbances into the chamber. A decrease in the decrement δT, a reduction of the relative oscillation amplitude A at a noise level, or a shortening of the relaxation time τ_r of damped oscillations indicate that the stability in the actual chamber has improved.

The trend in the change of stability in the model chamber can be inferred from the change of the boundaries for spontaneous excitation and hysteresis of oscillations. In this case a measure of the stability margin can be a dimensionless quantity R characterizing the distance of the working region of operating parameters in the actual chamber from the boundaries of the predicted instability region and hysteresis (Fig. 10.1). A value of R is determined from the same operating parameters X and Y used as coordinates for presentation of experimental and hysteresis boundaries:

$$\overline{R} = \sqrt{R_x^2 + R_y^2}$$

where

$$R_x = \Delta X/X_A, \qquad R_Y = \Delta Y/Y_A$$
$$\Delta X_b = X_b - X_A, \qquad \Delta X_{\rm hyst} = X_{\rm hyst} - X_A$$
$$\Delta Y_b = Y_b - Y_A, \qquad \Delta Y_{\rm hyst} - Y_A$$

X_A, Y_A are the values of coordinates of some point A belonging to the working region of the actual chamber and located closest to the instability region boundaries and X_b, Y_b, $X_{\rm hyst}$, $Y_{\rm hyst}$ are values at stability boundaries (b index) and in the hysteresis region (subscript hyst) of coordinates of the point, which is closest to the working point A.

Location of the coordinates of the simulated operating process in the hystersis region indicates that high-frequency oscillations can be excited from random

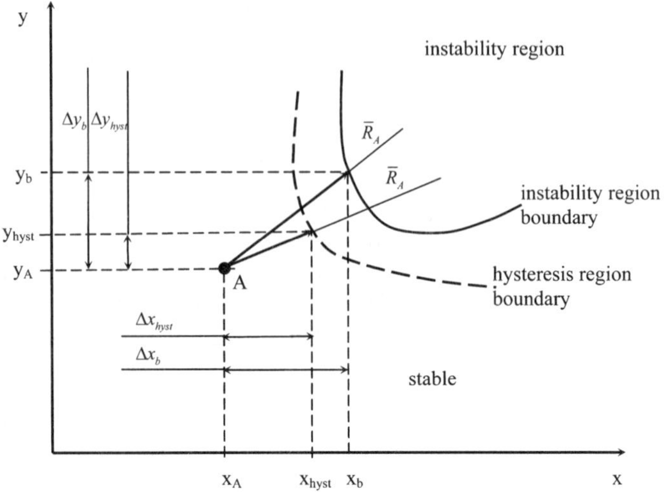

Fig. 10.1 Determination procedure of stability margin parametric index $\bar{R}$. (*A* is a working point for actual operating conditions.)

pulse disturbances. In comparative evaluation of different versions of injectors or mixing heads, that version for which the values of parameters $\bar{R}_b$ and $\bar{R}_{hust}$ are higher is considered to be more stable. Reduction of the instability region, or lowering of the level of relative self-oscillation amplitude $\bar{A} = \overline{P'}/P_{ch}$, even without changing the position of stability boundaries, also demonstrates the usefulness of the measure.

Variations of the parameters δT, $\bar{A}$, and τ_r based on tests of the actual chamber and of the parameters $\bar{R}_b$, $\bar{R}_{hyst}$, and $\bar{A}_{self}$ based on tests of the model chamber will indicate the qualitative similarity of phenomena defining the combustion process instability under actual and model conditions. Both a soft and hard excitation of pressure oscillations can be studied.

Quantitative comparison of operating parameters of the model and actual chambers at the stability boundaries must be made on the basis of a parametric stability criterion determined by other criteria. These criteria are required for two purposes: first, to predict the stability boundaries from model data of a motor; and second, to choose the operating conditions for development of the model. The general approach to the estimation of parametric stability criteria can be accomplished with the use of a power method. In the general case the combustion zone can be a source of not only driving but also damping oscillations. This is determined by the phase difference φ between the heat addition rate and pressure oscillations. The phase shift depends on a typical delay time of the processes in the combustion zone τ and the period of acoustic oscillations T:

$$\varphi = 2\pi\tau f = 2\pi/\Omega$$

where $\Omega = (\tau f)^{-1}$.

In a generalized form, the coefficient of oscillatory energy generation as a result of the processes in the combustion zone δ_{gen} can be represented as a product:

$$\delta_{gen} = |\delta_{gen}| \cos j = |\delta_{gen}| \cos\left(\frac{2p}{W}\right)$$

where $|\delta_{gen}|$ is the modulus of the generation coefficient characterizing the dissipation of oscillatory energy in the combustion zone.

In the general case for a selected oscillation frequency having period $T = 1/f$, the balance between the energy generated in the combustion zone and the energy dissipated for the same period of time is determined by the oscillation decrement

$$\delta T = \delta_1 T - \delta_{gen} T = \delta_1 T\left(\frac{1 - \delta_{gen}}{\delta_1}\right)$$

where δ_1 is the coefficient of passive losses.

Analysis of the expression for the oscillation decrement δT shows that closeness of δT to zero, that is, remoteness of the operating process from the instability region boundary, is completely defined by the difference $\Delta = 1 - \delta_{gen}/\delta_1$, which depends on the relation

$$N = \frac{\delta_{gen}}{\delta_1} = |\delta_{gen}|/\delta_1 \cos(2\pi/\Omega) = N_* \cos(2\pi/\Omega)$$

Hence the parameter $N = N_* \cos(2\pi/\Omega)$ is a generalized expression for the stability criterion. The parameter N includes two other dimensionless parameters: $N_* = \delta_{gen}/\delta_1$ and $\Omega = (\tau f^{1})$.

In a physical sense the parameter N_* is an energy (or amplitude criterion of stability) and represents a modulus of the magnification (amplification) ratio $|\delta_{gen}|$ divided by the coefficient of acoustic losses δ_1. The parameter Ω is a phase criterion of stability and characterizes the phase condition of oscillation excitation. The operating process will be stable to acoustic oscillations when the following inequality is met:

$$N = \left|\frac{\delta_{gen}}{\delta_1}\right| < 1 \quad \text{or} \quad N = N_* \cos(2\pi/\Omega) < 1$$

For a stable process $\delta T > 0$, $N < 1$, $\cos \varphi$ can be positive or negative. At the oscillation excitation boundary $\delta T = 0$, $N = N_* \cos(2\pi/\Theta) = 1$, and $\cos(2\pi/\Theta) > 0$.

In general, presentation of the model and test data for locating stability boundaries should be performed in the coordinates $\Omega - N_*$. However, analysis of the conditions for oscillation at the stability boundary ($\delta T = 0$) has shown that for a number of actual and model conditions a phase criterion written in the form of a dimensionless delay time $\Omega = (\tau f)^{-1}$, can be used as the stability criterion. If during simulation the frequency of oscillation under model and actual conditions is the same, the parametric stability criterion Ω (within a constant of the order of the oscillation period) is completely defined by a delay time τ sensitive to variations of the combustion processes caused by pressure oscillations. It follows that simulation of the conditions for oscillations in a liquid rocket engine must be based on similarity of the phase relations defining the relative homochronous property of acoustic and combustion processes. Not claiming complete confidence in the data obtained, such an approach makes it possible to provide the identical mechanisms of combustion instability under model and actual conditions of LRE and to obtain data for further analysis and generalization. A great variety of combustion instability phenomena require a differentiated approach in determination of phase similarity conditions as applied to one or another pattern of mixing and combustion processes.

Certainly the cases of small amplification ratios $|\delta_{gen}|$, for which the value of the energy criterion $N_* = |\delta_{gen}|/\delta_1$ is close to unity, must not be completely ruled out. Significant increase in the coefficient of passive losses δ_1 can result in complete elimination of the instability region. Hence it is evident that providing constancy of the phase criterion at the boundary and in the instability region $\Omega = \text{const}$ (or $\tau = \text{const}$ with $f = \text{const}$) is the first stage of approximate simulation. The second stage is introduction of corrections on the basis of calculation and experimental evaluation data to compensate distortions occurring under model conditions as a result of ignoring the similarity of amplitude relations.

The structure of the dimensionless $\Omega = (\tau f)^{-1}$ or dimensional τ^{-1} parametric stability criteria in terms of operating and design parameters is determined from physical considerations. This is done on the basis of logical analysis of regularities in the process being studied. To simplify the problem, the process is generally reduced to some typical scheme, for which the calculation or semi-empirical relations has already been developed. Therefore, an explicit form of an expression for the dimensionless parametric stability criterion $\Omega = (\tau f)^{-1}$ or dimensional parametric stability criterion τ^{-1} can be determined. The nature of the characteristic time τ is clarified in each case. We discuss this problem with the following example.

The problem of high-frequency instability is more commonly encountered during improvement of large-size chambers and gas generators used in modern powerful LRE. As a rule, the operating process then takes place under pressures P_{ch} exceeding the value of the critical pressure P_{cr} for the liquid-propellant components. The temperature of these components is also close to the critical temperature T_{cr} or exceeds it. As a rule, under such conditions the jets of liquid components approach jets of dense gas in their physical properties. For approximate estimations of total delay time of the processes in the combustion zone τ, the time of liquid phase transformations (spraying time τ_{spr} and evaporation time τ_{evap}) can then be excluded from consideration. In addition, in most cases a modern LRE is made as a closed circuit; oxidizer or fuel are gasified in the gas generator and fed to the combustion chamber for afterburning in the gaseous state.

Under the high pressures and temperatures typical for LRE, the time of chemical reactions τ_{chem} also must not have a noticeable effect on the delay of processes in the combustion zone. Therefore, from quality considerations confirmed by estimations, it follows that under actual operating conditions in modern powerful LRE the inertia of the processes in the combustion zone must be determined primarily by the mixing time τ_{mix}. That is, the rate of the mixing process is the main limiting stage of the combustion process.

In first approximation, the characteristic time of combustion, identifiable in this case with the mixing time ($\tau = \tau_{mix}$), can be represented as the residence time of the gaseous component in the flame, that is, as some time of transport lag:

$$\tau^{-1} = \frac{U}{l}$$

where l is the representative size of the flowfield and U is the representative velocity.

Consider the flow of jet 1 into the concurrent flow 2. A longitudinal dimension of the combustion (mixing) zone is proportional to the equivalent initial diameter d_{equiv}, of the jet, related to the geometrical initial diameter as $d_{equiv} = d\sqrt{n}$; $n = \rho_2/\rho_1$ is the ratio of the densities of the concurrent flow ρ_2 and the jet ρ_l. The value of U must be identified with some effective velocity representing the rate of mixing of propellant components U_{mix} or the propagation speed of various disturbances U_d (for example, vortex formation) along the jet zone. In the general case the speeds U_{mix} and U_d depend on the density ratio $n = \rho_2/\rho_1$ and speed ratio U_2/U_1. However, studies have shown that for the first approximation it is sufficient to use as a representative speed U the speed of the more rapidly flowing component U_σ, which has a higher volumetric flow Q_σ. Taking into account the foregoing, the parametric criterion Q will be proportional to the following expression:

$$\Omega \sim \Omega_* = (U_\sigma\sqrt{n})\,\mathrm{d}f$$

The parametric criterion Ω in the form Ω_* is already suitable for extrapolation of model parameters at the stability boundary to the actual parameters. In this case, depending on the conditions of the model tests, the conversion criterion Ω_* can be used in a simpler form. Thus, for example, if during simulation an actual oscillation frequency is maintained (f = const) and actual mixing elements are used (d = const), with a speed ratio of $m \gg q$ the parametric criterion can be used in forms:

$$\Omega_{*1} = U_\sigma\sqrt{n} \quad \text{or} \quad \Omega_{*2} = Q_\sigma\sqrt{n}$$

where Q_σ is the volumetric flow of the more rapidly flowing component.

The structure of the expression $\Omega_* = (U_\sigma\sqrt{n})\mathrm{d}f$ possesses some universality. It formally coincides with the structure of the parametric criterion, for the cases under consideration, for a representative time τ representing 1) the time of turbulent mixing of a homogeneous mixture propagating in the stationary medium, or 2) the time of destruction of a liquid jet or a sheet carried by a high-speed gas flow.

The parameter $m_g(\rho_g)^{-0.5}$ used in the studies by American researchers for evaluation of the effect of the gaseous hydrogen temperature on the stability of oxygen-hydrogen LRE can also be reduced to the structure of the criterion $\Omega_{*1,2}$.

IV. Schematic Diagrams of Model Units and Test Conditions

As just discussed, stability in the model unit is controlled by changing the exhaust velocities V of components from the injectors by changing the volumetric flows Q of components. In this connection a schematic diagram of the model unit and test conditions will be defined by changing the volumetric flow of components. The volumetric flow of components into the chamber can be represented in the form

$$Q = \frac{\dot{m}}{\bar{\rho}} = \frac{\dot{m}}{\bar{P}}\,RT$$

where $\dot{m}$ is the total mass flow of propellant, $\overline{P}$ is the average pressure in the model chamber, and R and T are the gas constant and temperature of combustion products, respectively.

With a constant value of the oxidizer excess coefficient $\alpha = (\dot{m}_o/\dot{m}_f)/(1/L_o)$, the condition $RT = \text{const}$ is satisfied, and so we have $Q \sim \dot{m}/\overline{P}$. Hence, the change of volumetric flow can be achieved by two methods: 1) by changing $\dot{m}$ with $P = \text{const}$, or 2) by changing P with $\dot{m} = \text{const}$.

The first method can be realized if the pressure in the model chamber is selected to be practically equal to atmospheric pressure, $P \approx 1.0$ bar $\approx$ constant. In this case the chamber operates under subsonic exhaust conditions of combustion products from the nozzle. The change of volumetric flow of components is provided by changing the mass flow of components, $Q \sim \dot{m}$.

The second method can be realized when the model chamber operates under supersonic exhaust conditions of combustion products from the nozzle. In this case, the pressure in the chamber must be $P > 2.0$ bar and varies in the range of $P \approx 2$–6 bar. The change of Q is provided by changing P with $\dot{m}$ constant. Because P is equal to $m\beta/F_{cr}$, Q is equal to $F_{cr}(RT/\beta)$, that is, $Q \sim F_{cr}$ variable with $RT/\beta \sim$ constant. It means that in the range of $P = 2$–6 bar, to change the volumetric flow of components the area of the nozzle throat F_{cr} should be changed.

In accordance with the preceding two methods of changing the volumetric flows, typical units (for instance, with full-scale chambers of gas–liquid engines) can have the following features.

When using the first method ($P = 1.0$ bar constant and $\dot{m}$ = variable, see Fig. 3. 1), mass flows of gaseous oxidizer m_o and gaseous fuel m_g can be regulated with the help of special gas distributors (3), (8). The gas distributors are arranged so that gas generators (2), (7) producing the gaseous oxidizer and gaseous fuel for the model chamber can operate under constant conditions. If necessary, the unit makes possible the dilution of active components (oxidizer and fuel) with neutral gases (N_2, He) preheated in the heat exchanger (1).

When using the second method (P variable and $\dot{m}$ constant, see Fig. 10.2), the volumetric flows of gaseous oxidizer and gaseous fuel fed to the model chamber (5) can be regulated by moving the cooled cone (6) installed in the nozzle throat.

If a single-injector model chamber is selected for tests, for which the stability boundaries are defined by the first method of changing the volumetric flows, the schematic diagram of such a single-injector model unit can have an appearance represented in Fig. 10.3. Chamber (1) practically operates at atmospheric pressure. It can be equipped with a transparent insert to observe the combustion process. The chamber is installed vertically on a flat steel plate (2) simulating a mixing head and can be freely moved in the horizontal plane. The outlet section of the injector being tested (3) is flush with the plate plane (2). The injector itself is mounted near the chamber wall to make provision for excitation of tangential oscillation modes. The injector is fitted with headers for supply of oxidizer and fuel. In operation of a single-injector model unit under laboratory conditions, the gaseous oxygen is used as an oxidizer, which can be diluted with a neutral gas (for instance, with N_2 or the air) as the need arises. If necessary, the oxidizing gas can be preheated in the heat exchanger (5), which consists of a coil (7) placed into a muffle kiln. The gases such as CH_4, C_3H_8, H_2, NH_3 can be used in the unit as

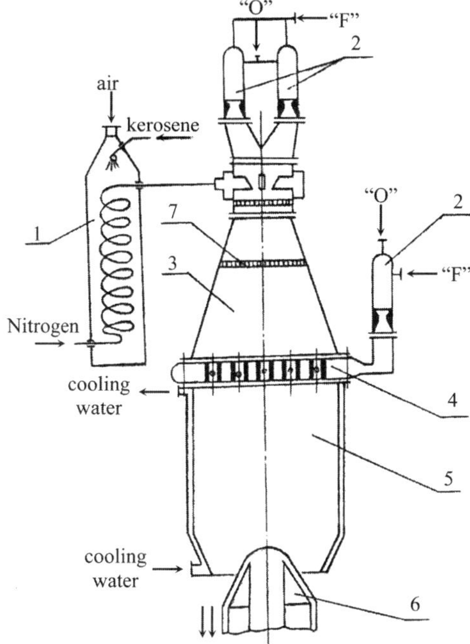

Fig. 10.2 Schematic diagram of a model unit with a full-scale chamber operating according to the method $\overline{P}'$ = variable, $\dot{m}$ = const: 1, heat exchanger; 2, gas generator; 3, gas passage; 4, mixing head; 5, combustion chamber; 6, regulating cone; and 7, honeycomb.

a fuel. If required, these combustion gases can be diluted with neutral gases (N_2, He). A flow of oxidizer and fuel can be controlled by throttle valves or special gas distributors.

The standard measuring devices for recording the flow rates, pressure, temperature, and pressure pulsations are used in all three units. In addition, special measuring tools can be used. They are optical devices for recording the constant and pulsating components of radiance of intermediate radicals of chemical reactions in the flame, high-speed photography of the flame and others. The internal walls of the chambers are cooled with water.

The pressure pulsation recording system in all units comprises a piezoelectric transducer, a wideband amplifier, a frequency counter, a voltmeter, and a personal computer.

During tests, this system enables visual check of a mode of vibration, determination of values of the frequencies and amplitudes of oscillations, as well as recording a signal from the pressure pulsation transducer for further reproduction, processing, and analysis.

The procedure for determining the instability region and hysteresis boundaries on the model units consists of increasing or decreasing gradually the

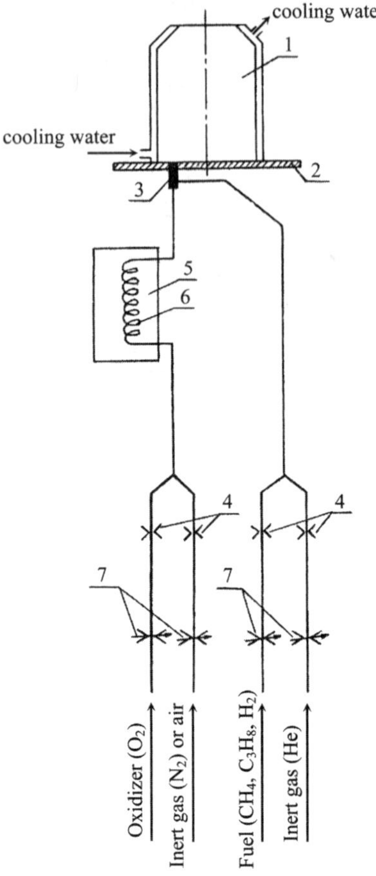

Fig. 10.3 Schematic diagram of a single-injector model unit: 1, combustion chamber; 2, plate; 3, tested injector with piping; 4, metering nozzle; 5, muffling kiln; 6, coil; and 7, throttle valve.

volumetric flow of oxidizer Q_o and fuel Q_f until the moment when oscillations occur. It is assumed that if possible a speed of reversal excludes the probabilistic excitation of high-frequency pressure oscillations (see Sec. III of Chapter 3).

The sequence of determining experimental points at the boundaries of stability and hysteresis is shown for illustration in Fig. 10.4, for the case when the flows Q_o and Q_f are controlled by the first method. The plot shown in Fig. 10.1 illustrates in the general case the approach to determining the stability margin R. The dimensionless value R is estimated from the difference between the simulated value of parametric criterion Ω to be determined (or its separate component, for example, Q) at a working point A and its value at the boundary of stability R_b and hysteresis R_{hyst}.

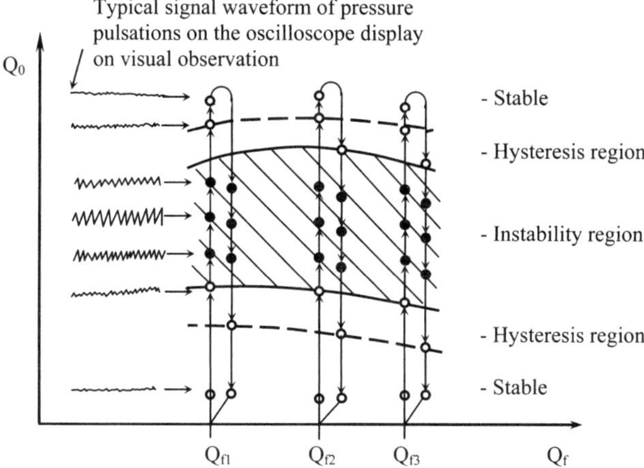

Fig. 10.4 Example of timeline in determination of instability boundaries for the model unit.

V. Examples of Using the Firing Simulation Methods for Studying Stability in a Combustion Chamber

Examples of using the firing simulation methods, especially on different modifications of a single-injector model unit, are presented in several sections of this book. Section III of Chapter 14 contains experimental results on increasing the stability margin by reducing the number of liquid holes from six to five in pneumatic gas–liquid injectors. Experiments have been carried out as a result of the need to increase the stability margin for LRE 4D75.

An influence of the generator gas velocity on stability as applied to bipropellant pneumatic injectors is discussed in Sec. VII of Chapter 13. Studies have been carried out on a full-scale model unit under a pressure in the chamber $P_{ch} \approx 0.1$ MPa.

As another example of a firing simulation, we examine the influence of the acoustic characteristics of the gas injector on the stability in the combustion chamber with afterburning of the generator gas. In this case the firing simulation is accompanied by acoustic simulation of a cold chamber without combustion, and by acoustic calculations. The procedure of acoustic simulation and approximate acoustic calculations is described in Chapter 7.

Tubes of a certain length, which frequently have restrictions at the inlet (a jet), are used as injectors. Gaseous components are fed to the combustion chambers of closed-circuit engines with afterburning of the generator gas. Pressure oscillations in the combustion chamber excite the same oscillation mode in the gas passage at an injector inlet. However, because the sound velocity in the generator gas is half or less than that in the combustion products, the frequency of the mode transmitted to the gas passage is much less than that in the combustion chamber. The resonance of the transverse mode in the gas passage is very sharp [43]. Taking into account the transformation of impedance, depending on the ratio of cross-sectional area of the gas passage to the cross-sectional area of injectors ($F_i/F_{gp} = 0.1$–0.3), the inlets of the channels present acoustically open ends ($Z \approx 0$).

From the side of the combustion chamber, the total flow area of injectors is 0.1–0.3 of the chamber cross-sectional area, and the wave impedance ρc in the generator gas exceeds the wave impedance in the combustion products in the chamber. So, from the side of combustion chamber, the gas injector channels can be also considered to first approximation as acoustically open. Therefore, the gas injector channels can be considered to first approximation as tubes with acoustically open ends. According to the formula (7.14), the resonant frequency of such tubes depends on the sound velocity c and an effective length of tubes L_{ef}:

$$f_{\text{ni}} = \frac{nc}{2L_{\text{ef}}} \tag{10.1}$$

where $L_{\text{ef}} = L_i + 0.8\,d_i$ is an effective length of the injector, m; L_i is the geometrical length of the injector, m; d_i is the injector diameter, m; c is the sound velocity in the generator gas, m/s; and $n = 1, 2$ ... is an integer number.

Choosing a length of injector channel with the condition that the frequency in the injector channel and the frequency in the combustion chamber are equal, $f_{\text{i.ch}} \approx f_{\text{c.ch}}$, it is possible to get an additional damping of pressure oscillations in the chamber. In this case the injectors are considered adjusted. The damping effect of the adjusted injectors is accounted for by losses caused by the hydraulic resistances of the channels. The main part of the hydraulic resistance is concentrated at the inlet (sudden contraction) and at the outlet (sudden expansion) of channels. These are the places where oscillatory velocity maxima are located.

The effect of injector channel length on the oscillation decrement at the first tangential-longitudinal mode is seen even in the simplest chamber model without flow. The model represents a cylinder with a diameter of 160 mm. From one side the cylinder is closed by a flat cover with vibration baffles made of pins (see Fig. 7.15), and from the other side it is closed by a cover with a slot. A tube injector with a diameter of 12 mm was mounted in one peripheral cell of baffles near the wall of the model chamber. The tube length was changed discretely. One end of the tube is mounted flush with the cover, and the other end is open into the room. The results of measuring the oscillation decrement for the first tangential-longitudinal mode are shown in Fig. 10.5.

With the increase in length of the injector channel from 94 to 124 mm, an effective length approaches the half-wavelength, which in this case is 136 mm, and the oscillation decrement attains the extreme value. Such a relationship is observed with three versions of baffles, but with the increase in length of baffles the waves showing the dependence of injector channel length on length of bafflers arise.

The damping effect resulting from the adjustment of injectors appears in the case of external mixing of propellant components. If fuel is injected inside the injector channel, with withdrawal (deepening) of an injection point from the exit section, the opposite result can be observed: the adjustment destabilizes the operating process. The destabilization occurs for the transverse oscillation mode with a frequency below critical, that is, when the oscillations are concentrated in the injector channels; in the combustion chamber the oscillation amplitude decreases exponentially from the mixing head to the nozzle. Therefore, before carrying out a hot firing and acoustic simulation of an actual combustion chamber, it is advisable to carry out an analytical study of the acoustic properties of the actual chamber and gas passage. Realistic dimensions of the injector and chamber must be

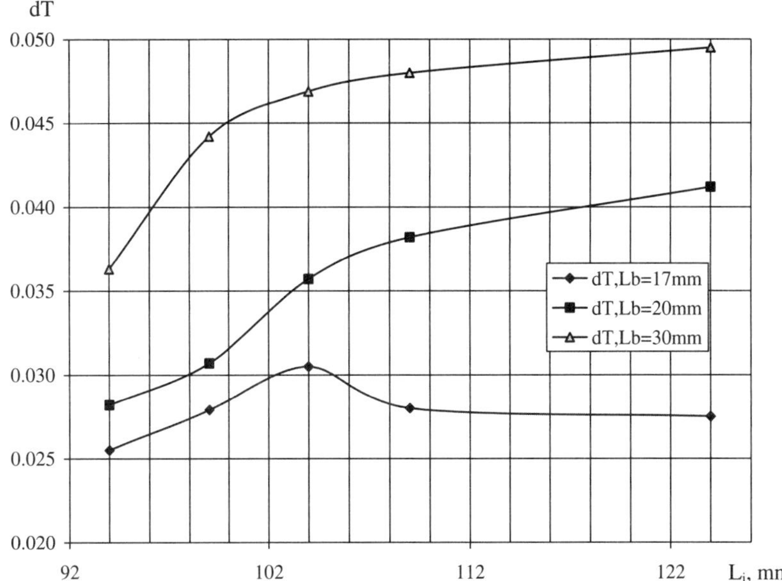

Fig. 10.5 Oscillation decrement in the first tangential-longitudinal mode as a function of a length of single injector with different height of baffles made from protruding pins.

used, as well as correct thermophysical properties of the generator gases and combustion products in the chamber. A design model of the combustion chamber for study of its acoustic properties is shown in Fig. 10.6.

As an example, let us perform an analytical study of the acoustical properties of the combustion chamber of engine RD-0208. The data for the dimensions of the combustion chamber and thermophysical properties of the combustion products are indicated in Table 16.1. The calculation formulas (7.3–7.17) given in Chapter 7 will be used.

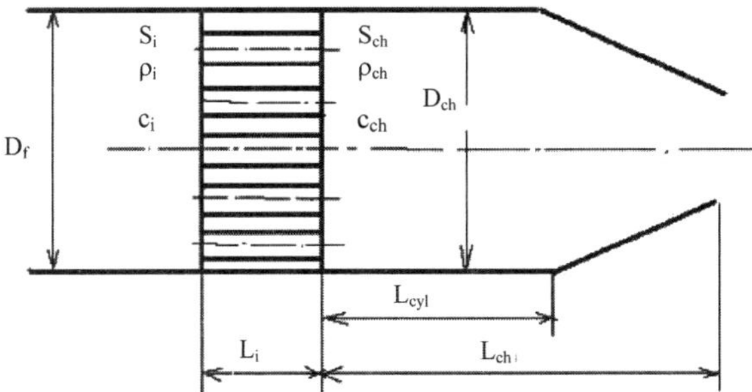

Fig. 6 Design model of combustion chamber.

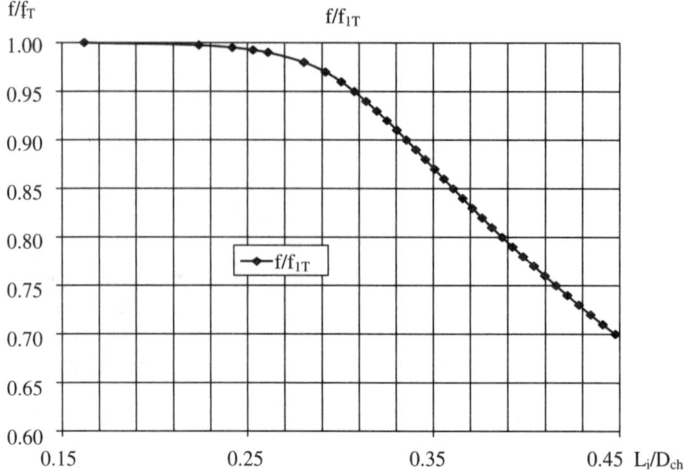

Fig. 10.7 Dimensionless frequency $\overline{\omega}_{1T} < 1$ vs dimensionless length of gas injector channels in the combustion chamber of engine RD-0208.

The equation for calculation of the frequency, which is lower than the frequency of the first tangential mode, sets the relation between the dimensionless length of the gas injector channels and the dimensionless frequency at given values of thermophysical properties of combustion products:

$$\overline{L}_i = \frac{\alpha_i}{4\nu_{mn}\overline{\omega}} \tan^{-1}\left[2\,\frac{S_i\rho_{ch}c_{ch}}{S_{ch}\rho_i c_i}\,\frac{\overline{\omega}}{\sqrt{1-\overline{\omega}^2}}\Bigg/\left(\frac{S_i\rho_{ch}c_{ch}}{S_{ch}\rho_i c_i}\,\frac{\omega}{\sqrt{1-\overline{\omega}^2}}\right)-1\right]+n\pi \tag{10.2}$$

Results of calculations using Eq. (10.2) are shown in Figs. 10.7 and 10.8.

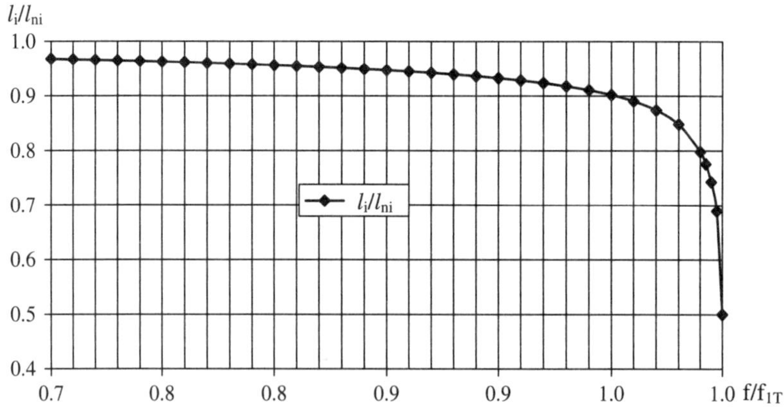

Fig. 10.8 Relation between the length of injector channels corresponding to the given frequency and a length of injector channels calculated by Eq. (10.1) for the tube open at both ends.

The dependence of frequency on injector length presented in Fig. 10.7 is suitable for any combustion chamber having the same permeability over the gas line and the same values of thermophysical parameters. To use the relation shown in Fig. 10.7, pressure pulsations in a particular combustion chamber, the dimensionless length of injector gas channels, and dimensionless frequencies of pressure pulsations in the amplitude spectrum are determined. In this case, the combustion chamber of the RD-0208 was taken as the example.

The dimensionless length of an injector is calculated as

$$\overline{L}_i = \frac{L_i + 0.8\,d_i}{D_{\text{ch}}} \tag{10.3}$$

All quantities in the formula (10.3) are known for a particular chamber, and determination of a dimensionless injector length does not present any difficulties. For the main injectors of engine RD-0208 chamber, $\overline{L}_i = 0.2101$.

Difficulties arise in determination of the dimensionless frequency, which represents the ratio of the frequency measured to the frequency of the first tangential mode. In the amplitude spectrum of pressure pulsations in the chamber with injectors extended to 50 mm, peaks are observed at the frequencies of 2070 and 2480 Hz (see Chapter 16). Because of boundary conditions (see Chapter 7), the oscillation modes with these frequencies cannot be purely transverse. They are transverse/longitudinal mixed modes. It can be assumed that for higher frequency of 2480 Hz the entire volume of the combustion chamber is involved in the wave motions. Experimental measurement of the distribution of the oscillation level along the chamber length in a standing-wave field has shown that it is similar to the distribution typical for a quarter-wave resonator. Assuming that the frequency of 2480 Hz corresponds to the first longitudinal-first tangential mode f_{1T1}, we get the expression for estimation of the effective sound velocity in the combustion products from formula (7.17):

$$c_{\text{ef}} = \frac{f_{1T1L}}{0.586} \Bigg/ D_{\text{ch}} \cdot \sqrt{1 + \left(\frac{1}{4 \,.\, 0.586 \,.\, \overline{L}_{\text{chef}}}\right)^2} \tag{10.4}$$

where $\overline{L}_{\text{chef}}$ is the ratio of sum of the length of the cylindrical part of the chamber plus two-thirds of the length of the nozzle entrance to the chamber diameter.

Using the data from Chapter 16, we get c_{ef} = 1086 m/s for the chamber of engine RD-0208. The frequency of the first tangential mode f_{1T} corresponding to this sound velocity is 2306 Hz. Hence, dimensionless frequencies recorded in the amplitude spectrum of pressure pulsations are equal to $\overline{\omega}_1 = 2070/2306 = 0.8976$ and $\overline{\omega}_2 = 2480/2306 = 1.0754$. According to the plot in Fig. 10.7, the dimensionless length of gas channels of injectors L_i = 0.335 should correspond to $\overline{\omega}_1 = 0.8976$, whereas their actual length is 0.2101. Hence the oscillations with frequency 2070 Hz are not purely intra-injector modes, but they involve, in addition to the injector channels, a region of combustion chamber near the mixing head, which is designated as a cold zone.

According to the relation shown in Fig. 10.8, a length of channels of the injectors adjusted to frequency of 2070 Hz should have been $444/(2 \cdot 2070) \cdot 0.94 = 0.1$ m or 100 mm instead of the actual length of 50 mm. Now let us estimate the effect of the cold zone.

We denote the length of the cold zone as L_{chc}, the density of combustion products in it as ρ_{chc}, and the velocity in the combustion products as c_{chc}. The length of the hot zone will be $L_{ch} = L_{chef} - L_{chc}$, the density ρ_{ch}, and the sound velocity c_{ch}. For the cold and hot zones, the frequencies of the purely tangential modes will be different. The length of zones and the parameters of the combustion products in them cannot be arbitrary. They are varied until the calculated frequencies become equal to the frequencies measured in the amplitude spectrum, 2070 and 2480 Hz. The sound velocity in the hot zone was taken constant and equal to 1170 m/s, which corresponds to the value for the measured specific pressure pulse. The calculations were made as follows.

The value of the sound velocity in the cold zone was preset; the plot of dimensionless frequency vs length of this zone was constructed. One such relation is shown in Fig. 10.9. Generally, the length of the cold zone is 5–10 diameters of the injector gas channel. Having assumed that its length is seven diameters, that is, 70 mm, we find from Fig. 10.9 that $f/f_{1T} = 1.3$. Using this value of dimensionless frequency, we find the corresponding dimension frequency in hertz:

$$f = \frac{(1.3 \times 0.586 \times 750)}{0.276} = 2070 \text{ Hz}$$

In this case the dimensionless frequency common for both zones is equal to $f/f_{1T} = 1.005$, which corresponds in hertz:

$$f = \frac{(1.005 \times 0.586 \times 1170)}{0.276} = 2496 \text{ Hz}$$

The natural frequency of the gas channels of the main injectors (50 mm) calculated by formula (10.1) is 3827 Hz, that is, they are not adjusted to any of the recorded frequencies.

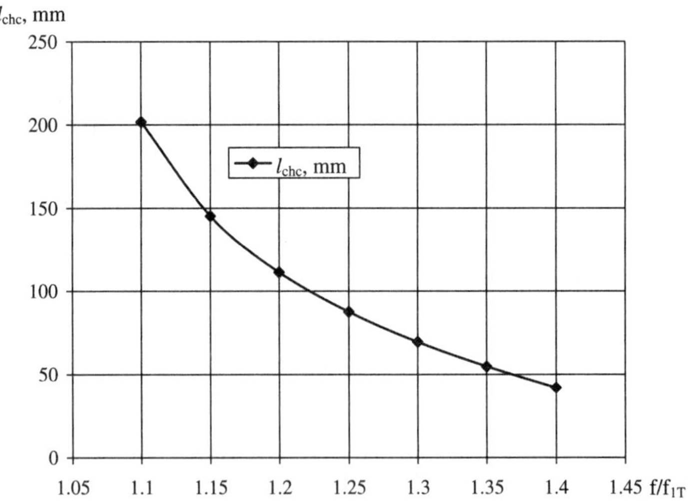

Fig. 10.9 Length of the cold zone vs dimensionless frequency; the sound velocity in the combustion products is 750 m/s.

Only the protruded injectors forming vibration baffles were adjusted to the frequency of ~2.4 kHz in the head of engine RD-0208: $f = 444/(2 \times 0.06) = 2467$ Hz.

It is emphasized in Chapter 16 that adjustment of all injectors on the mixing head could have increased essentially the stability characteristics. A great body of studies on the effect of adjustment of gas channels and jets at the injector inlet upon the stability characteristics has been carried out on single-injector model units similar to that shown in Fig. 10.2.

As an example, let us consider the study of the influence of jet diameter and injector diameter on damping in the model combustion chamber.

Emulsion injectors with a diameter of 12.2 mm and a length of 30 and 90 mm (see Fig. 10.10) were used in the experiments. Gaseous methane and oxygen were used as propellant components. Methane was fed to the injector through five holes

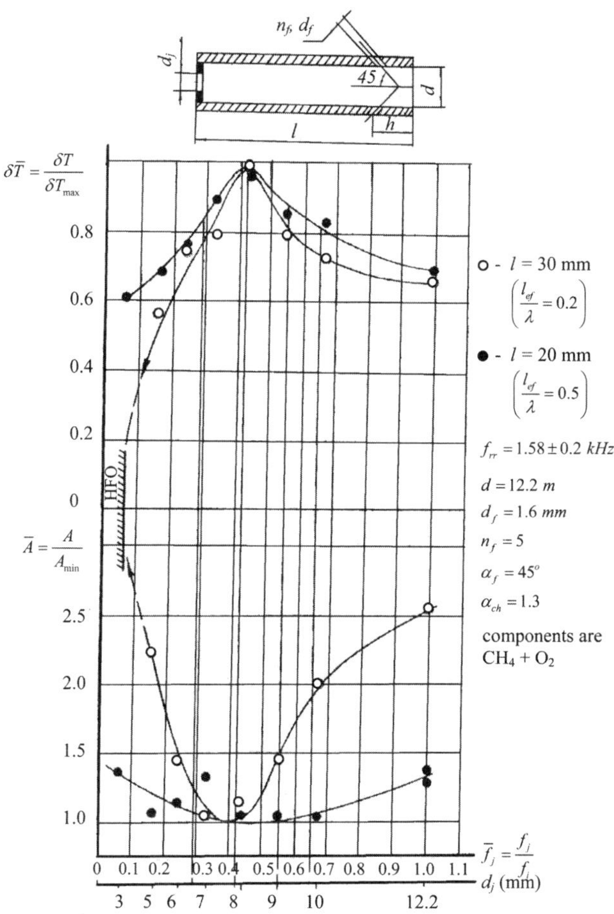

Fig. 10.10 Variation of the decrement determined from noise δT_n and height of resonant rise $\bar{A}$ as a function of jet diameter at the injector inlet.

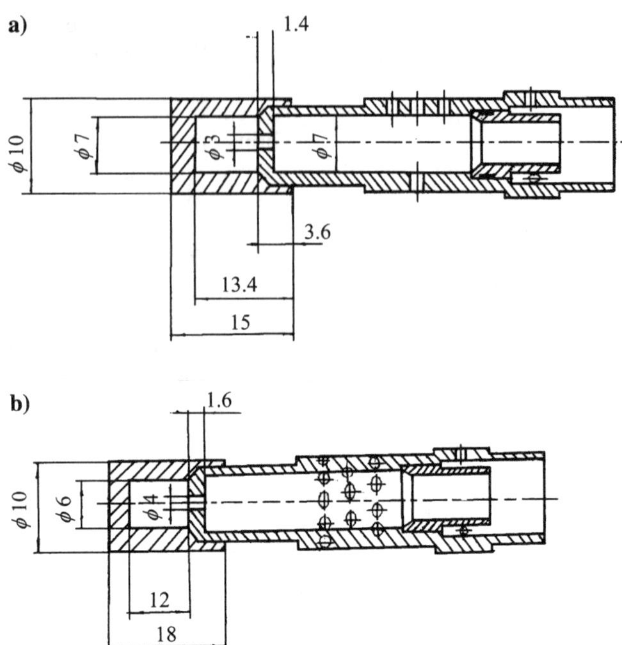

Fig. 10.11 Schematic diagrams of injectors with resonant acoustic dampers: a) a spray–centrifugal injector $f_r = 5.9$ kHz and b) centrifugal–centrifugal injector $f_r = 6.7$ kHz.

having diameter of 1.6 mm, inclined at an angle of 45 deg to the injector axis and located at a distance of 7 mm from the outlet section. A 30-mm-long injector corresponded to 0.2 of the wavelength; a 90-mm injector corresponded to a half-wavelength in gas at the frequency of the first tangential mode. The studies carried out on a single-injector unit gave results for the dependence on jet diameter of the decrement calculated from noise in the combustion chamber and the height of resonance rise (see Fig. 10.10). It is seen from the relations obtained that 1) the maximum values of decrements calculated from noise, and hence the minimum levels of pressure pulsation amplitudes, are realized with a relative jet area of ~0.45; 2) the damping for an injector having length of a half-wavelength exceeds the damping for an unadjusted injector with the same jets; and 3) a deviation of the area of a jet from the optimal one (~0.45) toward increasing values results in a reduction of the damping effect. This is because of the decrease in hydraulic losses in the jet, and a deviation toward decreasing values of jet relative area also results in a reduction of damping caused by the frequency detuning (transition from half-wave resonator to quarter-wave tube).

When the jet area of the injector with a length of 0.2 wavelength (30 mm) decreased to a value below 0.15, high-frequency oscillations occurred spontaneously. As is shown in Fig. 10.11, the higher oscillation damping in injector gas channels can also be attained by installation of resonant dampers in the injectors. Tests of combustion chambers with such injectors have shown a significant increase in stability.

Chapter 11

Estimation of Operating Process Stability from Pressure Oscillation Decrements

A WEALTH of experience has been acquired for several decades in the study of operating processes in LRE combustion chambers and gas generators. The accumulated experience makes it possible to solve specific problems and reach general conclusions. The results of studies on attaining the operating process stabilities in combustion chambers of liquid–liquid and gas–liquid engines are given in the subsequent sections of this book.

All of the data on estimation of operating process stability (i.e., stability margin) with respect to high-frequency pressure oscillations are considered on the basis of the characteristics given in Chapter 6, that is, decrement of pressure oscillations in the chamber and response of operating process to pulse disturbances.

Prediction of operating process stability in a combustion chamber and a gas generator should be made at the beginning of the design stage. During the conceptual design, an injector head mixing system is defined. The worldwide experience of LRE development has shown that with practically any mixing system it is possible to provide operating process stability. However, as known from practice, the development of different mixing systems require extensive time and financial resources. The success in development of a mixing system with a stable operating process for engines with specified criteria is defined by experience of engine designers, availability of engines of similar designs, results of studies carried out when new engine configurations are developed, and so on. The design specifications stipulating the operating process stability must be worked out in three directions (see Sec. I of Chapter 5):

1) The first step is selection of mixing system configurations and elements that are optimized in terms of stability, consistency, and economical efficiency.

2) The second step is use of structural means for suppression or prevention of operating process instability.

3) During development of the system for feed of propellants, it is necessary to prevent disturbances in the feed system from entering the combustion chamber (or gas generator). In some cases, one has to take into account pressure pulsations from the chamber cooling line.

It is advisable to foresee the disparity of frequency characteristics between the combustion chamber and the propellant feed system.

The main functions of a mixing system for providing operating process stability, steady-state performance with high combustion efficiency, and required service life can be formulated as follows:

1) Provide high combustion efficiency with distribution of energy release throughout the chamber volume at the expense of maximum permissible nonuniformity along the chamber length (i.e., an extended burning curve) and at the expense of nonuniformity in the radial direction for each circumference of an injector head. Such nonuniformity is achieved by appropriately manipulating the following: variation of propellant mixture ratio, introduction of injector rating, and orientation of injectors.

2) Provide mixing-system steadiness. The stability characteristics of each chamber should be kept within prespecified tolerance limits. The mixing system steadiness is achieved by the following:

a) Have strict requirements for manufacturing technology, that is, limited tolerance, precision of tools and accessories, accuracy of water-flow test stands, control of the effects of technological operations, such as head soldering on injectors, etc.

b) Provide stable injector sprays, for instance, through manufacturing the injector head bottom with countersinks. Making countersinks (cones) at the injector head serves two purposes: cooling of the combustor head end and stabilizing the spray cone of the injector.

c) Provide a well-defined gas flow near the combustor head end from the center to the periphery with a higher mixture ratio in the central region of the injector face plate.

d) Introduce injector tolerance rating (up to 15%) and a fixed arrangement at the face plate. The introduction of injector rating provides a higher definiteness in distribution of backflows for each chamber specimen. The lack of rating results in a chaotic (different) distribution of backflows for each chamber specimen.

It follows, however, from the data shown here and from the experience with LRE development, that in most cases modifications made in the design of the chamber mixing head and other engine components have an uncertain result with respect to their effect on operating process stability for high-frequency pressure oscillations. Therefore, the effectiveness of all design modifications made on a LRE, and determination of the operating process in a combustion chamber, must be evaluated on the basis of stability characteristics.

I. Main Design Characteristics of Tested Combustion Chambers and Injector Heads

The results of evaluation of operating process stability with respect to high-frequency pressure oscillations in chambers by using the pressure oscillation decrement as a stability characteristic determined from noise in the chamber are discussed in this section. The parameters of the tested chambers are given here: nominal total propellant flow rate ranged from ~3 to 146 kg/s, nominal pressure in the chamber ranged from 4.8 to 11.94 MPa, and chamber diameter ranged from 90 to 480 mm. Both combustion chambers of full-scale production engines and experimental chambers with injector heads for commercial engines were studied.

Table 11.1 Main design and operating parameters of combustion chambers

Combustion chamber index	P_{ch}, bar	G_{Σ}, kg/s	$\mathcal{K}$	D_{ch}, mm	D_a, mm	D_{cr}, mm	L_{cyl}, mm	f_{10}, Hz	N_i, pcs	δT at frequency f_{10} (nom. cond.)
T170-000	85	146	2.77	480	790	194.5	180	1200	967	0.24
T180-000	91	146	2.77	480	1260	188.4	180	1200	967	0.22
S5.1.0100	48	73.3	3.34	380	400	173	468	1600	265	0.10
S5.3.0100	69	41	3.60	267	365	111.3	176	2400	157	0.11
8030-600T	119.4	40.5	2.67	200	865	86	127	3400	145	0.13
4D28-0100	100	24	2.60	190	400	73	150	3600	127	0.11
U107-000	80	6.21	2.30	120	417	41.6	80	5800	61	0.34
U104-000	73	4.77	2.14	105	190	36.6	100	6000	89	0.29
U108-000	80	3.14	2.30	90	295	29.2	80	6500	37	0.26

Most of the studies were carried out on separate combustion chambers on benches with propellants supplied from cylinders. Some tests were performed on chambers installed in engines. The following were used as hypergolic propellants: oxidizer – nitrogen tetroxide, N_2O_4 (NTO); and fuel – unsymmetrical dimethylhydrazine, $(CH_3)_2N_2H_2$ (UDMH). The following hypergolic propellants were used only during tests of chambers S5.1.0100 and S5.3.0100: oxidizer – NA27I – a mixture of nitric acid HNO_3 and 27% nitrogen tetroxide with the addition of an inhibitor; and fuel – (TG-02) – a mixture of 50% aliphatic and 50% aromatic amines (triethylamine and xylidine). The main design parameters of the tested chambers and injector heads are given in Tables 11.1 and 11.2, where the following designations are used: P_{ch}, G_{Σ}, and K_m are the pressure, total propellant flow rate, and oxidizer-to-fuel ratio; D_{ch}, D_n, d_{cr}, and L_{cyl} are the chamber diameter, nozzle edge diameter, nozzle throat diameter, and length of the the cylindrical part of combustion chamber; f_{10} is the frequency corresponding to the first tangential mode; δT is the decrement of pressure oscillations under nominal operating conditions for the baseline design; ΔP_{oi} and ΔP_{fi} are the pressure drops across the oxidizer and fuel injectors, respectively; n_i is the total number of injectors; n_{oi} and n_{fi} are the numbers of oxidizer and fuel injectors, respectively; α_o and α_f are the angles of oxidizer and fuel spray cones, respectively; h_{vb} is the height of vibration baffles; and Δh is recess of nozzle edge of central injector (as applied to bipropellant injectors).

The studies made it possible to examine the influences of a number of design and operating parameters on the characteristics of operating process stability with respect to soft excitation of pressure oscillations.

II. Pressure Oscillation Decrements for Guaranteed Stability

A great number of factors affect the stability characteristics in response to soft excitation of high-frequency pressure oscillations in LRE combustion chambers. Conditionally, they can be grouped as follows: 1) operating parameters—changes

Table 11.2 Main design parameter of injector heads

CCh[a]	Type of IH[b]	ΔP_{oi}, bar	ΔP_{fi}, bar	$H_{p\pi}$, mm	n_i, pcs	n_{oi}, pcs	n_{fi}, pcs	$2\alpha_o$, deg	$2\alpha_T$, deg	Δh, mm	Note
T170	RPK-35	~8.0	13.3	35				65	80		Monopropellant
(180)-000	RPR-35	~8.0	–	35	967	652	315	100	105		injectors, RP on
	RPKN-35	5.6	9.3	35					110		extended injectors
C5.1.0100	–200	9.5	18.6	–	265	265	265	67	115	1.4–1.7	Bipropellant injectors
C5.3. 0100	RP-60			60							Bipropellant injectors
	Without RP	5.5	12.0	–	157	157	157	60	120	2	
	RP-45			45							
	RP-30			30							
8030-600T	6RP	8.5	8.2	20		145	145	74	125	0.5	Bipropellant injectors
	6	–	–	–	145	–	–	56	118	1.0	
	20RPπ	10	6.0	25	169	169	169	50	110	1.4–1.7	
	16	6.0	9.8	–	169	169	169	87	68	1.0–1.3	
	16RP	–	–	40		–	–	87	54		
4D28-0100	RP-25			25					115		Bipropellant injectors
	RP-25*	7.2	6.2	–	127	127	127		115		*RP do not reach
	Without RP							60	120	1.0	the wall
U107-000	$15D_3V$	18.5	18.0	10–16	61	61	61	58/67	112/119	1.05/0.65	Bipropellant injectors
	$15D_3PR$	22.3	26.0	10–16	61	61	61	58/67	112/119	1.05/0.65	*RP do not reach
	$15D_3$	22.3	26.0	10–16	61	61	61	58/67	112/119	1.05/0.65	the wall
	15DR	22.3	23.0	10–16	61	61	61	58/67	112/119	1.06	
	$15D_2R$	22.3	23.0	10–16	61	61	61	58/67	112/119	0.6	
U104-000	200L	6.7	6.0	15–10	89	36	53	65	65	–	Monopropellant
	2001	6.7	6.0	15–10	89	36	53	65	55	–	injectors
									75		
U108-000	$18R_1$	8.2	9.3	16	37	37	37	60	117	1.1/0.1	Bipropellant injectors
	$18R_2$	8.2	9.3	10	37	37	37	60	117	1.1/0.1	
	0.9	8.4	8.2	–	37	37	37	60	120	–1.5/0.5	

[a]Combustion chamber. [b]Injector head.

of chamber pressure, flow rate, oxidizer-to-fuel mixture ratio, and propellant temperature; 2) design characteristics of combustion chamber, mixing head, and injectors—geometry and dimensions of combustion chamber, presence of vibration baffles, and the number, type, and arrangement of injectors on the mixing head; 3) dynamic characteristics of the system—dynamic characteristics of injectors, dependence of average characteristics of the mixing system on pressure oscillations in combustion chamber and pre-injector cavities, combustion-chamber vibrations, and turbulence in oxidizer and fuel flows at the head inlet; and 4) steadiness of the mixing system, that is, its stability to disturbances.

As a rule, several factors exert influence on the operating process simultaneously. It is thus impossible to obtain exact quantitative dependence of stability characteristics on the preceding parameters. In some cases discussed next, however, it is possible to determine the effects of individual factors on the stability characteristics.

The stability with respect to soft excitation was estimated from the decrement of pressure oscillations. Numerous results of liquid–liquid chamber firing tests carried out for evaluating of their operating process stability to soft excitation of high-frequency pressure oscillations have shown that for combustion chambers with stable operating processes, the pressure oscillation decrements lie in the range of $\delta T = 0.1$–0.3.

III. Determination of Operating Process Stability to High-Frequency Pressure Oscillations and Its Dependence on Propellant Flow Rates

When a problem concerning the influence of the total propellant flow rate on oscillation decrement was studied, the test results for combustion chambers U107-000 and U108-000 were used, in which the propellant flow rate was changed with a fixed oxidizer-to-fuel ratio. Chambers U107-000 and U108-000, for which the data are presented in Table 11.1, are part of engines 11D411 and 11D412 [12]. Engine 11D411 is the main component of engine pack 11D410 of the moon ship RKK-N1. The engine was designed for soft lunar landing and injection of the moon ship into a lunar satellite orbit. This is a single-chamber two-mode engine with deep throttling in thrust and is equipped with a turbopump feed system. The propellant components are nitrogen tetroxide and unsymmetrical dimethylhydrazine:

$$P_{\text{vbm}} = 2050 \text{ kgf (20.1 kN)}; \qquad J_{\text{vbm}} = 315 \text{ s}$$

$$P_{\text{vtm}} = 858 \text{ kgf (8.41 kN)}; \qquad J_{\text{vtm}} = 285 \text{ s}$$

where the subscript bm denotes the basic mode and tm the throttling mode. Engine 11D412 (combustion chamber U108-000) is a standby component of engine pack 11D410 of the moon ship RKK-N1. This is a two-chamber engine with a pump-feed system.

The relative spectra of pressure oscillations and the oscillation decrement for different propellant flow rates with a fixed mixture ratio ($K_m = \text{const}$) for chamber U107-000 of engine 11D411 are shown in Figs. 11.1 and 11.2. With increasing propellant flow rate, the pressure oscillation decrement corresponding to the first

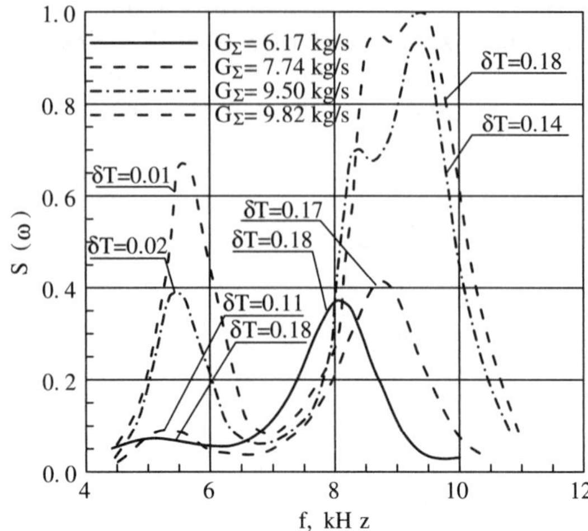

Fig. 11.1 Effects of propellant flow rate G_Σ on the pressure oscillation power spectrum in the combustion chamber and oscillation decrement (f = 5.5 kHz, the first tangential mode; D_{ch} = 120 mm, P_{ch} = 8.0 MPa; cylinder feed system for NT and UDMH).

tangential mode at the frequency of 5.5 kHz decreases, becoming nearly zero with a flow rate of ~8.8 kg/s. The dependence of the pressure oscillation decrement on the propellant flow rate in the combustion chamber U108-000 with 18R-type of injector heads is shown in Fig. 11.3. Results have been obtained for two different oxidizer-to-fuel ratios (K_m = 2.55 and 2) and for two different vibration baffle heights of 16 and 10 mm (see Table 11.2). As in the case of chambers of U107-000-type, with an increase in the propellant flow rate the pressure oscillation decrement decreases. Figure 11.3 also shows that an increase in the vibration baffle height from 10 to 16 mm did not markedly affect value of decrement.

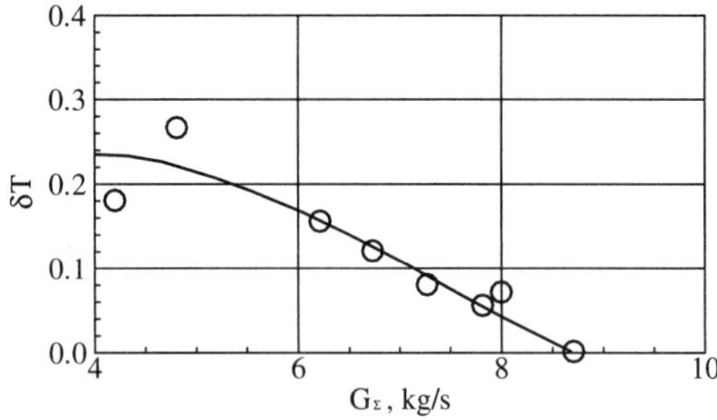

Fig. 11.2 Effects of propellant flow rate G_Σ on the pressure oscillation decrement for the combustion chamber U107-000, K_m = 2.7.

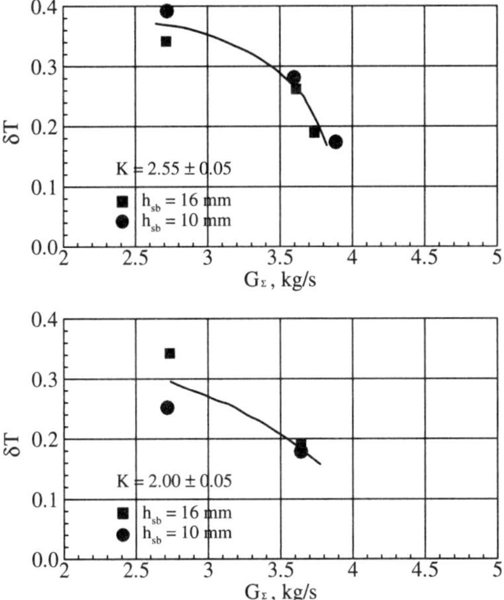

Fig. 11.3 Effects of propellant flow rate G_Σ on the pressure oscillation decrement for the chamber U108-000 with 18R-type injection heads.

IV. Dependence of Pressure Oscillation Decrement on Oxidizer-to-Fuel Ratio

For some combustion chambers, in particular, for combustion chamber U104-000 of engine 8D69M, the oxidizer-to-fuel ratio is a primary parameter defining the operating process stability. The four-chamber control engine 8D69M [12] is designed to control the flight of the second stage of rocket R-36. Propellant components are nitrogen tetroxide and unsymmetrical dimethylhydrazine. The thrust and specific impulse are

$$P_v = 5.53 \text{ tf } (54.23 \text{ kN})$$

$$J_v = 280.5 \text{ s}$$

The combustion chamber of engine 8D69M has a flat injector head with monopropellant centrifugal injectors of three different types for fuel and of two different types for oxidizer. The tests of combustion chambers with two different types of injector heads, U104-2001 and U104-200L1, were carried out. Injector head U104-200L1 differs from head U104-2001 in the presence of vibration baffles and in the type of central injectors (16 oxidizer injectors). On head U104-200L1 these are screw-type injectors, and on head U104-2001 they are tangential.

The influence of the oxidizer-to-fuel ratio on the operating process stability in combustion chambers was studied over a wide range of oxidizer-to-fuel ratios (from $K_m = 0.7$–2.2), with the flow rates close to its nominal quantity ($G_\Sigma = 4.5$–4.8 kg/s). The frequency of high-frequency oscillations recorded during tests of chambers U104-000 is 5800 Hz. The stability characteristics were estimated from the values of pressure oscillation decrements at the first tangential mode frequency, although in the energy spectrum of pressure oscillations under different test conditions, two outstanding peaks at frequencies of 6000 Hz and ~8000 Hz were observed for both combustion-chamber designs.

It has been found that with the change of oxidizer-to-fuel ratio within the prespecified limit the energy spectrum of oscillations changes. When K_m decreases, an essential change of the oscillation decrement of the first tangential mode is observed, whereas the pressure oscillation decrement for the frequency corresponding to the second harmonic either changes slightly (for chambers U104-200L1, see Fig. 11.4) or even increases (for chambers U104-2001, see Fig. 11.5) as the oxidizer-to-fuel ratio decreases.

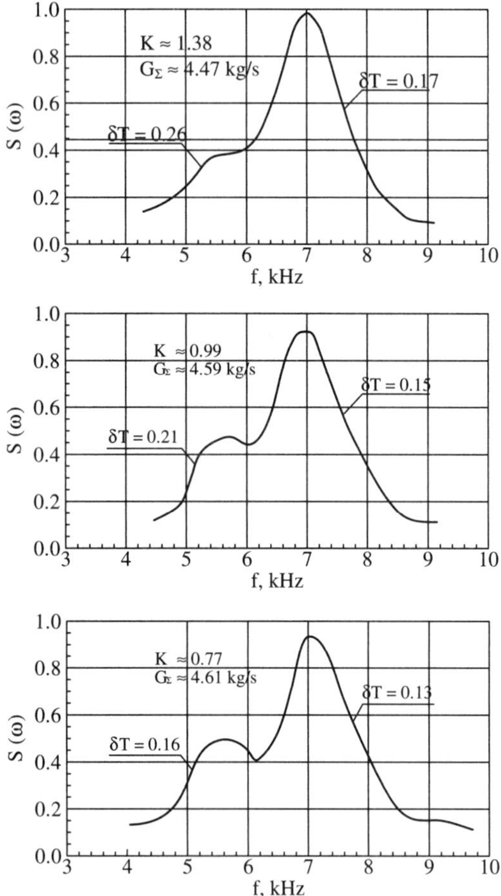

Fig. 11.4 Effect of operating conditions on pressure oscillation power spectrum for the combustion chamber U104-200L1.

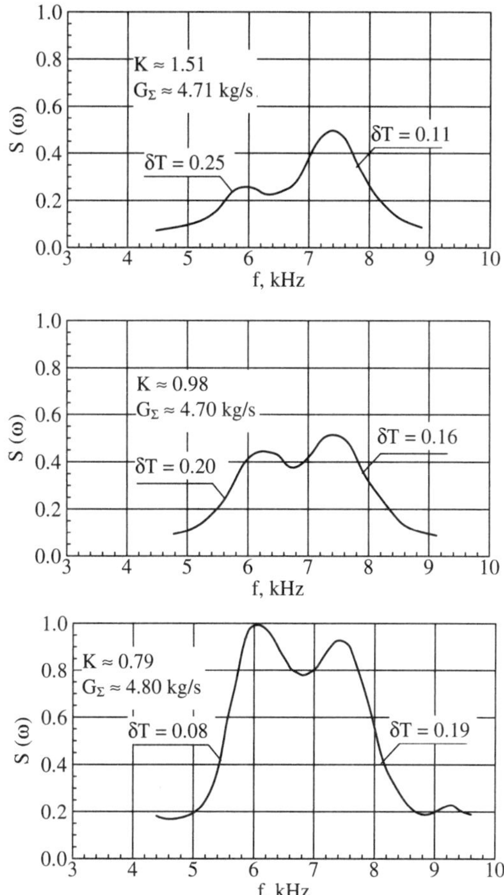

Fig. 11.5 Effect of oxidizer-to-fuel ratio on pressure oscillation decrement for the combustion chamber U104-2001.

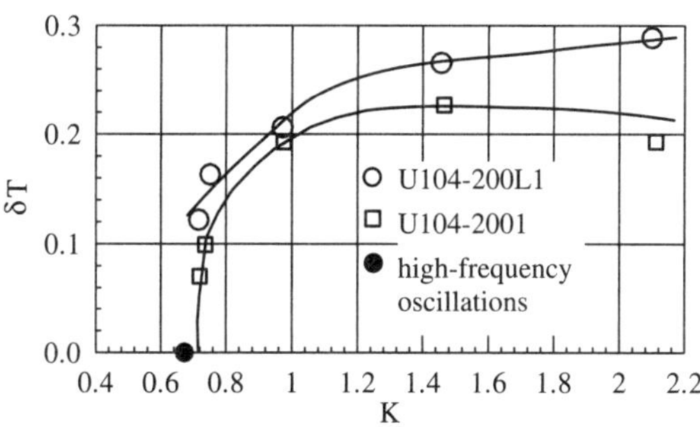

Fig. 11.6 Effect of oxidizer-to-fuel ratio on pressure oscillation decrement—curves (G_Σ = 4.5–4.8 kg/s).

The change of pressure oscillation decrement with varying oxidizer-to-fuel ratio for combustion chambers U104-2001 and U104-200L1 is shown in Fig. 11.6. As the instability boundary is approached, the decrement falls rather smoothly and reduces to zero. Thus, near the stability boundary the oscillation amplitudes increase with frequencies close to a dangerous value, and the oscillation decrement decreases. Similar relationships between the decrement and the oxidizer-to-fuel ratio were obtained for combustion chamber U108-000 with injector heads $18R_1$ and $18R_2$ (see Tables 11.1 and 11.2). Similar to chambers U104-000 and U104-2001, the oscillation decrement decreases with reduced oxidizer-to-fuel ratio, while the total flow rate remains approximately constant, as shown in Fig. 11.7.

V. Influence of Injection Pressure Drop on Stability

Although in some cases the change of operating parameters defines the operating process stability, such a method cannot be used for improving the stability characteristics in the basic mode or for expanding the chamber stability these parameters are assigned by the main engine characteristics. One of the main factors that has a remarkable effect on stability and can be changed by design variations is the injection pressure drop. The injection pressure drop in many ways defines the operating process and its stability as this parameter determines the propellant distribution and hence the location of the combustion zone in the chamber volume.

As an example of the influence of the injection pressure drop on the operating process stability for soft excitation, the test results of combustion chambers T180-000 (see Tables 11.1 and 11.2) with injector heads RPK-35, RPKN-35, and combustion chambers with injector heads $15D_3$, $15D_3$ AR, $200V_5VV$, and $200V_5NN$ are discussed next. (The parameters of heads $15D_3AR$, $200V_5VV$, and $200V_5NN$ are not shown in Table 11.2). The injection pressure drops of fuel and oxidizer for these injector heads are shown in Table 11.3.

The reduction in the injection pressure drop across injectors with IH RPKN-35 by 30% as compared to IH RPK-35 resulted in the essential increase of the pressure oscillation decrement (see Fig. 11.8). The stability characteristics in the accelerated mode at reduced pressure drops improved considerably for chamber U107-000 (see Fig. 11.9).

Table 11.3 Pressure drops of oxidizer and fuel injectors

Chamber	Injector heads	Oxidizer injector pressure drop, bar	Fuel injector pressure drop, bar
T180-000	200 RPK-35	$\Delta P_{o.i.} = 8.0$	$\Delta P_{f.i.} = 13.3$
	200 RPKN-35	$\Delta P_{o.i.} = 5.6$	$\Delta P_{f.i.} = 9.3$
U107-000	$15D_3R$	$\Delta P_{o.i.} = 22.3$	$\Delta P_{f.i.} = 26.0$
	$15D_3AR$	$\Delta P_{o.i.} = 15.0$	$\Delta P_{f.i.} = 13.0$
	$200V_5VV$	$\Delta P_{o.i.} = 16.5$	$\Delta P_{f.i.} = 16.0$
	$200V_5NN$	$\Delta P_{o.i.} = 12.7$	$\Delta P_{f.i.} = 12.2$

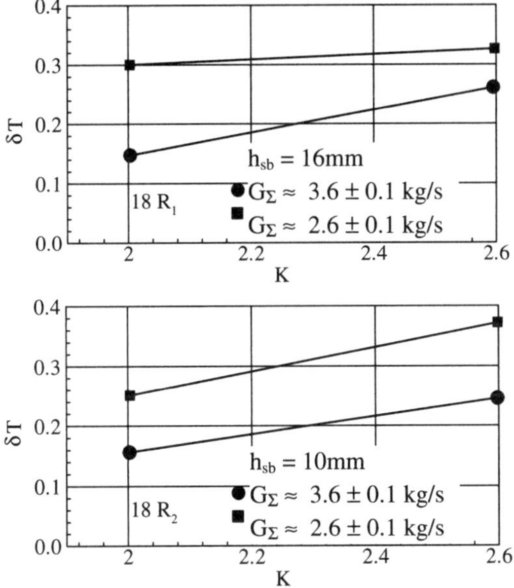

Fig. 11.7 Effect of oxidizer-to-fuel ratio on pressure decrement for chamber U108-000 with injector heads $18R_1$ and $18R_2$.

Similar results have been obtained for chamber U107-000 with injector heads IH $200V_5NN$ and $200V_5$ VV. For a high-pressure drop injector head $200V_5$ VV, the stability boundary was achieved with a total flow rate $G_\Sigma = 8.8$ kg/s. For a low-pressure drop head, unstable operating conditions did not occur even at a flow rate of 9.5 kg/s.

Therefore, for the just-mentioned combustion chambers, the reduction in the injection pressure drop with other conditions remaining fixed expanded the region of stable operation of combustion chambers with respect to high-frequency oscillations. The conclusion obtained, however, is not absolute. In practice,

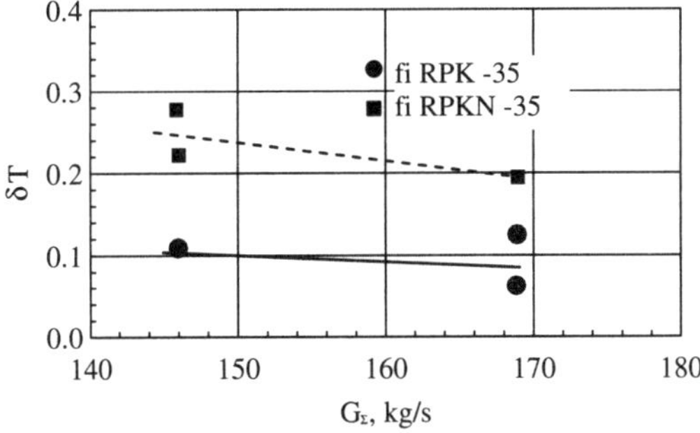

Fig. 11.8 Influence of injection pressure drop on pressure oscillation decrement for the combustion chamber T180-000.

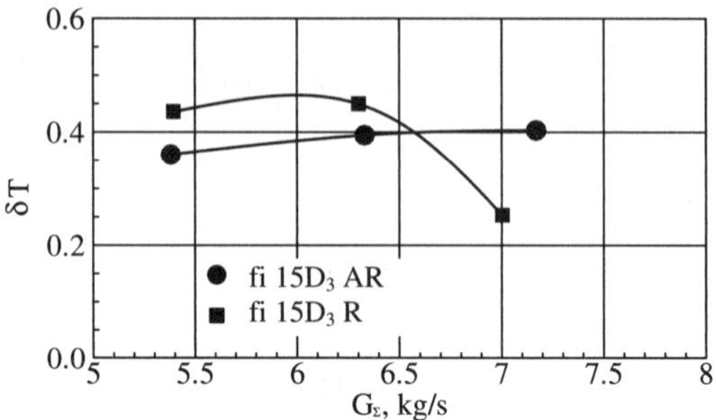

Fig. 11.9 Influence of injection pressure drop on the behavior of the pressure oscillation decrement in combustion chamber U107-000.

there existed cases in which a reduction in the injection pressure drop caused a deterioration in the operation process stability. Such phenomena were observed for chambers with vibration baffles, probably because of the displacement of the combustion zone vulnerable to disturbances beyond the limits of vibration baffles.

VI. Influence of Recess of Injector Nozzle Edge on Stability

The recess length Δh of the centrifugal injector nozzle with respect to the edge of the external injector nozzle, shown in Table 11.2, is one of the main parameters. From the experience acquired in the development of liquid–liquid combustion chambers with bipropellant injectors, this design parameter affects essentially stability of high-frequency pressure oscillations.

As an example, the test results for combustion chamber U107-000 with injector heads 15DR and $15D_2R$ are shown in Fig. 11.10. The injector heads have the same design parameters, except for recessing the central nozzle. On injector head 15DR, Δh is 1.06 mm, and on injector head $15D_2R$, it is 0.6 mm. On both heads there are vibration baffles made in the form of a circle in the center with a height of 16 mm

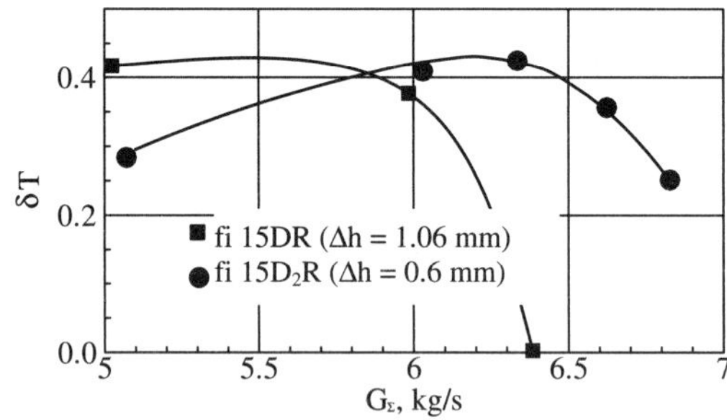

Fig. 11.10 Effect of central injector nozzle recess on pressure oscillation decrement for chamber U107-000 with injector heads 15 DR and $15D_2R$.

and six radial blades with a height of 10 mm. The increase of Δh from 0.6 to 1.06 mm reduced stability (see Fig. 11.10).

VII. Influence of Dynamic Characteristics of Mixing System on Stability

The practice of engine development has shown that occurrence of high-frequency pressure oscillations in pre-injector cavities, vibrations, and other external factors, as a rule, have an adverse effect on stability. For example, during the development of engines S2-72 (the engine designed for winged rockets, $P_g = 1200$ kgf (11.8 kN), it was found out that in terms of the flow rate the stability boundary of the combustion chamber with TPU propellant feed lay 5–40% lower than the boundary of the same combustion chamber with pressure feed of the propellants.

During the development of the combustion chambers of engines 8D74 and 8D75, it was found out that in four-chamber engines the probability of the occurrence of high-frequency instability is higher than that in two-chamber engines. This example has shown that the effect of one chamber upon the other, probably through additional vibrations, can result in deterioration of the stability characteristics.

Development tests of engine 11D24 have shown that in ground tests the stability of high-frequency pressure oscillations depends on the divergence of the nozzle exit. Engine 11D24 has a single chamber and is designed for providing thrust and control of rocket stage 11$\mathcal{K}$68 over all orthogonal channels [12]. The propellant components are nitrogen tetroxide and unsymmetrical dimethylhydrazine. The engine parameters are $J_v = 317$ s; $K_m = 2.10$; $P_{\text{ch}} = 8.88$ MPa; and $P_n = 5.2$ kPa. Without the uncooled diverging part of a nozzle, the measured amplitudes of pressure oscillations in the oxidizer cavity and vibration accelerations under steady-state operating conditions were $2A_{\text{o.i.}} \approx 2$ bar and $V \approx 50$ g, respectively, and in the chamber with a full-size nozzle tested under ground conditions $2A_{\text{o.i.}} \approx 9$ bar and $V \approx 200$ g, respectively. Probably, the increase of vibration loading was caused by flow separation phenomenon in a high-altitude nozzle, when tested under ground conditions.

A single case of the occurrence of high-frequency pressure oscillations under operating conditions far from the operating square was recorded in the chamber without an uncooled part of the nozzle. All other cases including conditions close to the operating square occurred in engines with chambers having a full-size nozzle. Under this situation the effect of vibrations on the change of unstable combustion tendency must not be considered as a direct effect through the flow rate feedback mechanism, but rather as an indirect effect resulting in the change of propellant spraying characteristics because the vibration frequency differs markedly from the chamber frequencies.

A comparison of the results of independent tests of combustion chambers U107-000 with the test results of chambers with a pressure-feed system and chambers as part of engine 11D411 has shown that in the basic mode the magnitude of vibration along the chamber axis changed markedly. In independent tests carried out under nominal conditions, the amplitude of vibration was ≈60 g; in tests of chambers as part of the engine, the amplitude of vibrations was ≈95 g; and

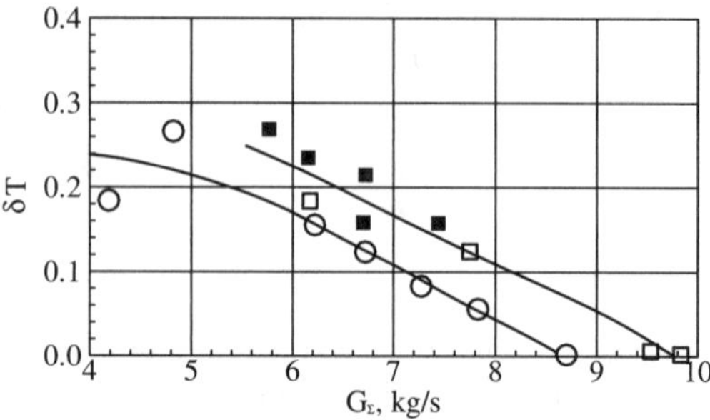

Fig. 11.11 Effect of propellant flow rate on pressure oscillation decrement (the first tangential mode): ○, testing of combustion chamber as part of engine; and ■□, independent testing of combustion chamber (cylinder feed system).

in the accelerated mode, it became up to 130 g. A noticeable change of the vibration level was observed in the deep throttling mode too. The following change of stability characteristics was recorded.

A comparison of test results of combustion chambers U107-000 with pressure feed and TPU feed as part of the engine is shown in Fig. 11.11. It can be seen from the results that in tests of the chamber as part of the engine the pressure oscillation decrement is considerably lower. In tests with a pressure-feed system, with $K_m = 2.7$, the soft excitation boundary is achieved with a propellant flow rate $G_\Sigma = 9.7$ kg/s. In tests with the propellant feed system as part of the engine, the soft excitation boundary is achieved with a lower flow rate, $G_\Sigma = 8.8$ kg/s.

The results of stability estimation for combustion chambers of some engines also have shown that with an increase of chamber vibrations the stability characteristics get worse.

From the results obtained, it follows that only stability characteristics determined from testing of an actual engine fully characterize the margin of operating process stability with respect to high-frequency pressure oscillations.

VIII. Influence of Mixing-System Unsteadiness on Stability

One of the factors responsible for a tendency to unstable combustion is unsteadiness of the mixing system in a LRE combustion chamber. Let us discuss some results of the estimation of stability characteristics of combustion chambers of engines 8D69, where unsteadiness of the mixing system was noted.

When control engine 8D69 in the engine cluster of the second stage was tested, during the startup of the cruise engine, the oxidizer flow rate fell down as a result of a sharp drop in pressure at the component extraction point. The duration of pressure undershoot was 0.1 s, and the reduction in the oxidizer flow rate was ~44%. Simultaneously with the reduction in the oxidizer flow rate and the combustion-chamber pressure, the fuel flow rate increased by 30% as a result of the backpressure decrease in the fuel line and relative increase of TPU rotational speed (with a reduced oxidizer pump load). During special tests of combustion

Table 11.4 Influence of mixing system unsteadiness on stability

Head	Before introduction disturbances	After introduction disturbances
β, s	138.7	161.2
δT_1	0.26	0.19
δT_2	0.12	0.14
f_1, Hz	5580	6300
f_2, Hz	7410	8000

chambers of engine 8D69 with simulated pressure undershoot, failure of pressure recovery after undershoot in the chamber was recorded under some operating conditions. This phenomenon was followed by the change of pressure oscillation decrement of the first tangential mode of up to 25% and the change of specific impulse of up to 10–25%. Actions on the mixing system in chamber U104-000 of engine 8D69 were conducted to check its steadiness by a sharp change of propellant flow rates and by introduction of disturbances into the combustion chamber. In this case unsteadiness of mixing was recorded, the operating process characteristics also changed.

The following results were obtained, as shown in Table 11.4, when the disturbances were introduced into the chamber.

Here δT_1 is the pressure oscillation decrement of the first tangential mode; δT_2 is the pressure oscillation decrement of the second tangential mode; and f_1, f_2 are the corresponding oscillation frequencies. The improvement of combustion efficiency and increase of pressure oscillation frequencies of the first and second tangential modes are indicative of an abrupt change in the mixing system and an essential reduction of the cold zone near the injection head. In other words, the mixing system is unsteady.

The questions of unsteadiness of propellant conversion to combustion products have already been discussed in Chapter 4. The example of chamber U104-000 of engine 8D69 is another confirmation of the necessity to check the steadiness of stability characteristics. It was pointed out in the preceding sections of the book that it is necessary to check the steadiness of the pressure oscillation decrement before and after introduction of disturbances into combustion chambers (or gas generators).

Chapter 12

Test Results for Pulsing Liquid–Liquid Chambers

THE results of studies obtained with the use of the procedures described in Chapter 6 are given in this chapter. Studies of the influence of design and operating parameters on the stability of combustion chambers to hard excitation are summarized. Tests were carried out on actual and experimental combustion chambers, with baseline and experimental injector heads. Disturbances were introduced into the combustion chambers from shock tubes and in-chamber disturbance devices as described in Chapter 9.

Stability characteristics at the first stage of the study were correlated using the parameters n and n_{cr} defined as

$$n = \frac{A_0}{2A_{nm}} \qquad \text{or} \qquad n_{cr} = \frac{A_{cr}}{2A_{nm}}$$

where A_0 is a disturbance amplitude, A_{cr} is an amplitude when undamped pressure oscillations are excited, and A_{nm} is the maximum value of noise amplitude in the combustion chamber in the time span preceding disturbance introduction.

Average values of decrements for the excited oscillations were determined. A chamber was considered to be stable if, with $n > 2$, undamped high-frequency pressure oscillations were not excited. Results of the estimation of stability by this procedure are given in this chapter.

Later, the following norm was set to improve the accuracy of the estimation of stability to hard excitation: a pressure pulse must provide

$$15 \leq n = \frac{A_{max}}{A_{mr}} \leq 25$$

where A_{max} is the maximum amplitude after disturbance introduction. The mean rectified value of the amplitude of random pressure oscillations on time interval t_r is A_{mr}:

$$A_{mr} = \frac{1}{n}\sum_{i=1}^{n} |Y_i|$$

The process in the combustion chamber was considered to be stable if after onset of free pressure oscillations a relaxation time (the time, in which the amplitude decreases e times) did not exceed $\tau = 0.015$ s (see Sec. VIII of Chapter 9).

Average values of decrements shown in this chapter are related to a relaxation time τ_r by

$$\delta T_\theta \approx \frac{1}{f\tau_r}$$

where f is the basic frequency of oscillations excited in the chamber when an artificial pulse is introduced.

In subsequent chapters the results are classified according to combustion-chamber circuits. Because in engine development hard excitation was observed for the first time in liquid–liquid combustion chambers, the discussion will be started with the test data for these combustion chambers.

I. Estimation of Operating Process Stability to Hard Excitation

Actual combustion chambers T170-000, T180-000, S5.1.0100, S5.3.0100, 8030-600T, 4D28.0100, U107-000, U108-000, and U104-000 with baseline and experimental injector heads were studied. The main design parameters and nominal operating conditions of combustion chambers are shown in Table 11.1, and design parameters of the injection heads are shown in Table 11.2. The investigation of actual combustion chambers T170-000, T180-000 and experimental chambers with injector heads RPK-35, RPKN-35, RPR-35, and so on were carried out in connection with development of engines for the rocket R-36.

RD-251 engine of the first-stage rocket R-36 consists of three engines RD-250. The engine RD-250 has two chambers T170-000. The engine of the second-stage rocket R-36 has two chambers T180-000 with high-altitude nozzles [3]. Cases of high-frequency oscillations occurred during bench tests.

Mounting vibration baffles on an injector head was studied as one of the methods to improve stability in the combustion chambers T170-000 and T180-000. Baffles were made from protruded fuel injectors connected by plates [44]. The protruded fuel injectors form a hexagon in the central part of the head. Whereas the RPK version comprises six beams of protruded injectors passing from the hexagon angles, the RPR version comprises six beams of protruded injectors passing from the midpoint of the hexagon edges. Chambers with an injector head of the RPK-type with baffles of heights 25, 35, 50, and 70 mm were studied.

Combustion chambers fitted with vibration baffles of height 70, 50, and 35 mm had the injection pressure drop reduced by 30%. In contrast to other chambers, these chambers were tested not only at normal propellant temperature ($t_{\text{prop}} <$ 20°C) but also with the propellant temperature increased to 22–30°C. Before the stability studies both actual and experimental combustion chambers were modified as follows:

1) Nipples $d_c = 12$ mm for introducing pulses from shock tubes at a distance of ~42 mm (in some cases also at a distance of ~72 mm) from an injector head, tangentially to the circle with a diameter of 2/3 D_{ch}.

2) Four sockets for cooled transducers measuring pressure pulsations were mounted at a distance of 57 mm from the head.

The test results for injector heads with a height of baffles of 35 mm only are discussed in this chapter. During tests of versions RPKN-50 and RPKN-70, the reduction of the oscillation decrement δT_θ was observed: with baffles

RPKN-35, $\delta T_\theta = 0.13$; with baffles RPKN-50, $\delta T_\theta \approx 0.10$; and with baffles RPKN-70, $\delta T_\theta \approx 0.06$. The increase in relaxation time with increase of baffle height was indicative of deteriorating stability. In addition, under certain conditions in chambers with 50- and 70-mm baffles, cases of spontaneous excitation of high-frequency pressure oscillations were observed.

The peculiar variations of stability characteristics for chambers T170 (T180) probably arise from special conditions of the combustion processes with this type of vibration baffle. When more than 25% of the fuel flow enters the chamber through the extended injectors and is burned within the chamber volume behind the baffles, stability is apparently reduced.

In the period of studying designs of new mixing systems, engines RD-251 and RD-252 were in serial production. Thus, changes were limited to the reduction of natural disturbances (see Chapter 1) [3]. Additional resistance was introduced into the hydraulic line of RD-250 engine chambers, which provided extended ignition in the combustion chamber. The stability of engine RD-252 was improved by decreasing the fuel preheating temperature, achieved by applying a thermal-protective coating to the internal wall of the combustion chamber nozzle [18]. For comparison of the stability of other chambers to final disturbances, the test data for actual chambers with 35-mm high vibration baffles are shown in this chapter.

Chamber S5.1.0100 is of interest for researchers in view of its spherical shape and quite large diameter. This chamber is part of engine S5.1 [12]. Engine S5.1 was developed in the 1950s for the S-25M rocket of a land antiaircraft rocket system: $P_v = 167$ kN (17 tons force).

The results of a hard excitation study for chambers having diameters less than 200 mm are shown in Tables 12.1–12.3. Engines 11D411, 11D412, and 8D69M with chambers U107-000, U108-000, and U104-000 are described in Chapter 11.

Chamber S5.3.0100, which has a high reliability, was used to study the influence of the height of vibration baffles on stability. Chamber S5.3.0100 is a part of engine S5.3.0000-0 for the rocket R-21 with underwater ignition.

The combustion chamber S5.3.0100 (propellant components are NA-271-TG-02, given Tables 11.1 and 11.2) has a cylindrical shape and a flat injector head. One-hundred fifty-seven bipropellant centrifugal injectors are mounted on the injector head; the edges of the fuel injector nozzles are submerged with respect to oxidizer injectors by 2 mm.

To improve stability, an essential nonuniformity of flow on a radius of the injector head is introduced. Fifty-five high-flow injectors surrounded by two concentric rows of 66 low-flow injectors are arranged in a honeycomb pattern in the central part of the injector head.

Internal wall cooling of the chamber is provided by a single concentric row of injectors (36 injectors) of the chamber having oxidizer-to-fuel ratio $K_w = 1.81$. An injector head of baseline design has six radial vibration baffles with a height of 60 mm, made from stainless-steel plates with a thickness of 2 mm (Fig. 12.1). In addition to the injector head of baseline design, versions of the injector heads with 45- and 30-mm high baffles, as well as without baffles, were tested during the study (see Tables 12.1–12.3).

At the early stages of S5.3.0000-0 engine development, the combustion chamber had no vibration baffles. It operated steadily over a wide range of propellant flow rates with cylinder propellant feed system during bench tests as well as

Table 12.1 Characteristics of stability to hard excitation for different types of combustion chambers (nominal operating mode)

Combustion chamber	Type of injector head	Conditions G_Σ, kg/s	Conditions K	Number of tests	Number of tested combustion chambers	Charges G, g	Maximum excited amplitudes	δT_{ef}	A_{cr}, bar	Maximum natural disturbances ($2A_n$ max), bar	Stability margin, n
T170.000	RPK-35	147	2.7	4	2	1.32–6.20	2.2–8.1	0.05–0.14		0.98	>8.3
	RPR-35	147	2.8	3	2	1.95–6.30	1.71–8.2	0.08–0.12	~7.2	1.0	~7.2
	Without RP	148	2.8	5	4	0.73–1.64	——	——	h.f.	k—	——
T180.000	RPK-35	147	2.7	2	2	1.95–6.20	2.5–9.2	0.09–0.13	——	1.1	>8.4
	RPKN-35	147	2.8	3	2	2.23–6.60	4.9–8.2	0.10–0.18	——	1.1	>7.4
	RPR-35	147	2.7	2	1	1.45–6.20	2.7–8.52	0.06–0.13	——	0.88	>9.7
S5.1.0100	——	73.4	3.5	7	5	1.12–1.63	1.2–1.5	0.09–0.10	~1.5	0.55	~2.7
S5.3.0100	RP-60	42	3.5	16	10	1.39–12.5	3.5–12.8	0.18–0.24	——	0.64	>20
	RP-45	41.5	3.5	3	1	1.39–12.5	4.2–12.0	0.10–0.19	——	0.43	>28
	RP-30	41.5	3.6	8	5	1.2–6.2	2.0–12.2	0.08–0.13	~7	0.41	~17
	Without RP	41.5	3.6	7	6	0.69–1.95	1.3–8.0	0.03–0.08	~5.8	0.36	~16

8030.600T	6RP	40.4	2.7	3	3	1.1–4.25	1.25–13.2	0.12–0.18	~14	0.9	~15
	6	41	2.6	12	9	0.72–1.45	0.7–1.50	0.08–0.17	~2	1.0	~2
4D28.0100	RP-25	24	2.6	10	9	1.1–6.1	4.6–48.0	0.1–0.25	——	0.51	>94
	RP-25*	24	2.6	3	3	3.1–6.1	11.7–14.2	0.09–0.15	——	0.67	>21
	Without RP	24	2.6	4	4	1.1	6.0–8.6	0.09–0.10	——	1.37	>6.3
U107–000	$\Delta l = 0$	6.3	2.3	13	2	1.23–3.85	15–23.0	0.04–0.08	——	1.3	>18
	$15D_3R$ $\Delta l > 0$	6.2	2.3	10	2	3.81–4.05	25–29.5	0.09–0.10	——	1.4	>21
	$\Delta l = 0$	6.3	2.3	16	7	0.69–3.85	1.7–26.5	0.08–0.17	——	1.2	>17
	$15D_3R$ $\Delta l > 0$	6.3	2.3	26	5	1.32–1.95	17.4–20.6	0.06–0.08	——	1.2	>17
	$15D_2R$	5.1	2.3	5	2	0.69–1.39	1.9–8.1	0.12–0.15	——	1.8	>4.5
U104–000	200LI	4.1–5.9	2.1–2.6	20	6	3.85–3.93	50.7–62.3	0.05–0.09	——	1.9	>33
	200I	4.1–5.9	2.1–2.6	30	7	0.76–3.93	13.1–31.5	0.06–0.11	——	1.8	>17
U108–000	18 R_1	3.2	2.3	9	7	0.69–3.85	9.0–22.2	0.07–2.15	——	0.65	>34
	18 R_2	3.2	2.3	5	2	1.27–3.04	12.6–30.1	0.07–0.10	——	0.78	>39
	0 g	3.2	2.3	10	3	0.4–1.39	3.5–35.5	0.04–0.07	——	1.65	>21

Table 12.2 Characteristics of stability to hard excitation for different types of combustion chambers (minimum power mode)

Combustion chamber	Type of injector head	Conditions		Number of tests	Number of tested combustion chambers	Charges G, g	Maximum excited amplitudes bar	δT_{ef}	A_{cr}, bar	Maximum natural disturbances ($2A_n$ max), bar	Stability margin, n
		G_Σ, kg/s	K								
T170.000	RPK-35	96	2.5	3	2	1.39–5.1	2.3–11.0	0.05–0.13	1.95 3.34	—	—
	RPR-35	67	1.5	1	1	1.95–6.3	2.1–3.8	0.07–0.12	5.28	—	—
	Without RP	—	—	—	—	—	—	—	—	—	—
T180.000	RPK-35	—	—	—	—	—	—	—	—	—	—
	RPKN-35	96	2.5	3	2	1.35–6.2	3.9–10.3	0.02–0.12	~9	1.1	~8.3
	RPR-35	—	—	—	—	—	—	—	—	—	—
	Without RP	96	2.5	1	1	0.69	—	—	h.f	—	—
S5.3.0100	RP-60	31.5	3.5	4	4	1.42–12.5	1.6–13.0	0.19–0.22	—	0.39	>33
	RP-45	31	3.5	2	1	1.42–12.8	6.0–9.4	0.13–0.17	—	0.28	>34
	RP-30	30.7	3.5	1	1	1.39–2.05	8.6–13.1	0.11–0.12	—	—	—
	Without RP	—	—	—	—	—	—	—	—	—	—

	RP-25	16	2.6	5	5	2.1–3.1	9.0–37.0	0.14–0.20	——	0.42	>88
4D28.0100	RP-25*	16	2.6	2	2	3.1	13.0	0.15	——	0.54	>24
	Without RP	16	2.6	3	2	1.1–2.1	——	——	——	——	——
	$\Delta l = 25$ 15D_3B	1.6–5.3	1.9–2.5	11	2	1.40–3.85	13–49	0.04-0.08	——	0.69	>27
	$\Delta l = 50$	1.7–4.0	1.6–2.4	14	4	3.85	19	0.09	——	0.69	>27
U107–000	$\Delta l = 0$ 15D_3R	1.7–4.0	1.7–2.4	2	2	0.69–1.32	7.0–15.0	0.08–0.10	——	0.8	>19
	$\Delta l = 25$	1.7–5.3	1.5–2.5	12	4	1.32	14.7	0.1	——	0.45	>32
	15D_2R	1.9	1.7	1	1	——	——	——	——	—0	——
U104–000	200LI	3.2–4.5	0.8–2.1	8	4	3.85–4.05	3.44–54.9	0.06–0.10	——	1.68	>33
	200I	3.2–4.5	0.8–2.4	10	4	3.85–3.95	26.0–44.6	0.05–0.10	——	1.45	>31
	18 R_1	2.7–2.9	2.0–2.3	8	4	1.32–4.1	8.0–21.3	0.08–0.19	——	0.70	>30
U108–000	18 R_2	2.7	2.1	12	2	1.4–3.86	9.6–31.0	0.1–0.15	——	0.52	>60
	0.9	2.7–2.9	1.9–2.6	8	2	1.32	22	0.06	——	1.43	>15

Table 12.3 Characteristics of stability to hard excitation for different types of combustion chambers (full-power mode)

Combustion chamber	Type of injector head	Conditions G_{Σ}, kg/s	Conditions K	Number of tests	Number of tested combustion chambers	Charges G, g	Maximum excited amplitudes, bar	δT_{ef}	A_{cr}, bar	Maximum natural disturbances ($2A_n$ max), bar	Stability margin, n
T170.000	RPK-35	171	3.2	4	2	1.45–6.20	6.6–11.0	0.06–0.16	~6	0.75	~8
	RPR-35	166	3.23	1	1	2.08–6.20	2.1–3.24	0.06–0.12	—	0.85	>3.8
	Without RP	170	3.12	1	1	—	—	—	—	—	—
T180.000	RPK-35	171	3.1	2	2	1.95–6.3	5.1–9.6	0.06–0.12	—	0.83	>12
	RPKN-35	175	3.1	3	1	1.45–6.2	2.5–8.2	0.04–0.1	—	0.97	>84
	RPR-35	166	3.2	2	2	2.36–4.6	2.9–4.8	0.1–0.18	~5	0.81	~59
	Without RP	—	—	—	—	—	—	—	—	—	—
S5.3.0100	RP-60	47.5	3.6	4	4	1.39–12.5	2.4–10.7	0.2–0.23	—	0.71	>15
	RP-45	47.5	3.5	1	1	1.39–12.5	4.2–12.0	0.14–0.16	—	0.37	>32
	RP-30	47.5	3.6	4	3	1.39–6.3	3.4–12.1	0.08–0.13	~7	0.43	~17
	Without RP	47.7	3.6	1	1	0.69–1.95	1.3–3.1	—	—	0.3	>10

4D28.0100	RP-25	29	2.6	5	5	1.1–6.1	7.0–30.0	010–0.19	——	0.85	>35
	RP-25*	29	2.6	2	2	3.1–6.1	8.8–14.0	0.09–0.15	——	0.59	>24
	Without RP	29	2.6	3	2	1.1	6.5	0.1	——	1.5	>4.3
U107-000	$\Delta l = 25$ $15D_3B$	7.0	2.6	4	3	1.21–3.85	15–46	0.06–0.09	——	1.5	>31
	$\Delta l = 50$	7.8	2.2–2.5	7	4	0.69–3.90	2.6–23.9	0.08–0.1	——	1.1	>22
	$\Delta l = 0$ $15D_3R$	7.8	2.6	6	5	0.69–3.85	16.4–26.5	0.02–0.1	——	1.8	>15
	$\Delta l = 25$	7.0	2.5–3.0	5	2	0.69–3.85	9.0–28.2	0.06–0.13	——	1.2	>23
	$15D_2R$	6.8	2.0–2.6	5	2	0.69	10.5	0.14	——	2.1	>5
U104–000	200LI	7.2	2.0–3.0	5	3	3.85–4.0	4.00–46.1	0.08–0.11	——	1.75	>26
	200I	7.2	2.0–3.0	7	4	3.85	26.4–42.4	0.05–0.06	——	1.4	>40
U108–000	18 R_1	3.6–4.0	2.0–3.1	7	4	1.32–4.1	5.7–30.6	0.06–0.16	——	0.76	>40
	18 R_2	3.6	2.0–3.1	5	2	1.29–3.94	11.0–37.5	0.08–0.09	——	0.60	>62
	3.4–4.0	1.9–2.5	5	2	1.39	29.0	0.11	——	——	——	——

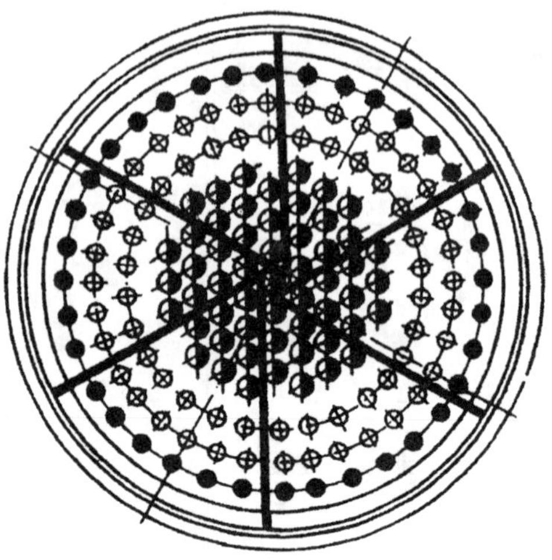

Fig. 12.1 Injector head of chamber S5.3.0100 with RP.

during tests of the chambers as a part of the engine. However, during ground tests of the engine installed in the rocket, cases of the breakdown of some combustion chambers were observed upon starting as a result of the occurrence of high-frequency pressure oscillations. The studies have shown that under conditions promoting emulsification of propellant entering the combustion chamber during startup (when long sections of feed lines are not filled) high-frequency instability is regularly excited. Chamber S5.3.0100 of baseline design with vibration baffles provided stability under these conditions as well.

Study of combustion chambers 8030-600T, being part of engine 15D13 with injector heads of type 6 (Fig. 12.2) and type 6RP (Fig. 12.3) with vibration

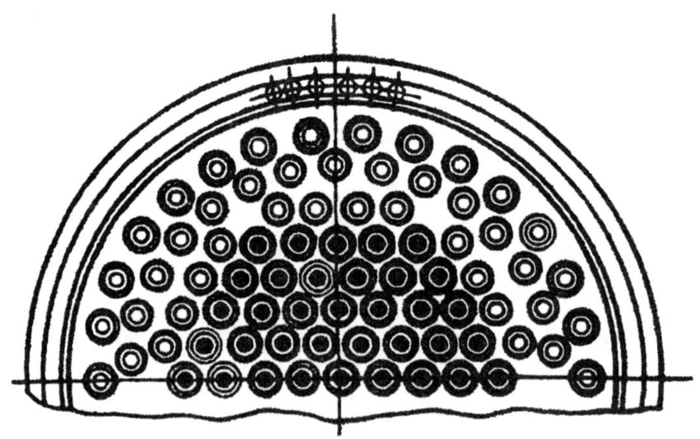

Fig. 12.2 Schematic diagram of injector head of type 6.

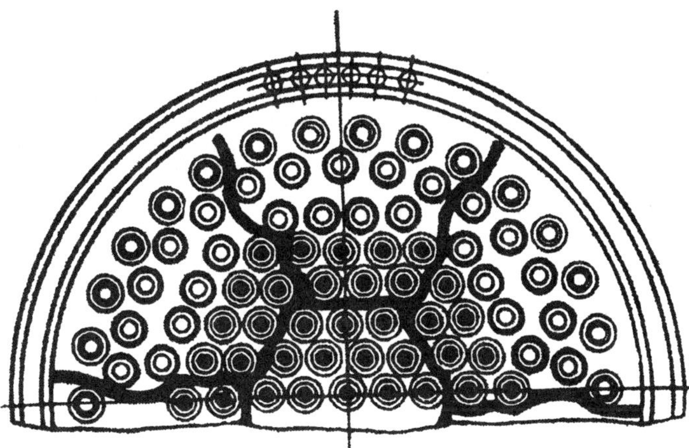

Fig. 12.3 Schematic diagram of 6-RP injector head (bipropellant centrifugal injectors similar to injector head type of 6).

baffles, is of special interest. Engine 15D13 is installed in the second-stage rocket UR-100.

In combustion chambers with injector heads of type 6, high-frequency pressure oscillations occurred from time to time, especially during switchover from preliminary stage operation ($G_{pr} \approx 0.7\ G_{nom}$) to nominal operating conditions. The occurrence of oscillations had a probabilistic nature. High-frequency pressure oscillations had a frequency $f \approx 3500$ Hz corresponding to the first tangential mode and amplitude $A \approx 5.5$–7.0 MPa. More detailed data on investigations of these chambers are given in Sec. II.

Stability characteristics obtained for hard excitation with indication of the actual operating modes of this unit are shown in Tables 12.1–12.3. In these tables the characteristic high-frequency pressure oscillations are shown for three operating modes: nominal, minimum, and maximum total flow rate of propellant. In addition, the following data are shown in the tables: the weight of charges used, peak values of amplitudes of excited oscillations, average values of oscillation decrements, and maximum amplitudes of natural noise in the chamber until the disturbance is introduced. In some cases it was possible to obtain the values of the critical disturbance amplitude, which are also shown in the tables.

Index h.f. in column of A_{cr} in Tables 12.1 and 12.2 for chambers T170-000 and T180-000 denotes that high-frequency pressure oscillations were excited at the minimum possible artificial pressure pulse.

The estimation data for chambers U107-000 with different lengths of the cylindrical part of the chamber are shown in Tables 12.1–12.3. The effect of chamber extension of up to 50 mm for different designs has not been found.

The values of explosive weight in grams, which caused high-frequency oscillations, is shown in column A_{cr} of Table 12.2 for chamber T170-000 with injector heads RPK-35 and RPR-35.

Examination of the data makes it possible to determine the influence of some factors on stability to hard excitation of pressure oscillations.

II. Relationship Between Probabilistic Excitation of Instability and Critical Amplitude

As pointed out in Sec. III of Chapter 3, the cases of so-called probabilistic occurrence of instabilities were noted rather commonly. The peculiarity of a probabilistic occurrence of high-frequency oscillations is poor reproducibility of test data. This is caused by a mechanism of excitation of oscillations such that the amplitude of noise pressure oscillations in the combustion chamber exceeds the critical amplitude A_{cr} (see Sec. II of Chapter 6).

The minimum value of amplitude causing excitation of high-frequency pressure oscillations is denoted A_{cr}. A value of A_{cr} is determined by introducing successively increasing pressure pulses into the chamber.

When engine 15D13 (combustion chamber 8030-600T) was put in service, cases of high-frequency pressure oscillations were encountered. Initially engine I5D13 had an injector head of type 6 (see Fig. 12.2) [23]. The chamber and head characteristics are shown in Tables 11.1 and 11.2. Twelve bench tests were carried out for research purposes with a head of type 6; nine chambers with the cylinder propellant feed system were tested. In tests of three combustion chambers with injector heads of type 6 during switchover from preliminary stage operation ($G_{pr} \cong 0.7G_{nom}$) to the nominal mode, transverse high-frequency oscillations occurred. The frequency $f \approx 3500$ Hz, and amplitude $A \approx 5.5$–7.0 at a pressure in the chamber $P_{ch} \approx 12.0$ MPa.

Estimation of stability characteristics to hard excitation was made on five examples of combustion chambers. For this purpose, pressure pulses with amplitude $A_{max} = 0.7$–1.50 MPa were introduced with a mass $G_{expl} = 0.72$–1.45 g. The oscillations at the initial amplitudes of pressure disturbance of 0.7–1.50 MPa decayed.

When the disturbance was introduced with an explosive of mass of 1.45 g causing initial pulse amplitude $A_{cr} \approx 0.2$ MPa, high-frequency pressure oscillations were excited in the combustion chamber with a head of type 6. The amplitude of developed high-frequency oscillations was ~8.5 MPa at a frequency $f \approx 3500$ Hz of the first tangential mode. To increase the stability margin, vibration baffles with a height of 20 mm were installed on the head of type 6 (Fig. 12.3). Tests of combustion chambers with vibration baffles on the injection head of 6RP-type have shown that high-frequency instability develops in these chambers at much higher initial disturbance pulses ($A_{cr} \approx 1.4$ MPa, $G = 5$ g). As in chambers without vibration baffles, the frequency of excited oscillations was ~3500 Hz; however, the amplitude of the developed pressure oscillations was much less and made up 2.6–44.9 MPa. During tests of combustion chambers with heads of 6RP type, spontaneous development of high-frequency pressure oscillations was not observed.

To reveal the causes of probabilistic excitation of high-frequency oscillations during tests with heads of type 6, frequency analysis was made of the pressure pulsation amplitudes. That was done to search for the occurrence of amplitudes comparable with the critical amplitude A_{cr}. For this purpose, the number of pressure peaks and their values were determined for pressure pulsations recorded in the combustion chamber. This was done for an interval of 0.1 s following ignition, in the mode of switchover from preliminary stage to the nominal conditions, and under nominal conditions. Hard excitation in the combustion chamber depends on the shape of the exciting pulse, that is, on the pulse amplitude spectrum.

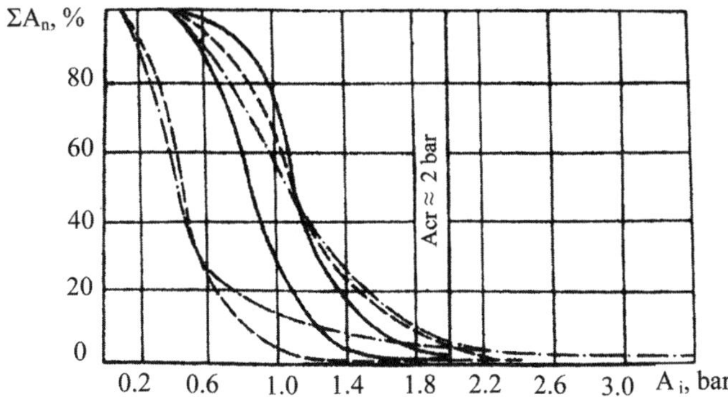

Fig. 12.4 Composition of pressure pulsation amplitudes in the combustion chamber with the injector head of type 6: —·—, startup; – – –, switchover; and ——, mode.

In processing of the pulsations, pressure rises with a durations of ~½ natural period of oscillations on the first tangential mode ($f \approx 3500$ Hz) were taken as one pressure peak (one natural pulse). Thereafter, a total number of pressure rises (pulses) in the selected area were determined, the number of pulses in intervals with the same amplitude were summed up, and the percentage of the total sum of pulses was calculated. The results of these calculations are shown in Figures 12.4 and 12.5. The values of fixed amplitude along the x axis, whereas their total percentage, that is, the density of their distribution $\Sigma A_n = f(A_i)$,were plotted along the y axis. Also plotted in Figs. 12.4 and 12.5 are values of the critical amplitude A_{cr} obtained in estimating the stability margin for the respective combustion chambers (see Table 12.1). The curves plotted for the conditions under study illustrate quite clearly the possibility of probabilistic hard excitation of high-frequency oscillations in the combustion chamber with an injector head, without vibration baffles (see Fig. 12.4). Also shown is the impossibility of the occurrence of such

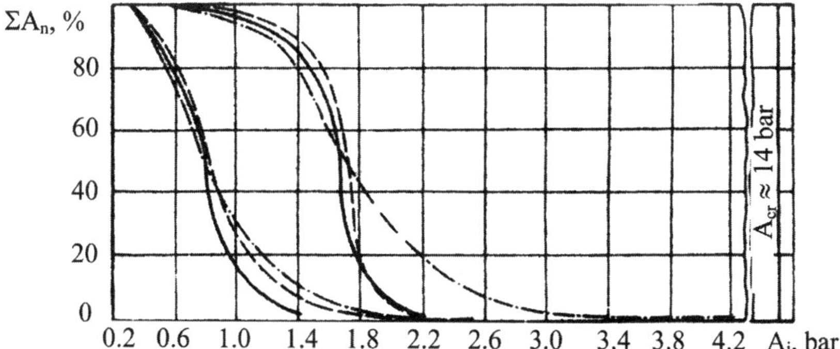

Fig. 12.5 Composition of pressure pulsation amplitudes in combustion chamber with the injector head of type 6 - RP: —·—, startup; – – –, switchover; and ——, mode.

oscillations at the measured level of natural disturbances in a chamber with an injector head equipped with vibration baffles (see Fig. 12.5).

It is also seen from the plots that under steady-state nominal conditions a level of maximum amplitudes in the combustion chambers with injector heads without baffles and with baffles is lower than the value of the critical amplitude A_{cr}. At the same time, during starting and switching over to the nominal conditions in the chamber without baffles, there are natural pressure oscillations at the injector head, with an amplitude exceeding the critical amplitude A_{cr}. This fact is in full agreement with test data for combustion chambers without baffles: high-frequency oscillations occurred during starting in some tests.

To provide sufficient margin for stability to the mechanism of probabilistic excitation of high-frequency oscillations, the quantitative stability characteristics must have appropriate values.

III. Estimation of Influence of Propellant Flow Rate and Fuel/Oxidizer Mixture Ratio

For each type of combustion chamber, the stability margin n_{cr} remains practically constant within a range of ±10% for the total flow rate and oxidizer-to-fuel ratio. When these parameters vary in a wider range, the stability margin starts changing.

The plots demonstrating the dependence of stability characteristics n_{cr} and δT on test conditions for combustion chambers S5.3.0100-0, T170-000, and U108-000 are shown in Figs. 12.6–12.9. It can be seen from the plots that the values of average decrements of artificially excited oscillations practically do not change for all considered types of combustion chambers, when the test conditions are varied within the specified limits. A similar conclusion holds for the minimum disturbance intensity A_{cr} causing the instability. The value of A_{cr} does not depend on test conditions, with consideration for the spread of its values within the specified limits.

Thus within the stability margin to hard excitation near the conditions for which this margin has been calculated, there exists some region in which the stability margin changes insignificantly. According to test data, this region overlaps the

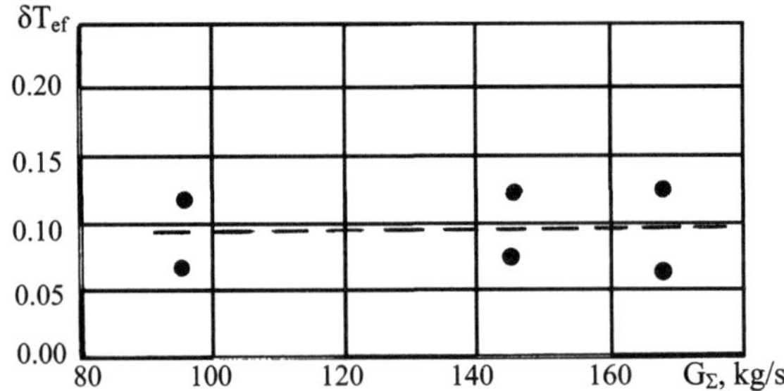

Fig. 12.6 Dependence of oscillation decrement on operating conditions in the combustion chamber T170-000 with injector head RPR-35.

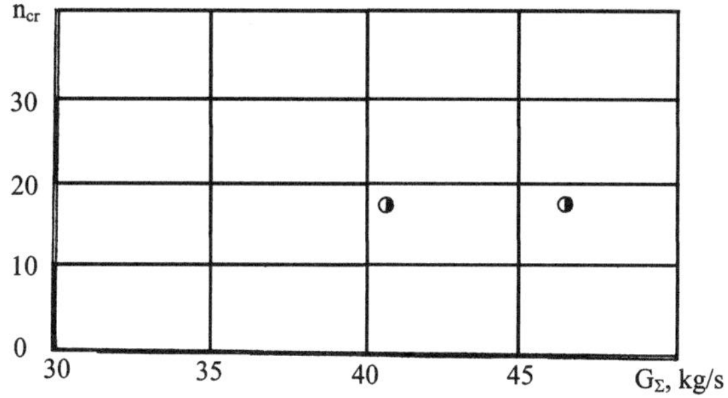

Fig. 12.7 Dependence of a value of n_{cr} on operating conditions in the combustion chamber S5.3.0100-0 with injector head RP-30.

square for checking the change of conditions (±10% of nominal pressure values in the chamber and of oxidizer-to-fuel ratio). The noticeable change of stability characteristics starts only when the operating parameters depart greatly from the values quoted.

IV. Application of Artificial Disturbance for Estimating Efficiency of Vibration Baffles

Test results for combustion chambers S5.3.0100-0, 8030-600T, and 4D28.0100 are shown in Table 12.4. The data compare stability characteristics in combustion chambers with and without vibration baffles at the head end under nominal conditions.

A definite dependence of stability characteristics for hard excitation, on the height of vibration baffles, is seen from Table 12.4, and Figs. 12.10 and 12.11. In all cases the combustion chambers with vibration baffles had higher stability characteristics to hard excitation than the combustion chambers without vibration

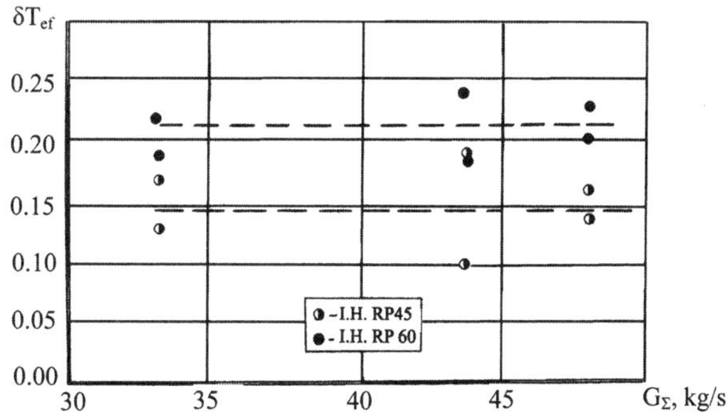

Fig. 12.8 Dependence of oscillation decrement on operating conditions in the combustion chamber S5.3.0100-0.

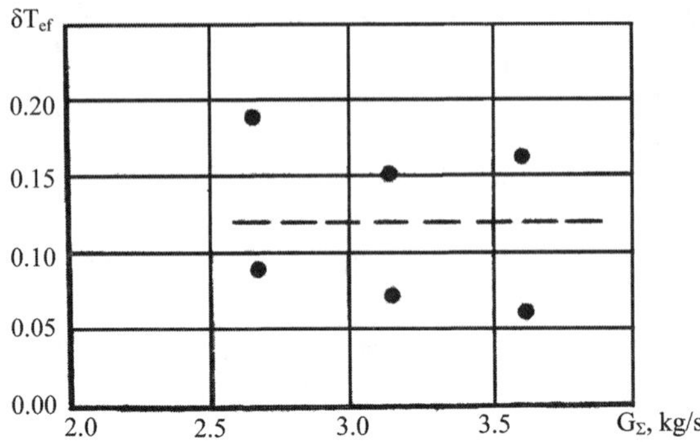

Fig. 12.9 Dependence of oscillation decrement on operating conditions in the combustion chamber U108-000 with injector head 18R$_1$.

baffles. This indicates that vibration baffles are an effective means for increasing the margin of stability to hard excitation.

Vibration baffles have an essential influence on the parameters of pressure oscillations excited by pulses from shock tubes as well. In combustion chambers with vibration baffles, the amplitude of high-frequency oscillations is much less than in chambers without baffles. When oscillations are excited in combustion chambers T170-000 with head RPR-35, the amplitude of oscillations was about 0.82 MPa at a frequency of 1200 Hz. In the same chambers without baffles, the amplitude of oscillations was about 6.2 MPa at a frequency of 1500 Hz. However, in some cases the installation of vibration baffles results in loss of stability to "soft" excitation of high-frequency oscillations.

Most representative in this respect are combustion chambers 8030-600T with heads 16 and 16RP (see Table 11.2). High-frequency pressure oscillations were excited in startup during tests of the chamber with head 16. After installing 40-mm high vibration baffles on the head, the oscillations also occurred during

Table 12.4 Stability characteristics of three combustion chambers

Type of combustion chamber	Type of injection head	Stability characteristics		Explosive charge (maximum), G, g	Parameters of excited oscillations	
		n	dT_{ef}		A, MPa	f, Hz
S5.3.0100	RP-60	>20	0.18–0.24	12.5	——	——
	RP-45	>28	0.10–0.19	12.5	——	——
	RP-30	~17	0.08–0.13	6.2	0.8–1.0	2000
	Without RP	~16	0.03–0.08	2	5.5–6.5	2400
8030-600T	6 RP	~15	0.12–0.18	4.25	2.6–4.92	3500
	6	~2	0.08–0.17	1.45	8.5	3500
4D8.100	RP-25	>94	0.10–0.25	6.1	——	——
	Without RP	>6.3	0.09–0.10	1.1	——	——

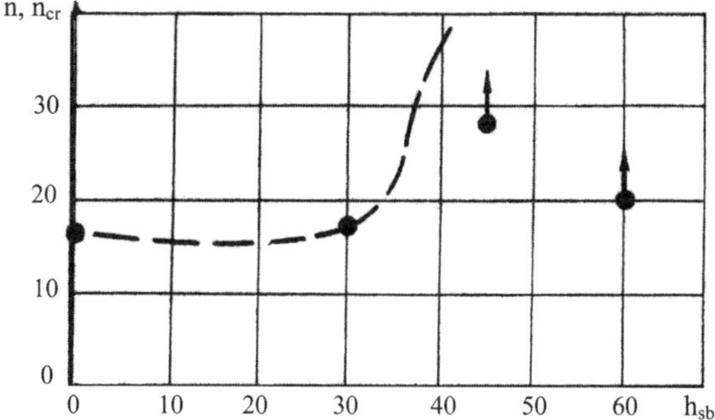

Fig. 12.10 Effect of baffle height on stability margin of combustion chamber S5.3.0100-0.

startup, but their amplitude increased smoothly and did not exceed ~1.0 MPa under nominal conditions.

V. Influence of Chamber Diameter and Relative Flow Ratio on Stability to Hard Excitation

The values of the stability margin for combustion chambers of various diameters under nominal, maximum, and minimum test conditions without vibration baffles are shown on the plot in Fig. 12.12. It can be seen from the plot that the chambers with a diameter of more than 200 mm have the lowest stability to hard excitation. The margin of stability to hard excitation of pressure oscillations drops sharply when the diameter increases beyond this value. Combustion chambers with smaller diameters (below 200 mm) have an increased margin of stability to hard excitation. This conclusion agrees with the available experience: with the increase of combustion diameter beyond some value, the stability to final disturbances worsens sharply.

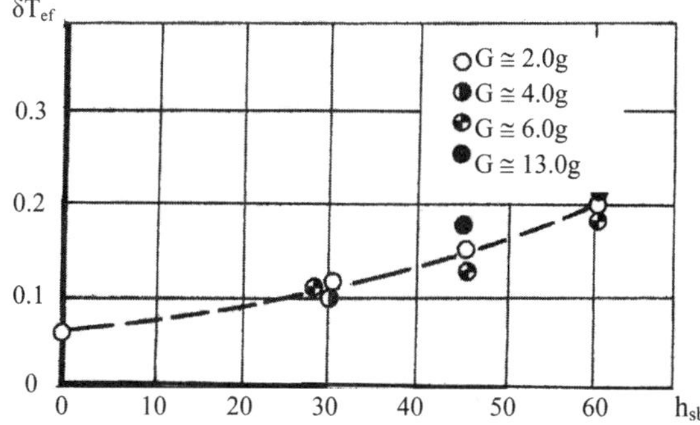

Fig. 12.11 Effect of baffle height on pressure oscillation decrement of combustion chamber S5.3.0100-0 (nominal operating mode).

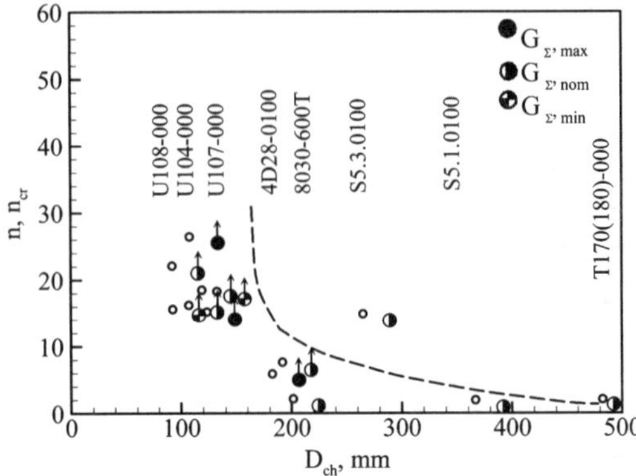

Fig. 12.12 Effect of combustion chamber diameter on stability margin (combustion chamber without separating baffles).

The combustion chambers under study had different mixing circuits. In Fig. 12.12 the cases when critical values of amplitudes were attained are marked by circles without arrows; cases when critical amplitude values were not attained are marked by circles with arrows.

As just stated, one of the main parameters characterizing the operating process in the combustion chamber is a flow ratio. Combustion chambers can differ significantly in dimensions, in propellant components, in pressure, oxidizer-to-fuel ratio, and so on. Therefore an integrating parameter such as flow ratio is used for correlation.

The flow ratio is determined in terms of total propellant flow rate G_Σ, cross-sectional area of combustion chamber F_{ch}, and chamber pressure P_{ch}:

$$q = \frac{G_\Sigma}{F_{ch}\, P_{ch}}$$

The dimension of the flow ratio is $[q] = 2/\text{s} \times \text{cm}^2 \times \text{bar}$. The total propellant flow rate is equal to the flow of combustion products through the critical section (throat) of nozzle F_{cr}:

$$G_\Sigma = \sqrt{kg\left(\frac{2}{k+1}\right)^{\frac{k+1}{k-1}}}\ \frac{P_{ch}\, F_{cr}}{\sqrt{R_{ch}\, T_{ch}}}$$

Substituting this expression into the formula for the flow ratio, we get the average flow rate across the chamber section:

$$q_{av} = \sqrt{kg\left(\frac{2}{k+1}\right)^{\frac{k+1}{k-1}}}\ \frac{F_{cr}}{F_{ch}\sqrt{R_{ch}\, T_{ch}}} = \sqrt{kg\left(\frac{2}{k+1}\right)^{\frac{k+1}{k-1}}}\left(\frac{D_{cr}}{D_{ch}}\right)^2 \frac{1}{\sqrt{R_{ch}\, T_{ch}}}$$

where k is an adiabatic index for combustion products; R_{ch} is the gas constant for combustion products, kg · m/(kg · deg); and T_{ch} is the temperature of combustion products, K. Numerical values of k, R_{ch}, and T_{ch} are determined by thermodynamic calculations. The flow rate q_h near the injection head depends on the uniformity of propellant flow across the head section; it can differ substantially from the average flow rate q_{av}.

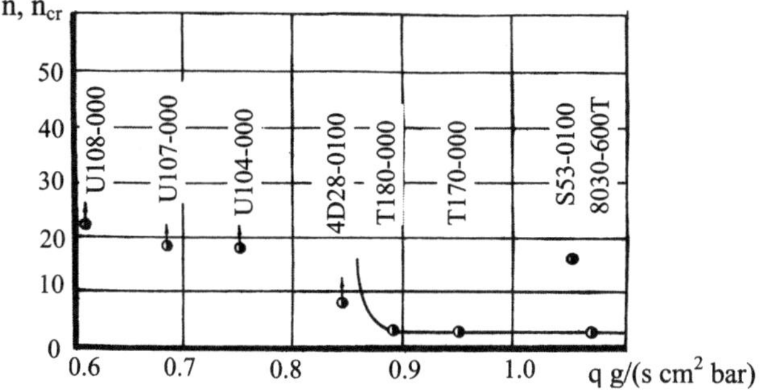

Fig. 12.13 Effect of relative flow ratio on stability margin (combustion chamber without separating baffles).

The dependence of the stability margin on the average value of relative flow ratio for the tested combustion chambers without vibration baffles is shown in Fig. 12.13. Values of n and n_{cr} are shown for nominal values of flow rates and oxidizer-to-fuel ratio.

Despite the essential differences in the arrangements of the combustion chambers studied a definite regularity could be observed in Fig. 12.13. Particular attention should be focused on a value n_{cr} for chambers 8030-600T and S5.3.0100. Combustion chamber 8030-600T has a diameter as small as 200 mm, and combustion chamber S5.3.0100 with a diameter of 267 mm has a high stability margin. The stability margin has been increased at the expense of nonuniformity of the flow rate across the injector head radius. However, for the both chambers large values of flow rates significantly reduce the stability margin.

It follows from Fig. 12.13 that the margin of stability to hard excitation decreases with increase of q_{av} beyond the values of 0.80–0.85. g/(s × cm^2 × bar). In all newly developed liquid–liquid combustion chambers, designers apply a value $q_{av} \approx 0.8$ g/(s × cm^2 × bar).

It will be shown in later sections of this book that for a gas–liquid combustion chamber and gas generators operating with these chambers the average flow rate can be increased to $q_{av} \approx 2$g/(s × cm^2 × bar).

Chapter 13

Stability of Gas–Liquid Combustion Chambers

I. Characteristics of Gas–Liquid Combustion Chambers and Injector Heads

THE results of development tests of engines RD-253 and RD-263 are a representative example of a pattern change of pressure-oscillation decrement determined from noise δT and its power spectrum density $S(\omega)$, depending on design and operating parameters. Engine RD–263 was developed in 1969–1973 for the first-stage IBR R–36M/36MU. Four LRE RD-263 form the engine unit RD-264 [12]. The propellant components are nitrogen tetroxide and unsymmetrical dimethylhydrazine. The engine operation and performance parameters are given here: $P_{ch} = 20.6$ MPa, $J_g = 293$ s, $P_g = 106$ tf (1040 kN), $J_v = 318$ s, and $P_v = 115$ tf (1130 kN).

Engine RD-253 was developed from 1961 to 1965 for the first-stage IBR UR-500 (CR Proton) [12]. The first flight of the carrier rocket Proton took place on 16 July 1965. A total of 247 launches of CR Proton were made with the engines RD-253. In recent years, upgraded engines RD-275 were used in the first stage CR Proton. The propellant components are nitrogen tetroxide and unsymmetrical dimethylhydrazine. The parameters of engine RD-253 are as follows: $P_{ch} = 14.7$ MPa, $J_g = 285$ s, $P_g = 150$ tf (1420 kN), $J_v = 316$ s, and $P_v = 167$ tf (1635 kN).

In addition, evaluation of the influence of combustion chamber design and operating parameters ($D_{ch} = 267$ mm) with use of pressure oscillation decrements δT and its maximum spectral density $S(\omega)$ was made on the experimental engine unit D-495. The combustion chamber operated over the range of pressure $P_{ch} = 1.5$–4.5 MPa. The relative flow ratio q could be varied over a wide range with the help of a movable cone in the nozzle throat. Unit D-495 comprises a combustion chamber with an adjustable throat, a mixing head, a gas passage, and a gas generator. The propellant flow rate, average pressure, pressure drops across the mixing head for the fuel and oxidizing gas line, and pressure pulsations were measured during tests of the unit. Unit D-495 is described in Chapter 6; its diagram is shown in Fig. 6.2.

The mixing heads had pneumatic internal mixing injectors. There were holes in the combustion bottom of the heads to feed some fuel for cooling of the combustor head-end. Some versions of mixing heads were equipped with pneumatic injectors with centrifugal fuel cascade. The mixing heads were different from each other in terms of the length of the injector gas channel, the number and diameter of holes for fuel feed, the permeability over the oxidizing line of injectors, and the number of injectors (see Tables 13.1 and 13.2). The total flow rate of propellants through unit D-495 was 35 kg/s. The oxidizer-to-fuel ratio (mixture ratio) in the combustion chamber varied in the range of 2.2–3.1 and in the oxidizing gas generator in the range of 20–30. The total flow rate of propellant components and the oxidizer-to-fuel ratio remained unchanged during each separate test.

A body of information on these matters was obtained using full-scale model units running on actual propellant components (NTO and UDMH) under considerably reduced pressures. The mixing heads with gas–liquid injectors were studied on the model unit, with "cold" reducing gas ($T = 70°C$) instead of liquid fuel fed into the liquid cavity of the head. During the simulation of the operating process at reduced pressure, fuel is fed to the injectors in the gaseous state to simulate closely the actual conditions by an aggregate state. In most cases, fuel in liquid holes in an actual engine's pneumatic injector is in a supercritical state, at which a liquid jet can be considered conditionally as a jet of a dense gas.

Let us consider some typical relationships for variations of the pressure oscillation decrement. The main design parameters of combustion chambers and injector heads are shown in Table 13.2. Their test results will be discussed next.

II. Stability, Chamber Pressure, and Oxidizer/Fuel Ratio

Let us consider variations of δT and $S(\omega)$ with respect to P_{ch} and mixture ratio K for two frequencies: $f_{1L} = 1.26–1.36$ kHz corresponding to the first longitudinal mode, and $f_{3T} = 3.76–4.02$ kHz corresponding to the third tangential mode for engine RD-253, a combustion chamber with bipropellant pneumatic injectors. Oscillations at the frequency f_{1L} have low intensity in the spectrum, whereas oscillations at the frequency f_{3T} have the lowest value of decrement in one of the regimes.

The decrement at the frequency f_{1L} with increasing chamber pressure P_{ch} and $K = 2.7$ is shown in Fig. 13.1a. With the pressure growth from 10 to 159 MPa, the decrement increases from 0.08 to 0.22. The δT vs $\mathcal{K}$ curve (with $P_{ch} = 13.2–13.7$ MPa) for the frequency f_{3T} is presented in Fig. 13.1b. With increase of mixture ratio from 2.0 to 3.7, the decrement at this frequency decreases from 0.24 to 0.14.

According to the firing test data, conditions with a reduced pressure and an elevated K are most dangerous for mixing heads from the standpoint of exciting high-frequency pressure oscillations. As it follows from the plots, decrements at

Table 13.1 Main design and operating parameters of gas–liquid combustion chambers

Engine index	D_{ch}, mm	d_{dr}, mm	$L_{subcr.}$, mm	P_{ch}, bar	G_{Σ}, kg/s	K	R, m
RD-253	430	279.4	733	~150	528	2.67	150
D-495	267	220	385	15–40	35	2.2 + 3.1	——
RD-263 exp. design	320	196.1	428	~200	344.8	2.67	100

Table 13.2 Main design parameters of injectors heads

Engine index	Type of injectors	Cond. index of inject. heads	n_i, mm	l_i, mm	d_{gas}, mm	n_1	d_1, mm	a_1, deg	l_1, mm	$\Delta P_{fi\,av.}$, bar	$\Delta P_{oi\,av.}$, bar
RD-253 baseline head	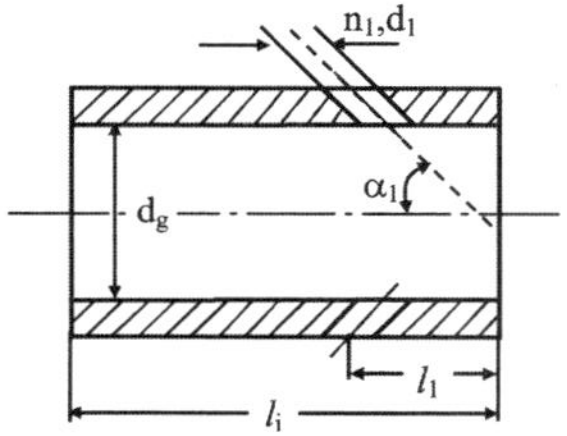	—	169	37	22.5	5	2.4–2.72	4.5	8.5	~11.5	~6
RD-263 experimental versions of injector heads	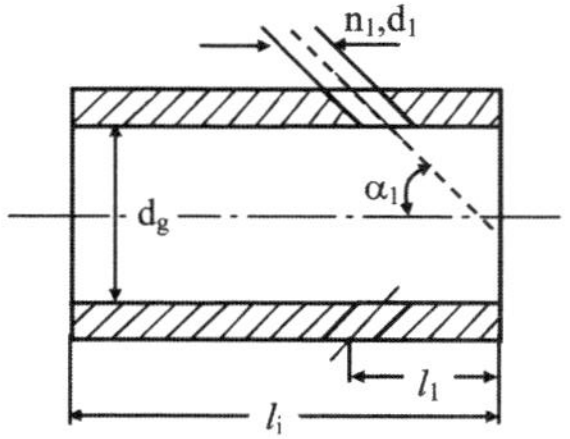	*a*	127	28	18.5	4	2.57–2.79	45	7.2	8.7	~9.5
		b	127	28	18.5	5	2.3–2.5	45	8.5	8.8	~9.5
		c	127	28	18.5	5	2.3–2.5	45	8.5	8.8	~9.5
		d	127	52.5–65.0	8.5	5	2.3–2.5	45	8.5	8.8	~9.5
		e	127	95	18.5	5	2.3–2.5	45	8.5	8.8	~9.5
D-495 experimental unit	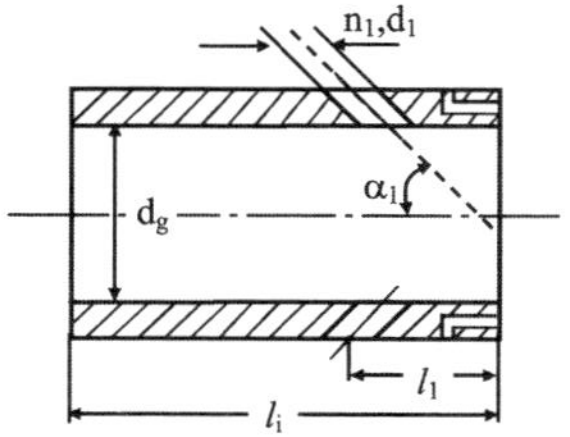	*a*	91	21	12.2	8	1.0	75	10	~2	3+6
		b	91	48	12.2	8	1.0	75	10	~2	3+6
		c	91	21	12.2	4	1.0	75	10	~2	3+6

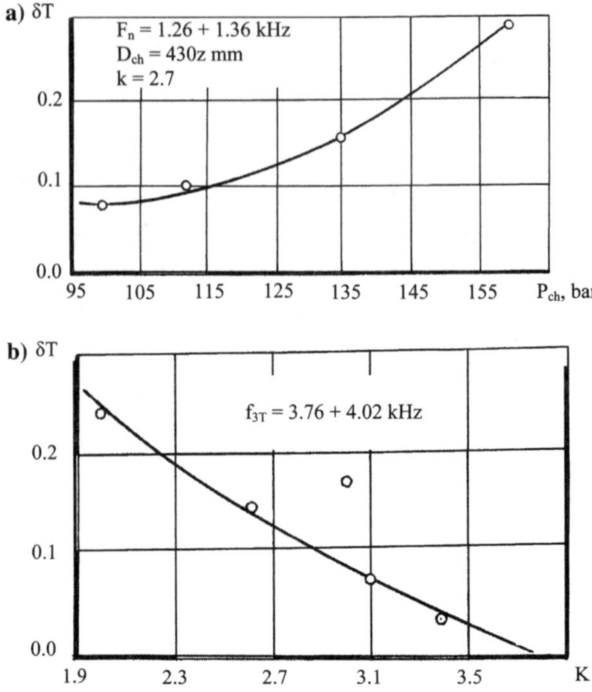

Fig. 13.1 Dependence of oscillation decrement on pressure P_{ch} and oxidizer-to-fuel ratio K in the gas–liquid combustion chamber.

frequencies f_{1L} and f_{3T} decrease, when engine operating conditions shift to the region of reduced pressure P_{ch} and high oxidizer-to-fuel ratio values K. In other words, the decrements tend to zero, when approaching the region of instability of high-frequency oscillations.

For unit D-495, using umbrella-type injectors, in the design of bipropellant pneumatic injectors, some fuel is fed through the swirl nozzles on the external injector wall to protect the combustion floor from burnout. Figure 13.2 shows δT as a function of K with a flow ratio $q = 2$ g/(s·cm^2·bar) (see injector head a in Table 13.2). The decrement at the frequency $f_{1T} = 2.4$ kHz of the first tangential mode of oscillations decreases with decrease of the mixture ratio in the combustion chamber. Apparently, this is connected with increasing fuel spray injection into the oxidizer spray under conditions of low K.

III. Influence of Relative Flow Ratio on Stability

On the experimental propulsion unit D-495, the value of q was varied over a wide range with the help of a cone moving in the nozzle throat. Therefore, it is reasonable to consider the influence of this parameter based on the unit test results. Figure 13.3 shows the curves of δT and $S(\omega)$ vs q at the frequencies $f_{1T} = 2.4$ kHz and $f_{2T} = 4.0$ kHz corresponding to the first and second modes of tangential

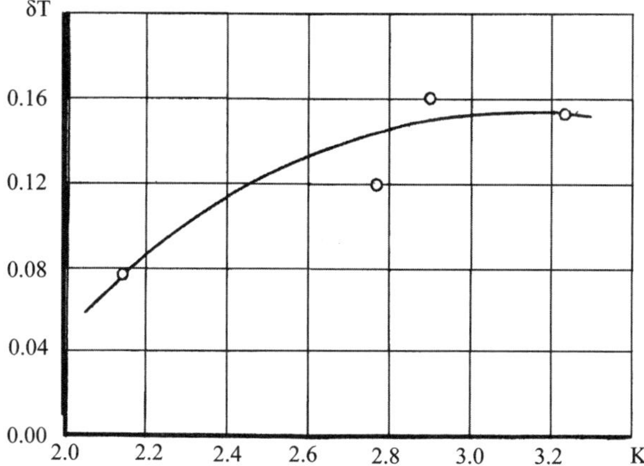

Fig. 13.2 Oscillation decrement δT vs oxidizer-fuel ratio K in gas–liquid combustion chamber (unit D-495, $G_\Sigma = 34.8$ kg/s, $K_{ggo} = 21.6$, and $f = 2.4$ kHz) (head *a* Table 13.2).

oscillations, respectively, for different combinations of chamber pressures and chamber and mixture ratios. It can be seen that with an increase of q the decrement decreases, and the relative value of energy spectrum maximum $S(\omega)^*$ increases. This result indicates a reduction in stability with an increase of the relative flow ratio in the combustion chamber, in which oscillations occurred at the frequencies f_{1T} = 2.4 kHz and f_{2T} = 4.0 kHz.

In modern gas–liquid combustion chambers, the value of $q = 2\text{g/(s}\cdot\text{cm}^2\cdot\text{bar)}$ is commonly considered to be the optimal value of flow ratio.

IV. Relationship Between Operating Process Stability and Number of Liquid Entry Holes in an Injector and Their Distances from Edge

The influence of the number of holes in an injector n_f on the oscillation decrement will be considered on the basis of the test results for engine RD-263 and experimental unit D-495. The decrement δT at the frequency $f = 2.1$ kHz of the first tangential mode of oscillation vs the number of fuel holes in the injector n_f with a constant pressure drop across the holes is shown for engine RD-263 in Fig. 13.4. It follows from Fig. 13.4 that with an increase in the number of holes from $n_f = 4$ to 5 the oscillation decrement decreases practically under all operating conditions studied so far.

The dependence of the decrement δT on the increase in number of liquid holes from $n_f = 4$ to 8 and simultaneous reduction in the pressure drop across liquid holes are shown in Fig. 13.4b. The oscillation decrement was determined for the first and second modes of tangential oscillations at $f_1 = 2.4$ kHz and $f_2 = 4.0$ kHz, at which self-excitation of high-frequency pressure oscillations was observed in some cases in unit D-495. Figure 13.4b shows that the decrement δT decreases with increasing number of liquid holcs.

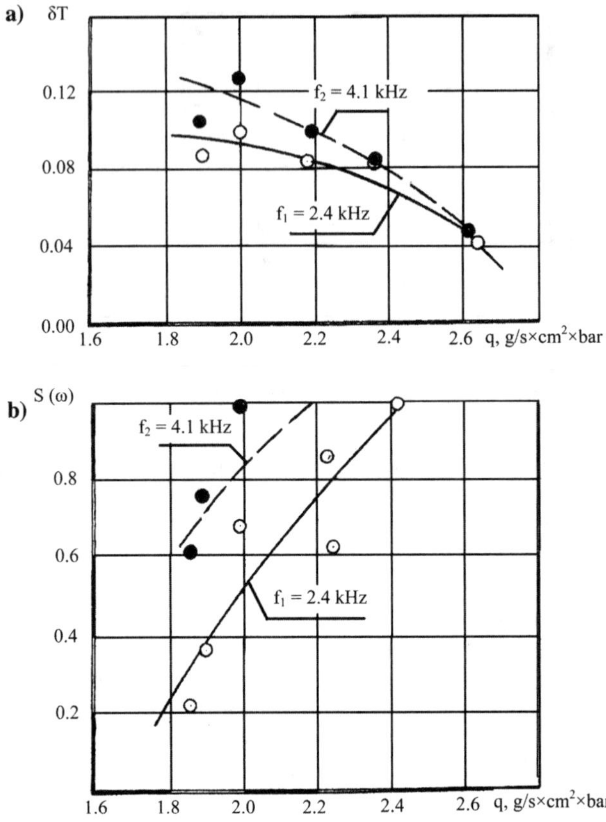

Fig. 13.3 Oscillation decrement δT and noise power spectral density $S(\omega)$ for first f_1 and second f_2 tangential oscillation modes vs relative flow ratio of the chamber q (unit D-495, $G_\Sigma = 34.8$ kg/s, $K_{ggo} = 21.6$, $f = 2.4$ kHz) (head b Table 13.2).

Using the decrements determined from noise, it was also possible to estimate the effects of design modifications [such as the distance of fuel holes to the edge of injector l_l (Tables 13.1 and 13.2) faced to the combustion chamber] on the stability characteristics for engine RD-263. Under conditions near the instability region ($\mathcal{K} < \mathcal{K}_{nom}$), the oscillation decrement at the frequency $f = 2.1$ kHz increased from 0.02 (with $l_l = 8.5$ mm) to 0.04–0.12 (with $l_l = 7.2$ mm), as shown in Fig. 13.5.

The statistical data on firing tests of the engine injector heads have shown that combustion chambers with injector heads having four fuel holes ($n_f = 4$) moved closer to the combustion bottom with a distance of $l_l = 7.2$ mm operate much farther from the instability region than with heads having $n_f = 5$ and $l_l = 8.5$ mm.

The overall data indicate that by changing the number of liquid holes, their distance from the injector edge, and the pressure drop across the fuel injector, it is possible to significantly modify the quantitative characteristics of stability in a gas–liquid engine combustion chamber with emulsion injectors.

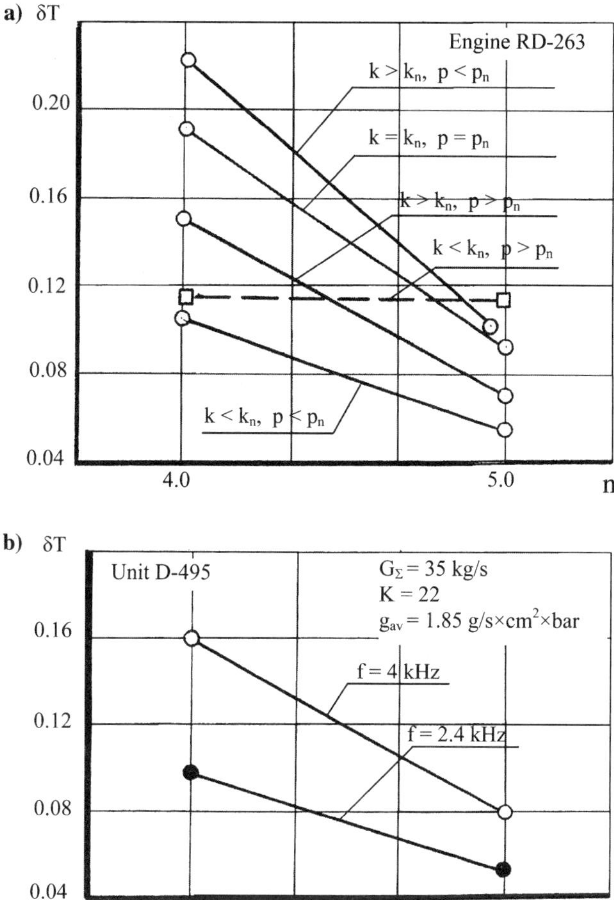

Fig. 13.4 Oscillation decrement vs number of fuel entry holes in the injector.

V. Effects of Injector Length on Stability

The results of enhancement of acoustic oscillation damping by adjusting the injector gas channel to acquire an appropriate oscillation frequency for gas-liquid engines are shown in Sec. V of Chapter 14. Such modification was first made in the design of combustion chambers of engines NK-15 (11D51), NK-31 (11D114), and NK-39 (11D113) developed by Samarsky Scientific and Technical Complex named after N.D. Kuznetsov public company. The engines were developed for the carrier rocket N-1 [43 and 45]. The possibility to utilize the modification of an injector length for increasing stability was studied for engine RD-263. The jets at the inlet of the injector gas channel were not used for increasing the oscillation energy losses in this engine.

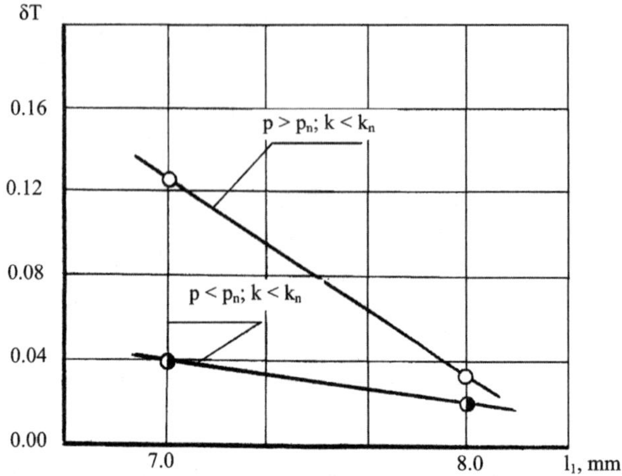

Fig. 13.5 Dependence of oscillation decrement δT on distance of fuel entry hole from the combustion bottom (engine RD-263, $f = 2.1$ kHz) (heads *b* and *c* in Table 13.2).

The oscillation decrement at the frequency of the first tangential mode $f = 2.1$ kHz vs the length of the injector l_i is shown in Fig. 13.6. Similar conditions as to P_{ch} and $\mathcal{K}(P_{chnom}, \mathcal{K}_{nom})$ were selected for comparison, and the injector length was the only design feature of the combustion-chamber heads being compared. As seen from the plot, the largest decrement was obtained for the injector head with the shortest injectors $l_i = 28$ mm. The statistical data of the tests also indicate that the combustion chamber with short injectors has the

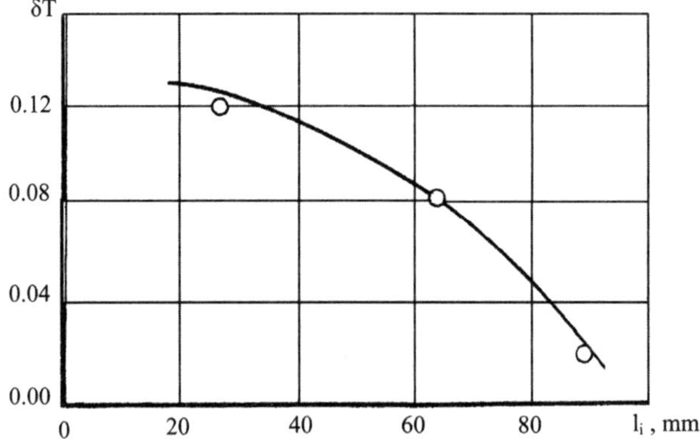

Fig. 13.6 Dependence of oscillation decrement δT on injector length (engine RD-263, $f = 2.1$ kHz) (heads *c*, *d*, and *e* in Table 13.2).

largest stability margin. In this case, the ratio of the effective length of the injector l_{ef} to the acoustic wavelength λ of the first mode of tangential oscillations has the value $l_{ef}/\lambda = 0.2$.

The data can be accounted for by the fact that at the inlet of the injector gas channel there were no jets that result in an increase of acoustic energy losses (see Chapters 15 and 16). With the increase in length of the injector gas channel without a jet, the losses are not so great as compared to the energy dissipation inside injectors and the Rijke tube effect [46]. Injectors with adjusted gas channels and jets at channel inlets were applied successfully in the development of the engine series RD-170, RD-172, RD-180, and so on.

VI. Influence of Fuel Temperature at Engine Inlet on Stability

The dependence of stability in a combustion chambers on the propellant temperature at the inlet of pumps has been found for a number of engines. Let us consider the influence of the fuel temperature at the inlet of engine RD-263 (see index *a* in Table 13.2). The deviation of engine operation was within ±10% in P_{ch} and κ from their nominal values.

The dependence of the decrement value at the frequency $f = 2.1$ kHz on the fuel temperature t_f at the engine inlet is presented in Fig. 13.7. The decrement value changes with the increase of fuel heating.

The preceding examples confirm once again the successful use of pressure oscillation decrement as a quantitative characteristic for evaluation of the influences of design modifications on the operating process stability with respect to high-frequency pressure oscillations.

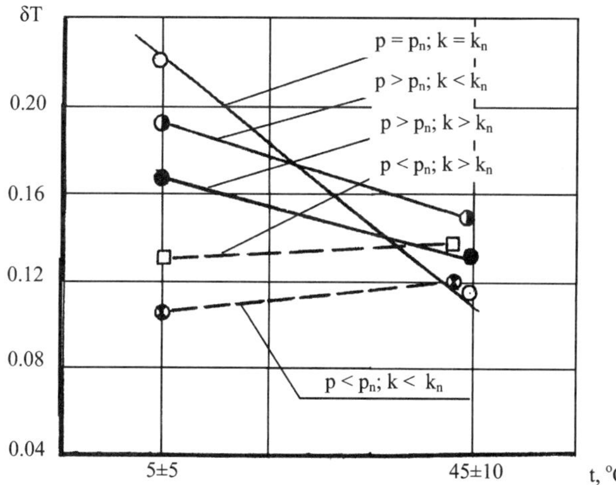

Fig. 13.7 Dependence of oscillation decrement δ*T* on fuel temperature at engine inlet (engine RD-263, *f* = 2.1 kHz); P_n and K_n are chamber pressure and fuel-to-oxidizer ratio under nominal conditions, respectively.

VII. Change of Stability Characteristics Near High Oscillation Region

It has been found experimentally that away from the stability boundary, the decrement determined from noise for a given type of mixing head on the average remains constant. In this connection, it is of interest to estimate the distance from the operating process stability boundary at which its reduction becomes notable. It is more convenient to make this evaluation based on the results of model tests of full-scale combustion chambers with gas–liquid heads. As a model unit, it is possible to change the mixing parameters over a wide range during the experiment (for example, the velocity of oxidizing gas at the outlet of injector U_o can be changed by eight to ten times).

As an illustration, let us consider testing of a model unit with an experimental head of engine 15D79 (see Table 14.1). The combustion-chamber diameter was $D_{ch} = 300$ mm, and the diameter of nozzle throat was $D_{cr} = 178$ mm. The injector head was equipped with a bipropellant pneumatic umbrella type of injector. The experiment was carried out with smooth approach to the self-excitation boundary with an excess oxidizer ratio $\alpha = 1.1$, starting with an operation located deep inside the stability regions. The boundaries of self-excitation and hysteresis regions were determined in this case. The frequency $f = 3.7$ kHz, corresponding to the second mode of tangential oscillations, was obtained at the stability boundary.

The boundaries of stability and hysteresis regions are shown in Fig. 13.8 within coordinates:

1) total flow rate G_Σ
2) relative flow ratio q
3) velocity of oxidizing gas U_o

} oxidizer excess ratio α

The initial U_o was ~20 m/s. Self-oscillations occurred at $U_o = 60$ m/s, that is, a three-fold change of the parameter U_o took place at the intersection of the stable operation region. In this case, the change of the stability index δT can be judged from the plot presented in Fig. 13.9. With increasing velocity U_o in the combustion chamber, the oscillation decrement at frequencies $f_1 = 2.2$ kHz and $f_2 = 3.5$ kHz, corresponding to the first and the second mode of tangential oscillations, respectively, varies about some average value. Any distinct tendency of the value change has not been observed. A clear tendency for δT to decrease at the frequency f_2 appears under conditions (numbers 22 and 23) quite close to the boundary of the hysteresis region, and the main change of this characteristic occurs inside the hysteresis region. In the region δT for the frequency f_2 decreases appreciably, and δT for the frequency f_1 increases significantly. Hence in the combustion zone, while the pressure oscillations at some frequency (in this particular case f_2) are enhanced, the oscillations at the other frequency (f_1) seem to be suppressed at the frequency f_1.

By varying the oxidizing gas velocity U_o, conditions 22 and 23 are 20–30% ($\mathcal{K}_{uo} = 0.2$–0.3) away from the self-excitation boundary. The statistical data of the model test have shown that the hysteresis region has just this width as to velocity U_o (in separate cases $\mathcal{K}_{uo}$ is as high as 0.5) for the experimental versions of gas–liquid engine heads.

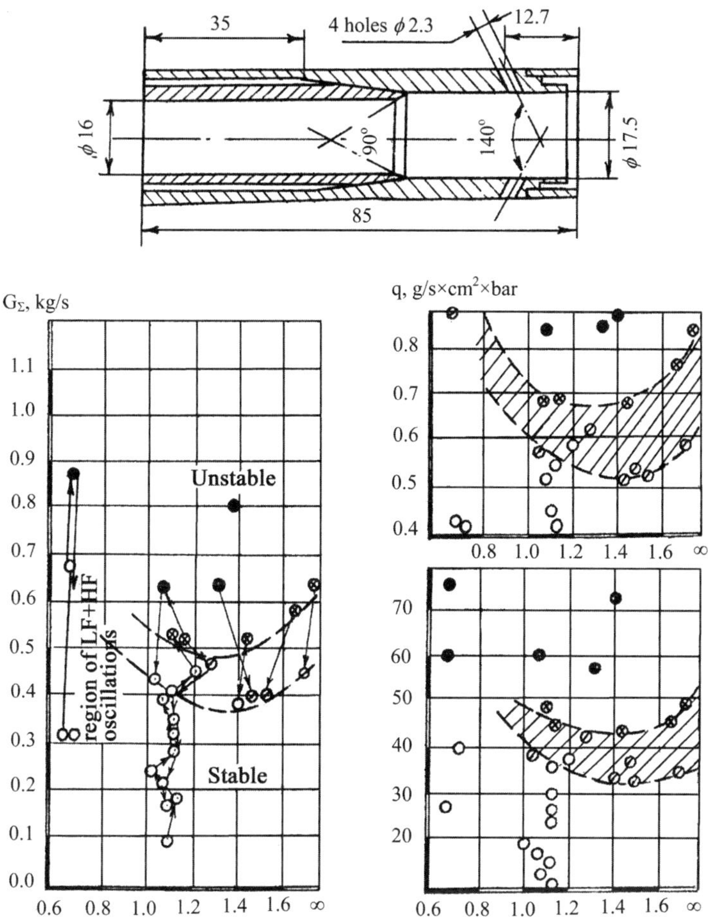

Fig. 13.8 High-frequency instability region and hysteresis zone (model test): ∘, stable; ⊗, HFO region boundary; ⌀, hysteresis region boundary; and •, developed oscillations.

The tendency toward the stability reduction with an increase in oxidizing gas velocity has been also obtained in the tests of 11D113 and 11D114 series of gas–liquid engines carried out by Samarsky Scientific and Technical Complex named after N.D. Kuznetsov [12 and 45]. The boundary for high-frequency operating process stability has been found experimentally during the development of combustion chambers of the engines 11D53 and 11D59 and of the prototype engines 11D113 and 11D114. The stability boundary is shown in Fig. 13.10 in terms of the mixture ratio at the chamber head and the oxidizing gas temperature at the inlet.

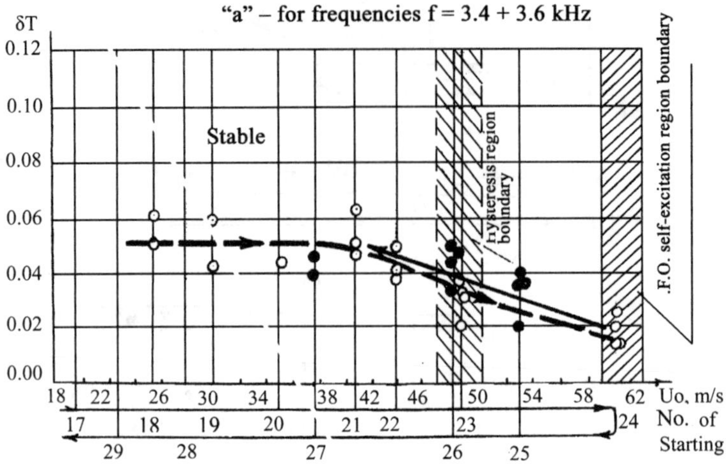

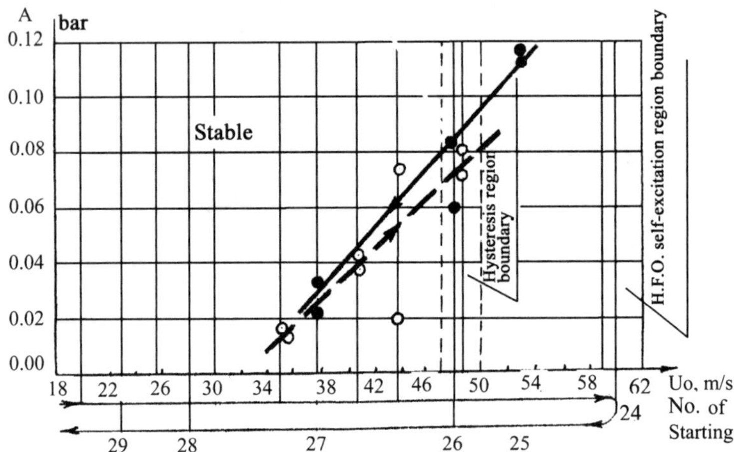

Fig. 13.9 Oscillation decrement while approaching high-frequency instability region: – –○→–, approaching the self-excitation boundary; and ——○←, receding from self-excitation boundary.

The increase in the oxidizing gas temperature at the head inlet results in an increase of the gas-generator velocity. The increase in the mixture ratio has the same effect. At the stability boundary, the amplitude of the high-frequency oscillation at the frequency of 1560 Hz of the first tangential mode was about 2.0–2.5 MPa. After increasing the stability in the chambers of engines 11D113 and 11D114, as compared to prototype engines 11D53 and 11D59, by increasing the length of the gas channels of the oxidizing gas spray injectors to 90 mm instead of 29 mm (see Sec. V of Chapter 10), the stability boundary changed.

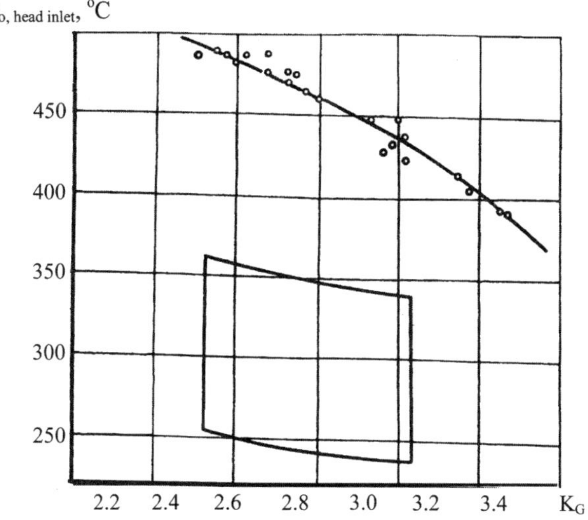

Fig. 13.10 Stability boundary for a combustion chamber of the first version.

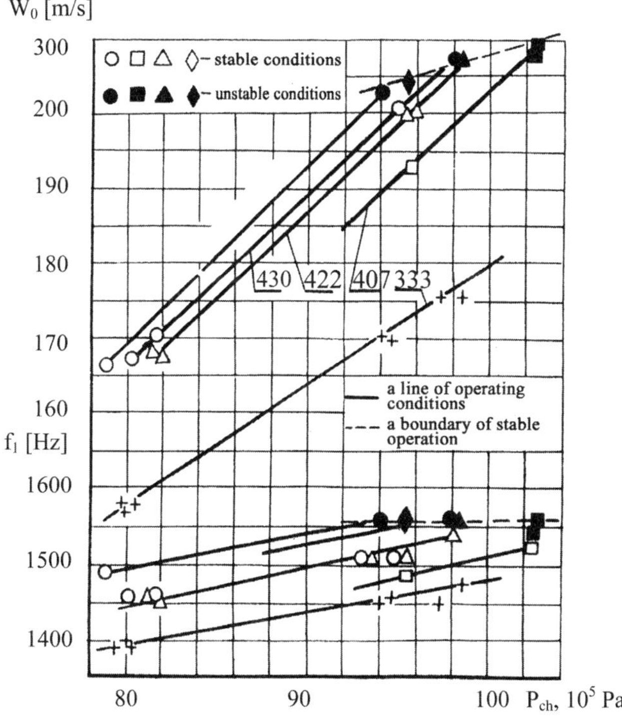

Fig. 13.11 Oxidizing-gas flow rate vs combustion-chamber pressure at different engine setting temperatures.

The stability boundary for engines 11D113 and 11D114 is shown in Fig. 13.11 in terms of the oxidizing gas velocity W_o and the chamber pressure P_{ch} for different oxidizing gas temperatures. High-frequency pressure oscillations with amplitudes $A_p > 2.0$ MPa and vibration loads $A_g > 180$ g at a frequency of 1540–1560 Hz were excited spontaneously at the stability boundary. All cases of spontaneous oscillations were observed under the conditions of $\mathcal{K}_{\text{c.ch.}} > 2.7$ and $t_{\text{setting}} > 410$°C when the oxidizing gas velocity attained the values $W_o > 200$ m/s.

Examination of the energy spectra near the unstable operation boundary has shown that with a change of operating conditions the spectra are deformed. The resonant rises of spectral density increase, as a rule, at the frequencies where the engine is excited. In this case, the oscillation frequency at the resonant condition also increases slightly.

Chapter 14

Gas–Liquid Combustion-Chamber Tests for Stability to Hard Excitation

I. Characteristics of Gas–Liquid Combustion Chambers

EXTENSIVE studies of hard excitation of high-frequency pressure oscillations carried out in actual and experimental gas–liquid combustion chambers allowed determination of a number of relations. The main design parameters and nominal operating conditions of the combustion chambers studied are listed in Table 14.1; the main design parameters of the injector heads are given in Table 14.2.

The following symbols are used in these tables:

D_{ch} = diameter of the cylindrical part of combustion chamber
D_{cr} = diameter of chamber nozzle throat
$L_{subcritical}$ = length of combustion chamber from the head-end to the throat
P_{ch} = pressure in the chamber
G_{Σ} = total propellant flow rate
K = oxidizer-to-fuel ratio
R = combustion-chamber thrust
n = number of injectors
l_i = length of injectors
d_{gas} = diameter of gaseous component hole
d_l = diameter of liquid component hole (a diameter of liquid hole)
n_l = number of liquid holes on the injector
α = geometrical angle of liquid hole axis inclination to the injector axis
$\Delta P_{oi}, \Delta P_{fi}$ = pressure drops on the injector over the oxidizer and fuel lines

The results of studies of the following engines are presented here.

1) *Engine RD-263*: The purpose and characteristics of this engine are presented in Chapter 13.

2) *Engine RD-268*: This engine was designed during 1969–1976 for the first stage of the IBR MR-UR-100/100U. The engine represents an augmented modification of engine RD-263 [12]. Propellant components are NT and UDMH. $P_v = 126$ tf (1236 kN), $J_v = 319$ s, and $J_g = 296$ s.

3) *Engine 15D79*: This engine was designed for the first stage of the rocket UR-100. Propellant components are TN and UDMH. $P_g = 75$ tf (765 kN), $P_v = 80$ tf (817 kN), and $J_g = 327$ s.

Table 14.1 Main design and operating parameters of gas–liquid combustion chambers

Engine index	D_{ch}, mm	D_{cr}, mm	$L_{subcr.}$, mm	P_{ch}, bar	G_{Σ}, kg/s	K	R_g, tf
RD-263	320	196.11	553	200	244.8	2.67	100
RD-268	320	196.1	553	230	~397	2.67	115
15D79	300	178	——	158	228.9	2.60	75
4D75	315.4	162.5	386.5	150	234.3	2.54	51.5
15D95 (96)	225	131	——	202	156.8	2.60	45
15D169 (15D12)	160	85	280	135	43.94	2.55	14.5
11D121	140	83.9	324.3	78	24.3	~2.55 ±0.3	8.5

4) *Engine 15D169 (RD-862)* [12]: This engine was developed on the basis of engine 15D12 and designed to provide thrust and control flight of the second stage of rockets 15A15, 15A16, and so on, over all the orthogonal channels. This is a closed-circuit engine with afterburning of the reducing generator gas. To control the thrust vector over pitch and yaw channels, the gas-dynamic method is used. It is based on reducing generator gas injection into the supersonic part of the combustion-chamber nozzle. The propellant components are NT and UDMH.

5) *Engine 4D75*: This engine was designed for the first stage of the rocket R-29 with underwater takeoff. This is a "drowned" engine, implying that its main part is located in the fuel tank. Propellant components are NT and UDMH. The other data are presented in Tables 14.1 and 14.2.

6) *Engine 11D121*: This engine was designed in 1969–1974 and installed in the first stage of CR N1 for control. The propellant components are liquid oxygen and kerosene [fuel is fed from TPU of cruise LRE NK-15 (NK-33)]. In 1964–1968, engine RD-58 (11D58) for the fifth stage of CR N1 and CR Proton was designed on the basis of this chamber. It was also used as a part of CR Energiya.

Thus, the evaluations of stability to final disturbances for the following engines are presented in this chapter: 1) engines running on nitrogen tetroxide and unsymmetrical dimethylhydrazine, as well as oxygen and kerosene; 2) engines with propellant flow rate from ~24 to ~400 kg/s; 3) engines with diameters of combustion chambers from 140 to 320 mm; and 4) engines with pressure in the combustion chamber from 78 to 230 bar (Table 14.1).

II. Hard Excitation Characteristics of Gas–Liquid Engine Combustion Chambers

Data for the characteristics of stability to hard excitation for the combustion chambers studied are given in Table 14.3. Limits of operating conditions in the combustion chambers are also included.

During development of these engines, the main device for creating the artificial disturbances was IDD (in-chamber disturbance device; see Chapter 9), which was used in the current series of studies. Shock tubes for creating the artificial disturbances (see Chapter 9) were used only for evaluation of stability to hard excitation

Table 14.2 Main design parameters of injector heads investigated

Engine index	Type of injectors	n_i	l_i, mm	d_{gas}, mm	n_1	d_1, mm	α_1, deg	l_1, mm	ΔP_{oi}, bar	ΔP_{fi}, bar
RD-263 (RD-268)		169	28	15.5	3	2.3+2.6	45	7.2	~9 + 10	8.7
15D79		61-center; 50-periphery	39.3	20.9	4	2.55	65	13.1	~7.2	~15
4D75		61	53	23.5	5	2.3	45	8.5	~3	12
15D95		48	27.6	19.8	4	2.55	45	13.5	5.5	14
15D12[a] (15D169)		37	$l_1 = 34$ $l_2 = 6$	$d_3 = 11.4$ $d_4 = 15.6$ $d_1 = 3.4$	$n_{d_2} = 12$ $n_{d_2} = 12$	$d_2 = 1.55$	—	—	11.2	7.6
11D121		Annular slots 6 - for gas 5 - for liquid	~6.2	—	—	0.9 + 1.15	45 60 90	6.2	7	8

[a]Engine 15D12 is made on closed circuit (staged-combustion cycle) with afterburning of reducing generator gas.

Table 14.3 Characteristics of stability to hard excitation of gas–liquid combustion chambers

Engine index	Conditions		Number of tests	Type of disturbance device	Weight of charge, g	Value of the first pressure pulse A_0, bar	Maximum amplitude of natural disturbances, bar	$n = \frac{A_0}{2A_{n\,max}}$
	P_{ch}, bar	K						
1	2	3	4	5	6	7	8	9
RD-263 RD-268	200–218	2.5–2.9	12	IDD	0.3–0.6	7–16	2–3	~1.7
15D79	150–235	2.1–2.8	8	IDD	0.3–1.0	8.4–22.8	~3	~2–2.4** >1.2–2.4
4D75	150	2.5–2.6	3	IDD	0.3–0.45	6.2–14.35*	~2–3.5	>5
15D95	105–222	2.1–2.6	5	IDD	0.45–1.0	9–27	3.4–8.1	>3.67
11D121	74–81	2.2–2.5	7	IDD	0.3–1.0	2–12.2	~0.8	>7.6
15D169 (15D12)	115–130	2.1–3.6	7	4ST	2.6–12.5	7.2–23.8	~2.88	>4.1

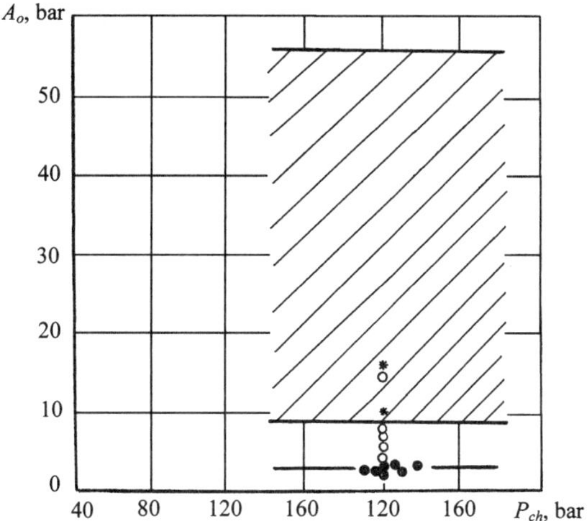

Fig. 14.1 Bifurcation diagram: •, level of natural noise in the chamber; ○, initial pressure pulse giving rise to damped oscillations; and *, initial pressure pulse giving rise to undamped oscillations.

of engine 15D169. The weight of charges used for creating the disturbances, and variation of the first pressure peak A_0 are shown in Table 14.3.

Stability evaluation, which existed at that time, was made according to the requirement that the relation $n = A_o/(2A_{\mathrm{m\ max}}) \geq 2$ should be met. It was also a requirement that the pressure oscillations should decay quickly. Sign ~ in column 9 of Table 14.3 represents excitation of high-frequency oscillations. As seen in the table, until the design modifications were introduced, these requirements were not met in some operating modes of engines RD-268, 15D79, and 4D75 (see Sec. III).

In evaluating stability to hard excitations of engine RD-268, the critical value of chamber noise intensity that initiates the occurrence of undamped pressure oscillations was found. During firing tests, the pressure pulses were initiated with the help of IDD using explosive charges of 0.3 and 0.6 g.

During some tests, pressure oscillations with large amplitude and a frequency of 2100 Hz were recorded after IDD actuation. According to calculations, the frequency was close to that of the first tangential mode. At a value of the initial pulse $A_0 = 3$ to 8 bar, damped oscillations occurred. At a value of $A_0 = 10$ and 16 bar, undamped pressure oscillations were excited.

An area on the bifurcation diagram, where a level of natural noise is 2 to 3 bar and a minimum value of pressure pulse exciting undamped oscillations lies within $8 < A_{\mathrm{cr}} < 10$ bar, is shown in Fig. 14.1. When high-frequency pressure oscillations are excited, the amplitude is as high as 55 bar.

III. Enhancement of Stability of Engine 4D75

The development of the combustion chamber for engine 4D75 by using quantitative stability characteristics is a striking example of hard excitation control in a gas–liquid combustion chamber. During development of this engine, after gathering a large set of statistical data, cases of high-frequency instability appeared, accompanied by accidents during firing tests. In each case the cause of instability was found. The causes were different in each case; they were either connected to design modifications or to the deviation of various parameters from specified limits. Before the appearance of instability, those deviations were quite admissible. In this situation, a conclusion was made about the necessity to take measures aimed at increasing the stability margin of the combustion chamber. It was decided to carry out firing tests for experimentally checking the three groups of measures. For this purpose three injector heads were developed. The heads were conditionally named a two-class head, a spray head, and a head equipped with injectors with five fuel feed holes instead of six holes in an initial version.

The two classes of injector heads were divided on the basis of the flow rates. This measure allowed use of the entire tolerance range for liquid flow through the injectors: $\pm 5\%$ instead of ± 2 %, which was in the manufacturing of the injectors of the initial head version.

A spray injector head differed from the initial design by having 18 spray injectors with a diameter of 2.2 mm installed between the main gas–liquid injectors. Injectors with five fuel feed holes were installed instead of the initial version, with any two adjacent holes being oriented along the overall radius of the head.

Studies have shown that heads with injectors having five fuel feed holes possess the largest margin of stability to hard excitation. This was expressed in high values of critical disturbances for these heads, as well as in the reduction of the level of natural noise in the chambers with such heads. The indices used to characterize stability to hard excitation for all injector head versions are shown in Fig. 14.2.

It follows from the data shown in Fig. 14.2 that the two-class version of the injector head did not give any notable positive results; only the level of noise in the combustion chamber decreased slightly. The head version with additional spray injectors has shown worsening of the margin of stability to final disturbances. It was impossible to determine the lower boundary of a pressure pulse causing instability because it is located lower than for the previous versions.

As mentioned earlier, the value of A_{cr} increased and the noise level in the combustion chamber decreased only for the injector head with five fuel feed holes. The decrement variation as a function of pressure in the combustion chamber is shown in Fig. 14.3. It can be seen that the chamber using a head with injectors having five fuel feed holes has the best characteristics.

To clarify the origin of the effect of changing to the version of injector head with five holes, experimental studies on a model unit with single actual injector were carried out (see Chapter 10). The chamber (a resonator) of the model unit was an uncooled cylinder ending with a blunt-nose cone. The chamber was installed vertically on a perforated plate-simulator of an injector head. The actual injector was located in the central hole of the plate.

In testing engine 4D75 injectors, the combustion chamber had a diameter of 143 mm. This size gave approximate coincidence of the frequencies of the transverse modes in the engine chambers and the model unit. Gaseous methane was used as a fuel; gaseous oxygen or a mixture of oxygen and nitrogen with 73% oxygen volume fraction of the mixture was used as an oxidizer.

The experimental results are given in Figs. 14.4 and 14.5, G_{Σ} is the total flow rate of gaseous components through an injector, and α is the oxygen excess ratio.

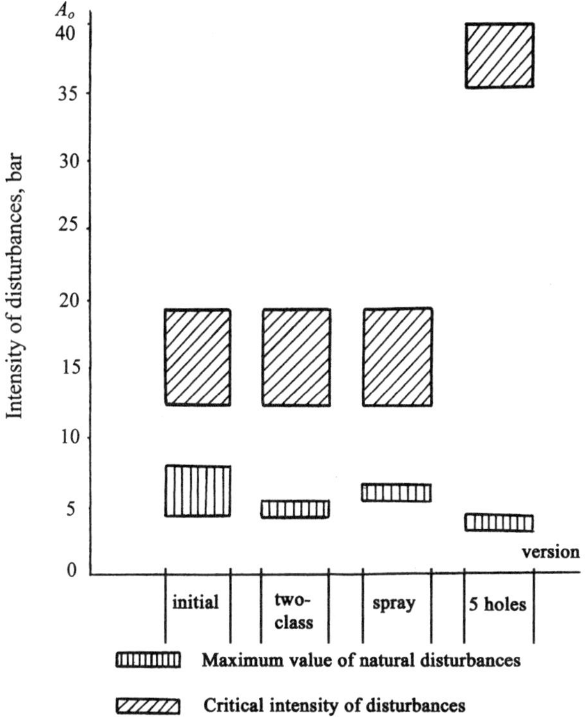

Fig. 14.2 Intensities of pressure oscillations in combustion chambers with different versions of injector heads.

The directions of mode variation are indicated by arrows at the stability boundaries. As the plots show, the region for instability of the injectors with five fuel feed holes is narrower, and the lower boundary of the region is located at higher flow rates.

Based on these data, it can be concluded that the change of the number of fuel feed holes from six to five holes resulted in a shift of the self-excitation boundary, which extended the stable region of combustion-chamber operation.

The injector model test results agree with the results of processing the oscillation decrements determined from noise. The significantly larger decrements for the injectors with five fuel feed holes compared with the decrements for other injector versions (see Chapter 13, Fig. 13.4) indicate that the self-excitation boundary moved away from the operating region. During testing of the injector head with five fuel feed holes, hard excitation did not occur even with excitation pulse $A_0 = 35$ bar (Fig. 14.2). This result also indicates a shift of the self-excitation boundary and expansion of the stable operation region in the chamber.

IV. Influence of Gaseous Oxidizer Velocity on Stability

The tendency for reduction of stability with increase in velocity of gaseous oxidizer in injectors is discussed in Chapter 13. To obtain the data on the influence of oxidizer gas velocity on the stability to final disturbances, a series of studies on

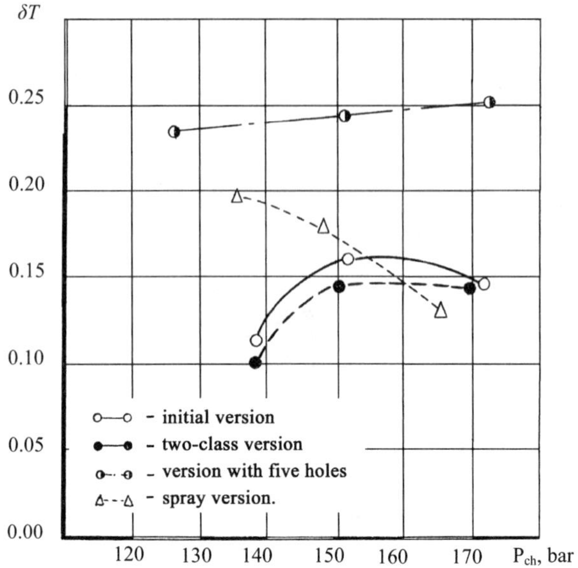

Fig. 14.3 Average values of oscillation decrement vs pressure in the combustion chamber at a frequency of 2000 ± 200 kHz.

hard excitation was carried out in the combustion chambers of experimental gas–liquid engines with afterburning of oxidizer gas. The experiments were carried out on units D-69 (propellant: NA-2471 + TG-02) and D118 (propellant: NT + UDMH). A schematic diagram of the unit is shown in Fig. 6.5.

The combustion chamber was essentially the same design as that of serial combustion chambers S.5.3.0100-0. Modifications were made to provide dismantling

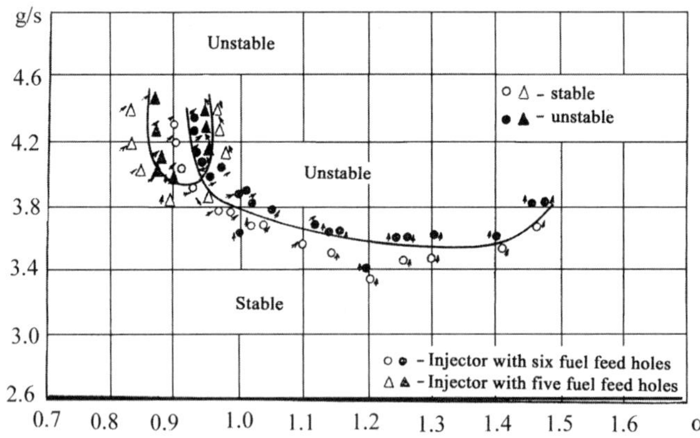

Fig. 14.4 High-frequency combustion instability regions at a frequency corresponding to the first tangential oscillation mode.

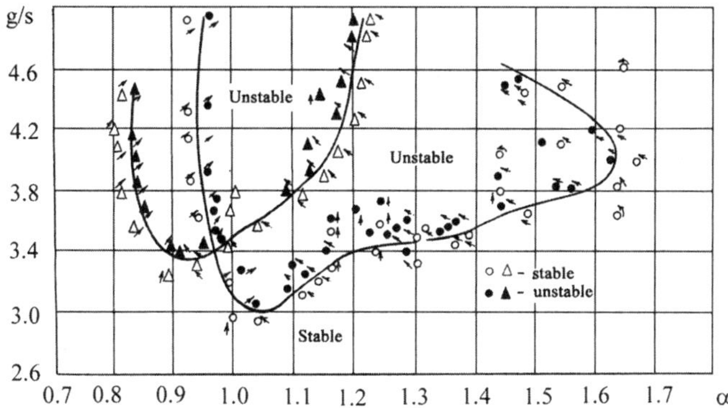

Fig. 14.5 High-frequency combustion instability regions at a frequency corresponding to the first radial oscillation mode: ○●, injector with six fuel feed holes; and ▲△, injector with five fuel feed holds (the direction of the change of conditions are indicated by arrows).

of the chamber with an injector head, shortening of the nozzle exit, and introduction of independent water cooling. The injector heads being tested were equipped with pneumatic injectors. In each injector the fuel was fed through one row of holes in the walls of the cylindrical channel, over which the gas was fed to the combustion chamber. In addition, on separate heads, fuel was fed through the swirl nozzles on the external wall of the injector (umbrella-type injectors). A general view of the head and structural dimensions of all tested versions are presented in Fig. 14.6.

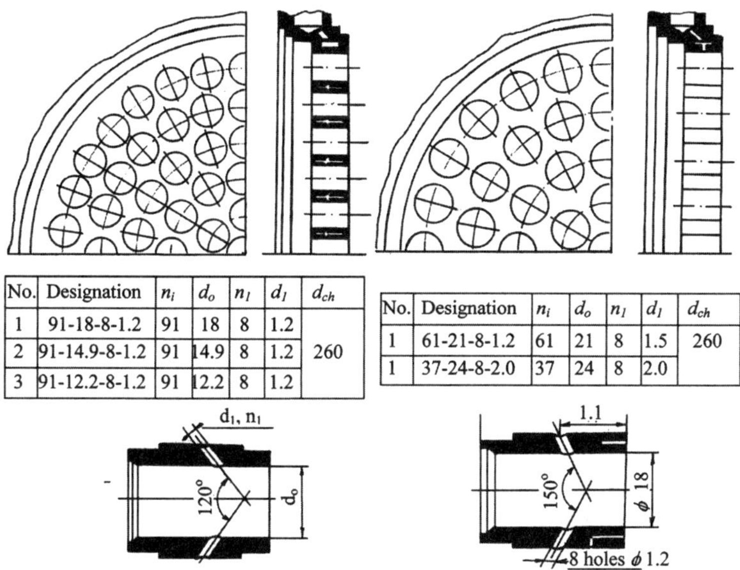

No.	Designation	n_i	d_o	n_l	d_l	d_{ch}
1	91-18-8-1.2	91	18	8	1.2	260
2	91-14.9-8-1.2	91	14.9	8	1.2	
3	91-12.2-8-1.2	91	12.2	8	1.2	

No.	Designation	n_i	d_o	n_l	d_l	d_{ch}
1	61-21-8-1.2	61	21	8	1.5	260
1	37-24-8-2.0	37	24	8	2.0	

Fig. 14.6 Design parameters of injector head.

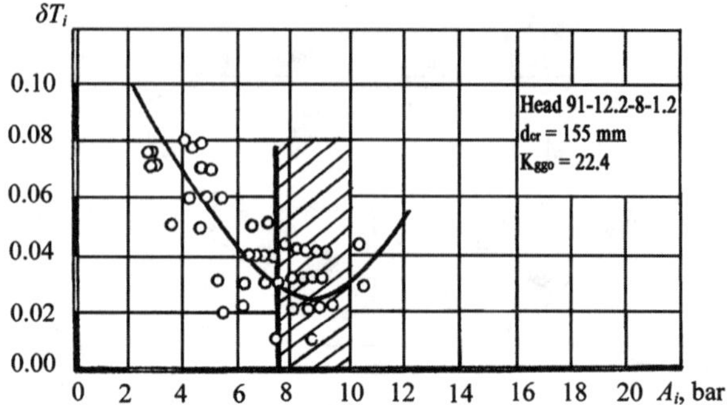

Fig. 14.7 Amplitude dependence of the oscillation decrement determined in each successive period of damped oscillations.

Disturbance pulses were initiated with the help of shock tubes. Damped and undamped high-frequency oscillations were excited in the combustion chamber; the frequencies corresponded to those of the first and second modes of tangential oscillations ($f_1 \approx 2400$ Hz and $f_2 \approx 4000$ Hz).

Some experimental results are presented. The average decrement of damped oscillations arising from the disturbances introduced into the combustion chamber is used here as an index of stability to hard excitation. According to the experimental data, reduction in the decrement with change of some parameter indicates that the system approaches the instability boundary.

The values of the decrement were determined for each successive period of the damped oscillations, as well as their average values, vs an initial pressure disturbance. Typical curves are shown in Figs. 14.7 and 14.8. These curves have minima as function of the area, $A_{cr} = A_{bound}$, where A_{bound} is the amplitude of oscillations occurring spontaneously in the given chamber.

The reductions in the average values of the oscillation decrements δT with decrease in oxidizer-to-fuel ratio K_{ggo} have been obtained for the experimental

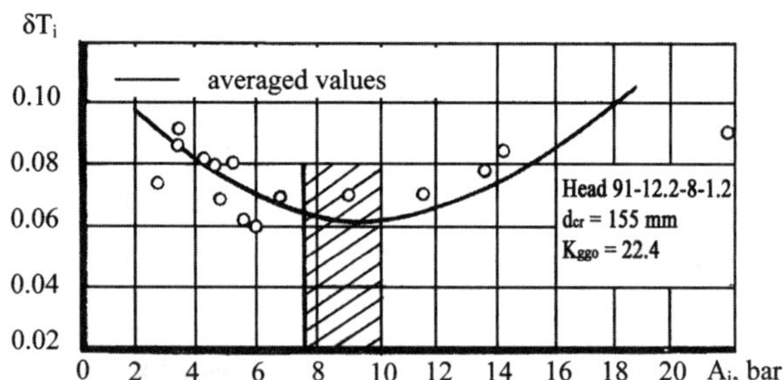

Fig. 14.8 Oscillation decrement determined as a function of initial pressure pulsation A_o.

unit D-118. This is clearly seen in the plot of δT vs the initial pressure pulse A_0 for modes with different K_{ggo} (Fig. 14.9). This effect is connected with the increase in temperature of the oxidizer gas and hence with increase in the injector outlet velocity. The decrease in K_{ogg} resulted in spontaneous oscillations with amplitudes of 6–8 bar.

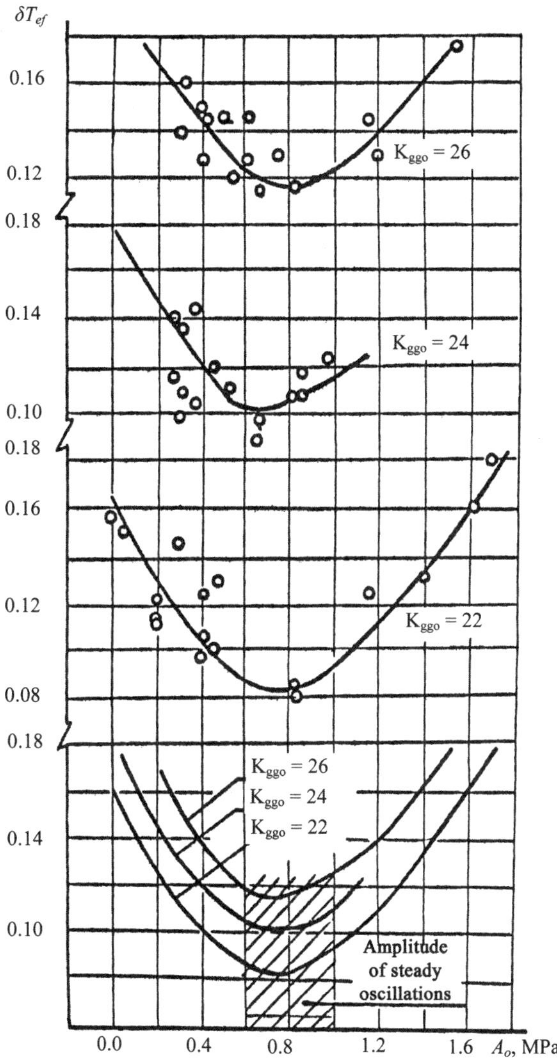

Fig. 14.9 Effective oscillation decrement vs initial pressure pulse A_0 for operating modes with various oxidizer-to-fuel mixture ratios in the oxidizing gas K_{ggo} (experimental gas–liquid combustion chamber).

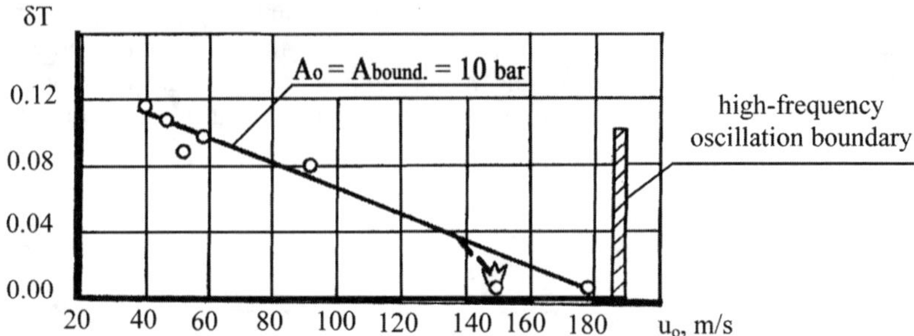

Fig. 14.10 Dependence of minimum oscillation decrement on chamber operating mode approaching the boundary of high-frequency oscillations (summary plot based on the unit D-118 actual test results).

The effect of changing oxidizer gas velocity, caused by the change of relative flow area of the heads over the gas line, upon stability is shown in Chapter 6 (Fig. 6.4). Variation of the minimum oscillation decrement δT as operation approaches the boundary of the high-frequency oscillation region is shown in Fig. 14.10.

Chapter 15

Injector Head for RD-170 Engine Combustion Chamber

I. Introduction

AS an example of evaluation of stabilities in LRE combustion chambers, the studies carried out during the development of the RD-170 engine and its modifications are described. The results of testing the injector head units 1UK, 2UK, and 2UKS are presented in Secs. II–VII, along with the selection of the optimal design based on the quantitative stability characteristics. The test results for the RD-170 combustion chamber as part of an engine unit are given in Sec. VIII.

The RD-170 engine was designed for the first stage of the Zenit rocket, and its high-power version was designed for the first stage of the Buran launch vehicle. The RD-170 engine employs a staged-combustion cycle with afterburning of oxidizing gases from gas generators. Each engine contains four combustion chambers, two gas generators, one turbo-pump unit, an automatic control system, and so on. The engine specifications are given next:

1) Propellant components are liquid oxygen for the oxidizer and RG-1 for the fuel.

2) Ignition of fuel in combustion chambers and gas generators is achieved with the help of starting fuel PG-2.

3) Nominal thrust is 740 tf at the sea level and 806 tf in vacuum.

4) Nominal mass flow rates of propellants through the engine combustion chambers are the following: oxidizer, 1732 kg/s, and fuel, 666.3 kg/s.

5) Pressure in combustion chambers under nominal conditions is 25.0 MPa.

6) Diameter of the cylindrical part of the combustion chamber is 380 mm.

7) Diameter of nozzle throat is 235.6 mm.

8) Length from the combustor head-end to the nozzle throat is 490 mm.

Development of such an engine according to its specifications required development of each assembly on special units. The development of combustion chambers was performed on units 2UK, 1UKS, and 2UKS. Those units were established on the basis of a serial engine operating with NTO and UDMH propellants. The engine was adapted for a new propellant combination: oxygen and kerosene. A variety of engine parts, such as TPU component seals, were replaced.

Table 15.1 Main characteristics of the unit 2UK combustion chamber

Parameters	Designation	Value
Pressure in combustion chamber, MPa	P_{ch}	18.0
Propellant mixture ratio	K_m	2.6
Relative flow ratio, g/(cm^2·s·bar)	q	2.06
Diameter of combustion chamber, mm	D_{cyl}	320
Diameter of nozzle throat, mm	D_{cr}	196.1

The dimensions of the main units remained completely unchanged. Provision was made for injection of starting fuel PG-2 (87% triethylborin and 13% triethyl aluminum) to achieve self-ignition with oxygen.

The units 2UK, 1UKS, and 2UKS differed in the arrangement, overall assembly, and attainable parameters. Units 1UKS and 2UK represent a LRE test bed adapted for multiple starts with a thrust of about 100 tf operating on oxygen and kerosene (RG-1) with afterburning of oxidizer-rich gases from the gas generator. The main characteristics of the unit 2UK combustion chamber are given in Table 15.1.

Unit 2UKS for the chamber development represented an engine with thrust of about 150 tf (at sea level) made on a scheme with afterburning of oxidizing gas and pump feed of propellants and pneumatic control. The conditions of the chamber tests of units 2UKS came up to ~80% of the nominal conditions for the RD-170 engine.

Design of the combustion chamber as part of unit 2UKS was regarded as the final stage. Any further significant design modifications of the combustion chambers were not required during tests of the combustion chambers for the RD-170 engine.

II. Separate Tests of RD-170 Engine Combustion Chambers in Special Units [47]

Evaluation of stability of the combustion chambers and gas generators with different versions of mixing heads under stable operating conditions was made in accordance with the methods presented in this book.

Measurements of pressure pulsations in the combustion chamber (PPIC-1) were made using two different recording channels: an insensitive channel with calibration of ±30 bar and a sensitive channel with calibration of ±5 bar. Pulsation transducer LH-608 was mounted on the chamber at 168 mm from the combustion end of the mixing head. A small amount of fuel was injected into the cavity underneath the diaphragm of the transducer socket from the cooling jacket to acquire additional protection of the diaphragm against burn-out.

Signal processing of the sensitive channel for recording pressure pulsations in the combustion chamber was performed automatically over the starting period in real time with the help of the instrument for recording the amplitude, decrement, and frequency of the filtered signal. A short name of the instrument is RADCh. The sample filter was tuned successively to the most typical (dominant in the spectrum) natural frequencies of the chamber: 1 kHz, 2.0–2.5 kHz, and 3.5–4.0 kHz.

Recording of the amplitude spectrum and determination of the following three parameters of a random stable process were performed with the help of the instrument RADCh: 1) oscillation decrement δT_n (by means of two different methods: the instantaneous period method in the time domain and the amplitude method in the frequency domain); 2) filtered signal amplitude A_s; and 3) frequency f corresponding to the maximum peak in the signal amplitude spectrum.

Evaluation of stability with respect to hard excitation was made in the tests of engine units 2UK by introducing external pulse disturbances from an MDD into the combustion chamber in the radial direction. The MDD (see Chapter 10) consists of a thick-wall cylindrical case with a diameter of 130 mm with four blasting chambers communicating through 6-mm-diam channels, along with a common outlet channel of the same diameter (6 mm). The charge units with electric detonators of the ED-202-type and detonating explosive charges are screwed into the blasting chambers. The weight of charges in the unit can be varied. The explosive initiation lag is less than 0.01 s. The blasting chamber cavities are covered with sleeves, whose mounting seats and contact conductor bushings are sealed by Teflon® gaskets and collars. The transverse dimension of channels ($d = 6$ mm) connecting the blasting chambers with the combustion chamber is dictated by the attempt to minimize modifications of the combustion chamber. The MDD was attached to the test article with four bolts connecting two plates of the MDD with two mating plates mounted on the cylindrical part of the combustion chamber. To prevent combustion products from entering the blasting chamber cavity, nitrogen purging with a pressure drop of ~5 bar was provided in the MDD outlet channel through a drilled hole with a diameter of 0.7 mm. Initiation of charges was controlled by a pressure indicator light and a clockwork mechanism.

The degree of stability with respect to hard excitation was evaluated from the rate of attenuation of oscillations excited by pulse disturbances. Data were processed for a period of $\Delta\tau = 0.017$ s in the vicinity of the introduction of pressure pulse in the pressure–time trace. A signal-processing program was developed for separating up to three frequencies with minimum oscillation decrements (i.e., three major maxima) from the spectrum in a specified frequency range. The magnitudes of spectral density with a frequency resolution of 5–9 Hz, and oscillation decrements and amplitudes of initial disturbances for the frequencies corresponding to the three major maxima, were printed in the table of processing results. By varying the frequency range whenever necessary, it is possible to determine the decrements and amplitudes at any frequency of interest.

In addition to oscillation decrements, the stability margin factors n were used as an additional stability property, expressed as a ratio of the first pressure peak A_0 of the resultant signal (initiated by an artificial disturbance) to the double maximum amplitude of noise $2A_n$ n the time period before applying the pulse disturbance $n = A_0/2A_n$.

When a combustion chamber was tested in unit 2UKS, the stability with respect to final disturbances was evaluated using the following formula: $n = A_0/2A_{nm}$, where A_0 is the magnitude of the first disturbance peak, and A_{nm} the maximum value of the noise amplitude in the time period of 5 ms immediately preceding the initiation of artificial oscillations.

The relaxation time of pressure oscillation was determined by

$$\tau_r = \tau_1 + \tau_e$$

where τ_1 is the time elapsed from the first peak A_0 of the disturbance to the maximum oscillation amplitude A_m and τ_r the time elapsed for the maximum amplitude A_m to decay by a factor of e. The decrement was determined from the relation

$$\delta T_{\text{ef}} = \frac{1}{\tau_r} \cdot f_i$$

where f_i is the frequency corresponding to the maximum in the amplitude spectrum of the process. The stability to hard excitation under transient conditions was evaluated using the same procedure as that under the main operating conditions.

III. Characteristics of Mixing Heads on Units 1UKS and 2UK

Determination of stability characteristics was carried out mainly for mixing heads, which operated under stable conditions with a duration of 10–30 s.

Mixing heads with emulsion injectors belong to the first group. Figure 15.1 and Table 15.2 show the design features of such an injector and the geometrical characteristics of mixing heads. The second version (I-2) differs in that the peripheral row consisted of injectors smaller than those in version I-1. Both versions have low flow fuel injectors to protect the combustion bottom from overheating.

In version II, mixing heads with coaxial-spray injectors (see Fig. 15.2 and Table 15.3) in which fuel was fed in parallel to the oxidizing gas flow through the central bushing. In version III, the fuel bushing was made flush with the edge of the injector gas channel, as shown in Fig. 15.3 and Table 15.4. In version IV, it projected 10 mm into the combustion chamber.

Injectors of the first and second groups (versions I and II) were short ($l_i = 28$–39 mm) and were not specifically adjusted to match the first tangential mode of oscillations ($\bar{l}_{\text{ef}} = l_{\text{ief}}/\lambda_{1T} = 0.19$–$0.23$).

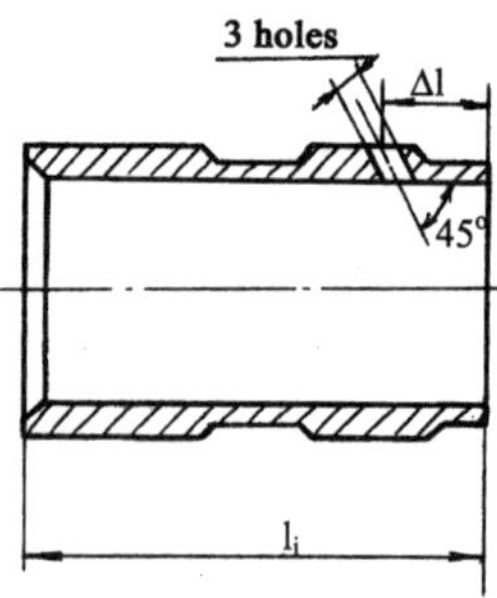

Fig. 15.1 Mixing heads with emulsion injectors.

Table 15.2 Mixing heads with emulsion injectors

Version no.	Number of main injectors n_i	Spacing between injectors t, mm	Length of injectors l_i, mm	Distance of holes from end-face Δl, mm	Permeability for oxidizing gas at outlet F_o, %	Dimensionless length of injectors $\bar{l}_{ef} l_{i_{ef}}/\lambda$
I-1	169	20.5	28	7.2	39.6	~0.2
I-2	= 54 = 91	= 25 = 19	-″-	7.2	-″-	-″-

Mixing heads with spray–centrifugal injectors belong to the third (the largest) group, as shown in Fig. 15.3 and Table 15.4. The oxidizing gas enters the combustion chamber through the central channels of injectors. The fuel first passes through the tangential holes, becomes swirled, and is injected into the combustion chamber. It spreads over the surface of countersinks of the combustion bottom. In the initial segment of the swirling cavity, directly at the location of fuel exit from the tangential holes, the bushing is mounted in the injector, which separates fuel from oxidizing gas. The position of the end-face of this bushing relative to the injector exit (i.e., depth of separating bushing undercutting Δl or recess length) defines the extent of preliminary mixing and combustion of inside the injector.

Injectors of this group (version III) can be further divided into the following three subgroups according to the design of the gas channel inlet. In the injectors of subgroup 3-A, the oxidizing gas is fed into the combustion chamber through cylindrical channels having the same flow area along their length. This subgroup (versions III-5 and III-6) was equipped with short injectors with a dimensionless effective length of the gas channel $\bar{l}_i$ around 0.24.

Here the wave length λ_{1T} was determined by the frequency corresponding to the first tangential mode of oscillation in the combustion chamber and the sound velocity in the oxidizing gas upstream of injectors. The recess of the separation bushing was $\Delta l = 0.5$ mm for version III-5 and $\Delta l = 2.0$ mm for version III-6. Hence, the mixing of propellants took place outside injectors themselves.

Versions III-7 and III-8 have a normalized injector length of $\bar{l}_{ef} \approx 0.5$, which, according to modern concepts, provides a maximum loss of acoustic energy from the combustion chamber. In version III-7, the recess of the separation bushing was

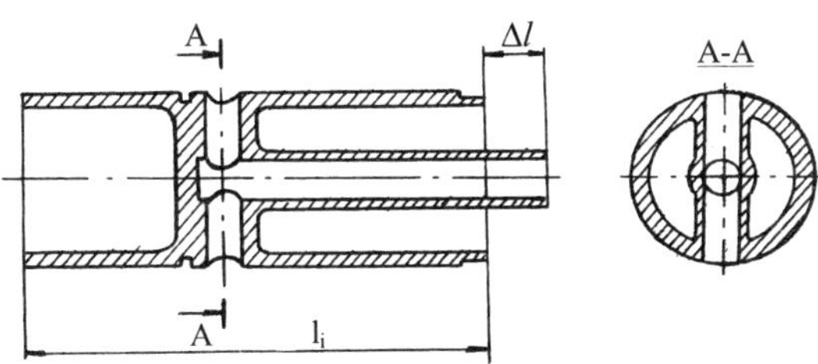

Fig. 15.2 Mixing heads with spray–spray injectors.

Table 15.3 Mixing heads with spray–spray injectors

Version no.	Number of main injectors n_i	Spacing between injectors t, mm	Length of injectors l_i, mm	Distance of holes from end-face Δl, mm	Permeability for oxidizing gas = 0.34 at outlet F_o, %	Dimensionless length of injectors $l_{ef}\, l_{i_{ef}}/\lambda$
II-3	271	17.35	39	0	35.2 (with increased permeability)	~0.25
II-4	-″-	-″-	-″-	10	-″-	-″-

$\Delta l = 2.0$ mm, and for version III-8 it was increased to $\Delta l = 8$ mm. It should be kept in mind that a successive increase of recess of the separation bushing will change the injection scheme from "external" mixing of propellants to "internal" mixing, for which propellants react and burn in the injector outlet channel.

The transition from one mixing scheme to the other has a significant effect on combustion instability for "adjusted" injectors because in the case internal mixing oscillations can be enhanced as a result of the resonance adjustment of two interrelated oscillatory loops: an injector and combustion chamber (see Sec. V of Chapter 10). Intensification of oscillation occurs and might not be compensated by the damping effect caused by the acoustic energy losses and ultimately can result in worsening of stability with respect to soft excitation, as compared to the case with no combustion inside the injector channel (i.e., external mixing).

The enhancement of stability as a result of increased damping of intra-injector oscillations by the jets at the inlet of the gas channel of an adjusted injector was obtained experimentally on full-scale and single-injector fire model units. This kind of adjusted injector ($\bar{l}_{ef} \approx 0.5$) is assigned to subgroup 3-B as shown in Fig. 15.4 and Table 15.5. Versions IV-9, IV-10, and IV-11 have an increased recess of the separation bushing of Δl = 10.5 mm. They differ in the jet diameter in the peripheral row of injectors. Version 9 was equipped with injectors with a jet diameter of d_j = 8 mm for both central and peripheral injectors. The jet diameter for the peripheral row is d_{jp} = 8.5 mm for version 10 and d_{jp} = 9 mm for version 11.

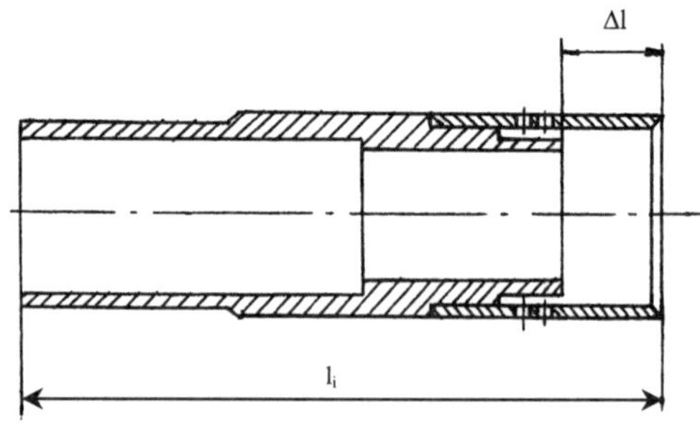

Fig. 15.3 Mixing heads with spray–centrifugal injector without jets.

Table 15.4 Mixing heads with spray–centrifugal injector without jets

Version no.	Number of main injectors n_i	Spacing between injectors t, mm	Length of injectors l_i, mm	Depth of under-cutting Δl, mm	Number of fuel holes in injectors n	Permeability for oxidizing gas F_o, %	Dimensionless length of injectors $l_{ef}\, l_{i_{ef}}/\lambda$
III-5	271	17.35	40.5	0.5	12	35	~0.25
III-6	-″-	-″-	-″-	2.0	-″-	-″-	-″-
III-7	-″-	-″-	95	-″-	-″-	-″-	~0.5
III-8	-″-	-″-	95	8.0	-″-	-″-	-″-

The change in the jet diameter in the peripheral row of injectors (that is, in the most critical part of the combustion zone with respect to excitation of the first tangential mode of oscillation), with the jet diameter in the remaining injectors kept unchanged, has two consequences: 1) the damping action of a jet changes (it decreases with increasing d_{jp}); and 2) the momentum-flux ratio of oxidizing gas to fuel, $\pi = \rho_o U_o^2/\rho_f U_f^2$, and the propellant mixture ratio, $K_w = G_o/G_f$, at the wall of the combustion chamber change (it increases with increase of d_{jp}).

As compared to version 9, which is the main version in the tests of unit 2UK combustion chambers, the recess of the separation bushing of version 12 increased to a maximum value of $\Delta l = 11.5$ mm. Version 13 injectors have in four peripheral rows (out of nine) gas channels adjusted to two frequencies simultaneously $f_{1T} = 2.0$–2.2 kHz and $f_{2T} = 3.5$ kHz, which correspond to the first and second modes of tangential oscillations according to acoustic calculations. Injectors with $\bar{l}_{ef} = l_{ief}/\lambda_{1T} = 0.5$ and $\bar{l}_{ef} = l_{ief}/\lambda_{2T} = 0.5$ were installed in four peripheral rows in an alternating manner.

Injectors with a contoured choked inlet through which the oxidizing gas is fed are assigned to subgroup 3-C, as shown in Fig. 15.5 and Table 15.6. Such a design of injectors acoustically separates the combustion chamber from the gas passage because the local velocity of the oxidizing gas at the nozzle throat equals the sound velocity. Taking into consideration the specific features of the RD-170 engine (a batch of four combustion chambers combined with a common gas passage), the injectors shown in Fig. 15.5 and Table 15.6 are of great interest.

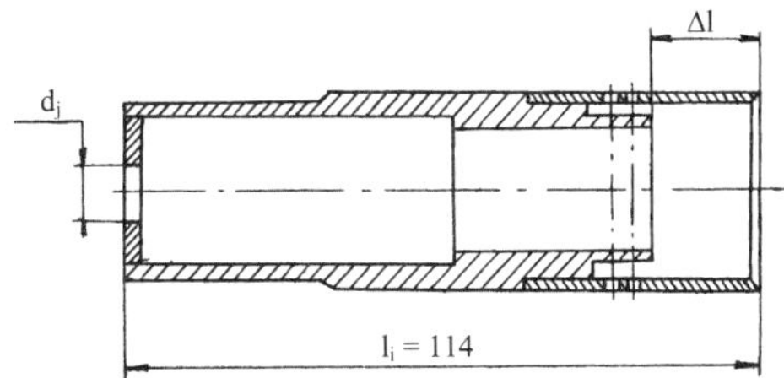

Fig. 15.4 Mixing heads with spray–centrifugal injector with jets.

Table 15.5 Mixing heads with spray–centrifugal injector with jets

Version no.	Number of main injectors n_i	Spacing between injectors t, mm	Length of injectors l_i, mm	Depth of under-cutting Δl, mm	Jet diameter in central and peripheral injectors d_{jc}/d_{jp}, mm	Number of fuel holes in injectors n_f,	Permeability for oxidizing gas $\bar{f}$, %,	Dimensionless length of injectors $\bar{l}_{ef}\, l_{i_{ef}}/\lambda$
IV-9	271	17.35	95	10.5	8/8	12	35	~0.5
IV-10	-″-	-″-	-″-	2.0	8/8.5	-″-	-″-	-″-
IV-11	-″-	-″-	95	-″-	8/9	-″-	-″-	-″-
IV-12	-″-	-″-	-″-	——	8/8	-″-	-″-	-″-
IV-13	-″-	-″-	——	——	8/8	-″-	-″-	~0.5

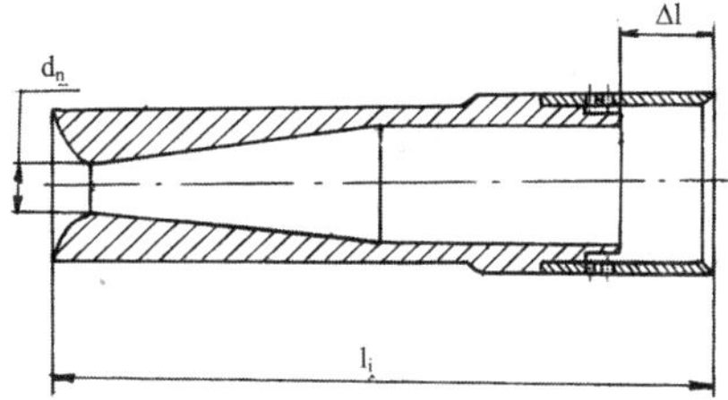

Fig. 15.5 Mixing heads with spray–centrifugal injector with sonic nozzles at inlet.

IV. Operating Process Stability for Units 1UKS and 2UK with Different Injector Heads

High-frequency oscillations, which occurred during starting at an initial stage of development of units 1UKS and 2UK, took place mainly at a frequency of $f \approx$ 2.2 kHz corresponding to the first tangential mode of oscillations f_{1T}. Thus, major consideration was given to evaluation of the stability characteristics at this frequency.

The spectral analysis has shown that there exist several harmonics close to the first tangential mode ($f \approx 2.2$ kHz). These harmonics in the amplitude spectrum of pressure pulsations in the combustion chamber have frequencies $f_{1T} = 1.95 \pm 0.05$ kHz and $f_{1T} = 2.15 \pm 0.05$ kHz. The former, as shown in Sec. V of Chapter 10, corresponds to the oscillatory system including a gas passage, injector gas channels, and part of the combustion chamber with a relatively "cold" zone near the mixing head. This frequency is less than that of a pure tangential mode. The second frequency corresponds to the oscillatory system comprising the just-mentioned regions plus the entire volume of the combustion chamber. This frequency corresponds to a longitudinal-tangential mode and is higher than that of a pure tangential mode.

Evaluation of the stability characteristics of the combustion chamber was made for these two frequencies separately. Figures 15.6 and 15.7 shows the stability characteristics of 16 different versions of mixing heads for the frequency $f_{1T} = 1.95 \pm 0.05$ kHz and $f_{1T} = 2.15 \pm 0.05$ kHz, respectively. Values of pressure decrements are plotted on the y axis, and an integrating parameter π proportional to the square root of the ratio between the momentum heads of the oxidizer to the fuel flows at the injector outlet is plotted on the x axis:

$$\pi^* = K_m \frac{\rho_f^{0.5} T_o^{0.5}}{(P_{ch} \cdot 10^4)^{0.5}} \approx \left(\frac{\rho_0 U_o^2}{\rho_f U_f^2} \right)^{0.5} = K_m \frac{\rho_f^{0.5} T_o^{0.5}}{(P_{ch} \cdot 10^4)^{0.5}} \frac{R_o^{0.5} F_f}{F_0}$$

In a physical sense, this parameter defines the mixing conditions for mixing heads of a specific geometry (F_o/F_f = const). It takes into account variations of the pressure P_{ch} and mixture ratio K_m in the combustion chamber, specific density ρ_f of fuel (during its preheating) and oxidizing gas temperature (with K_m = const).

Figure 15.6 shows that version 9 (see Fig. 15.4 and Table 15.5) with spray–centrifugal injectors (with recess of the separation bushing being $\Delta l = 10.5$ mm) adjusted to the maximum damping of oscillations at the frequency of the first tangential mode f_{1T} (i.e., $l_i = 95$ mm and $\bar{l}_{ief} = 0.5$), and having a jet diameter of

Table 15.6 Mixing heads with spray–centrifugal injector with sonic nozzles at inlet

Version no.	Number of main injectors n_i	Spacing between injectors t, mm	Length of injectors l_i, mm	Depth of under-cutting Δl, mm	Number of fuel holes in injectors n_f	Permeability for oxidizing gas $\bar{f}$, %	Diameter of sonic nozzle d_n, mm	Dimensionless length of injectors $\bar{l}_{ef}\, l_{i_{ef}}/\lambda$
V-14	271	17.35	95	10.5	12	35	6	~0.5
V-15	-″-	-″-	47.5	-″-	-″-	-″-	-″-	~0.25
V-16	-″-	-″-	-″-	-″-	8 (coarse spraying)	-″-	—	-″-

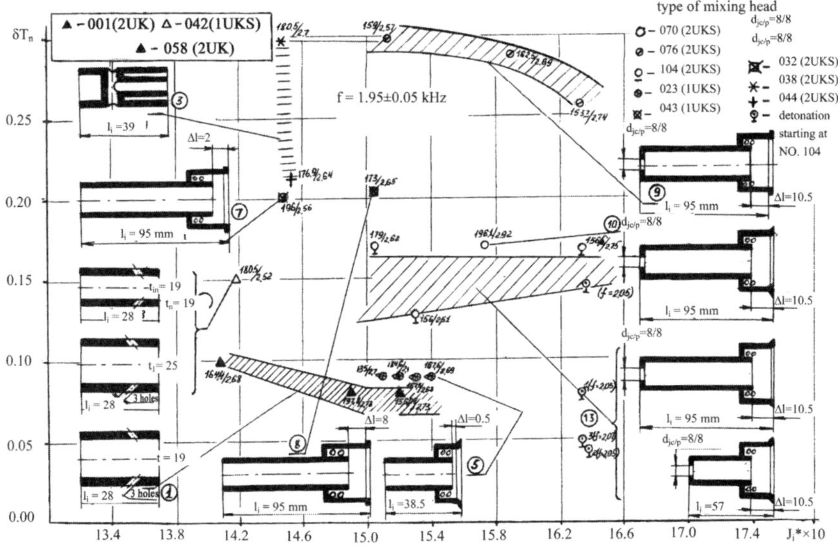

Fig. 15.6 Values of δT_n at frequency of $f = 1.95 \pm 0.05$ kHz, depending on parameter π^*.

$d_j = 8$ mm at the injector inlet (with $d_o = 11.5$ mm), has the highest stability with respect to self-excited oscillations at the frequency of the first tangential mode $f_{1T} \approx 1.95 \pm 0.05$ kHz. The oscillation decrements are $\delta T_n = 0.26 \pm 0.03$.

When the jet diameter in the peripheral row of version 10 (see Fig. 15.4 and Table 15.5) is increased to $d_{\text{jp}} = 8.5$ mm, the stability is slightly worsened. The oscillation decrement decreases and becomes $\delta T_n = 0.17$.

The oscillation decrements determined from the noise signal for version 13, in which some injectors in the four peripheral rows (out of nine) are adjusted to damp oscillations of the first tangential mode $f_{1T} = 2.0$–2.2 kHz, and other injectors are adjusted to damp oscillations of the second tangential mode $f_{2T} = 3.5$ kHz, have the same level of oscillation decrements as those for version 10 having $d_{\text{jc/p}} = 8/8.5$ mm. The δT_n for version 13 is = 0.13–0.17.

The lowest oscillation decrements determined from noise at the frequency $f_{1T} = 1.95 \pm 0.05$ kHz were obtained for version 1 with emulsion injectors (see Fig. 15.6) and for version 5 with short spray–centrifugal injectors ($\bar{l}_{\text{ef}} = 0.25$) as follows: δT_n (version 1) = 0.08–0.10; and δT_n (version 15) = 0.09.

Extension of the spray–centrifugal injectors to $\bar{l}_{\text{ief}} \approx 0.5$ in version 7 (see Fig. 15.3 and Table 15.4) with $\Delta l = 2.0$ mm improved the stability characteristics, with $\delta T_n = 0.2$. The increase of the separation bushing recess from $\Delta l = 2.0$ to 8 mm in version 8 practically did not affect the stability. The δT_n under this condition is 0.20 (see Fig. 15.6).

For injectors of the emulsion type (version 2), the combination of larger injectors in the center region and smaller injectors in the peripheral row has a positive effect on the stability. The oscillation decrement is 0.15, which is greater than that of version 1 by a factor of 1.5–2.0.

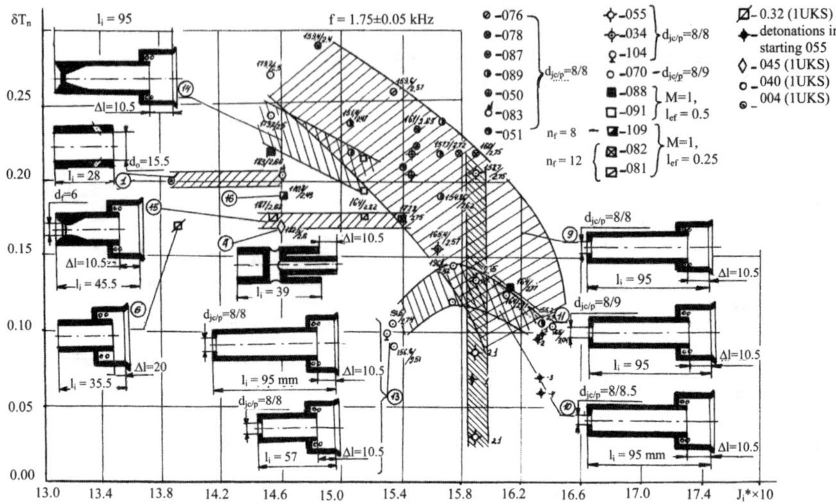

Fig. 15.7 Values of oscillation decrement δT_n at frequency of $f = 2.15 \pm 0.05$ kHz, depending on parameter π^*.

Version 3 equipped with injectors of the spray–spray type, in which the outlet of the central bushing for fuel is made flush with the exit of the injector gas channel, does not have bad stability characteristics. The oscillation decrement δT_n is 0.21–0.30 (see points + and * in Fig. 15.6).

Stability characteristics of the various versions of the mixing heads at the oscillation frequency $f_{1T} = 2.15 \pm 0.05$ kHz are shown in Fig. 15.7. For such given oscillation frequency (as well as for the frequency $f_{1T} = 1.95 \pm 0.05$ kHz), version 9 has the highest oscillation decrements determined from noise and is thus employed as the main version in the development of engine units 1UKS and 2UK (see Fig. 15.4 and Table 15.5). A tendency is observed of decreasing δT_n with increasing parameter π^*. The phenomenon occurs by reducing the pressure in the combustion chamber P_{ch} or by increasing the propellant discharge coefficient K_m, that is, with shifting to the "right lower corner" of the combustion chamber P_{ch}-K_m operating regime. Thus, with the change of parameter π^* from 14.5 to 16.5, the oscillation decrement determined from noise decreases from 0.20–0.29 to 0.11.

Such a reduction in stability with the increase of parameter π^* can be presumably accounted for by enhancing the interactions between the fuel shroud and oxidizing gas in the outlet section of the injector channel. The reduction in fuel kinetic head (i.e., momentum flux), the increase in oxidizing gas kinetic head, and the increase of the ejecting effect of the oxidizing gas spray with increasing π^* can result in separation of the fuel shroud from the injector wall and an earlier contact between the propellant components in the injector itself.

The majority of experimental points for version 9 have oscillation decrements $\delta T_n \geq 0.2$, determined from noise.

Increasing the jet diameter in peripheral injectors to $d_{jp} = 8.5$ mm (i.e., version 10) resulted in reduction in oscillation decrement δT_n to 0.09–0.15. The same

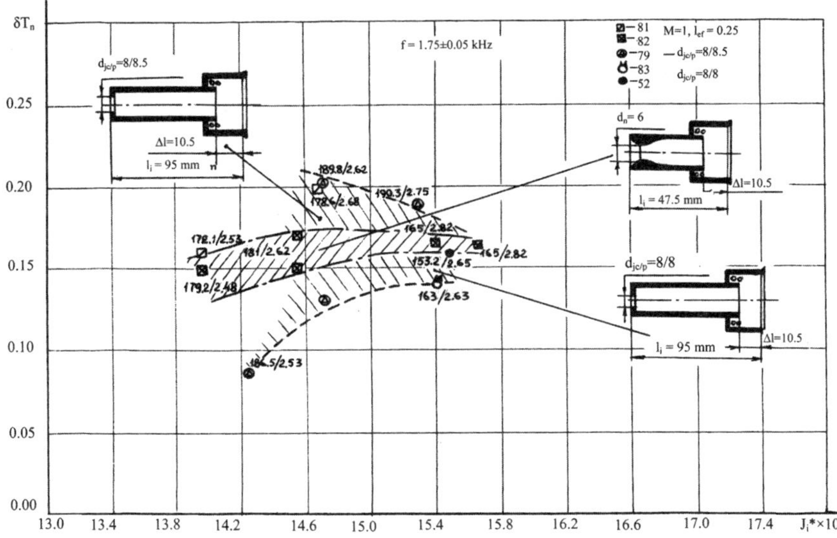

Fig. 15.8 Values of oscillation decrement δT_n at frequency of $f = 1.75 \pm 0.05$ kHz, depending on parameter π^*.

effect was also obtained in the case of using injectors of different height in version 13, for which $\delta T_n = 0.1$.

Version 14 has stability characteristics closest to the main version 9. It has injectors of the same relative length $\bar{l}_{\text{ief}} \approx 0.5$ with sonic nozzles at the inlet. The throat diameter d_n is 6 mm instead of the jet diameter $d_j = 8$ mm. For this injector head, the oscillation decrements δT_n fall in the range of 0.19–0.22 (see Fig. 15.7).

Version 15 with injectors of similar design but having a reduced gas channel of $\bar{l}_{\text{ief}} \approx 0.25$ has, in the same range of π^*, lower oscillation decrements of $\delta T_n = 0.175$ (see Fig. 15.7). The use of more coarse fuel spraying for this kind of mixing head by reducing the number of tangential holes for fuel feed (while maintaining the same total flow passage area with $F_f = \text{const}$) has a positive effect on stability. The oscillation decrement for version 16, in which the preceding modifications were made, is somewhat higher, with $\delta T_n = 019$.

Experimental studies have shown that in repeated tests of the same specimen of unit 2UK the stability characteristics are consistent. For example, in the first test of no. 087, $\delta T_n = 0.19$–0.20 for the frequency of 2.3 kHz, and in the fifth test of no. 087 of the same unit, $\delta T_n = 0.23$ for the frequency of 2.27 kHz.

V. Stability to Hard Excitation for Chambers Tested as Part of Unit 2UK

Pulse disturbances from a 4-g explosive charge were introduced for three different mixing heads under stable operating conditions in the combustion chamber of unit 2UK (see Figs. 15.8 and 15.9).

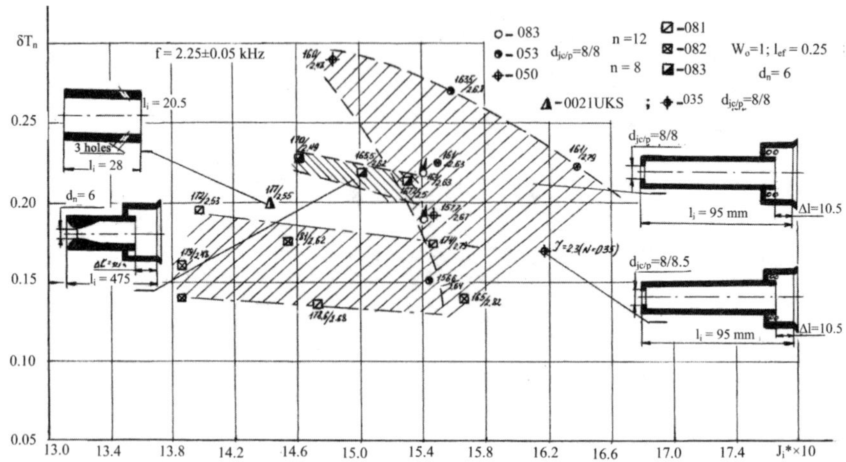

Fig. 15.9 Values of oscillation decrement δT_n at frequency of $f = 2.25 \pm 0.05$ kHz, depending on parameter π^*.

For the main version of head no. 9 with jets of $d_j = 8$ mm, the oscillation decrements from pulse disturbances at the frequency of the first tangential mode of oscillations, $f_{1T} = 2.16$–2.30 kHz, have values in the range of $\delta T = 0.06$–0.134. The maximum value of n made up ~0.98, and a relaxation time $\tau_r = 73.3$–7.5 ms.

For head version no. 12 with injectors of a different length, the decrements of damped oscillations from pulse disturbances have values of $\delta T = 0.045$–0.140. The maximum value n is greater than 0.54, and $\tau_r = 10$–3.2 ms. These stability indices were obtained by applying pulse disturbances under operating conditions for which the decrease of oscillation decrements determined from noise with reduced P_{ch} and increased K_m (i.e., the right lower corner of the P_{ch}-K_m operating regime) is typical.

In spite of the fact that the 4-g explosive charge was the maximum for the given type of disturbance device, the disturbance pulses for different heads were 3–5 bar below the pressure in the combustion chamber of 150–170 bar.

Under startup conditions ($P_{ch} = 50$–110 bar), A_0 reached 20 bar, and during tests of a combustion chamber equipped with 271 spray-centrifugal injectors with a length of 95/57 mm and jets with a diameter of 8 mm, high-frequency pressure oscillations were excited. In the case of $P_{ch} = 87$ bar and the temperature of oxidizing gas $T_0 = 10°$C, the initial pulse A_0 was 29.5 bar, and under the condition of $P_{ch} = 109$ bar and $T_0 = 80°$C, the initial pulse A_0 was 21 bar.

We consider the relation between stability characteristics under nominal conditions and combustion efficiency in the next section.

VI. Relation Between Stability Characteristics and Combustion Efficiency Under Nominal Conditions

Let us compare the stability data with the indices of operating process efficiency, in particular the flow-rate coefficient, $\varphi = \beta_m/\beta_t$ as shown in Fig. 15.10.

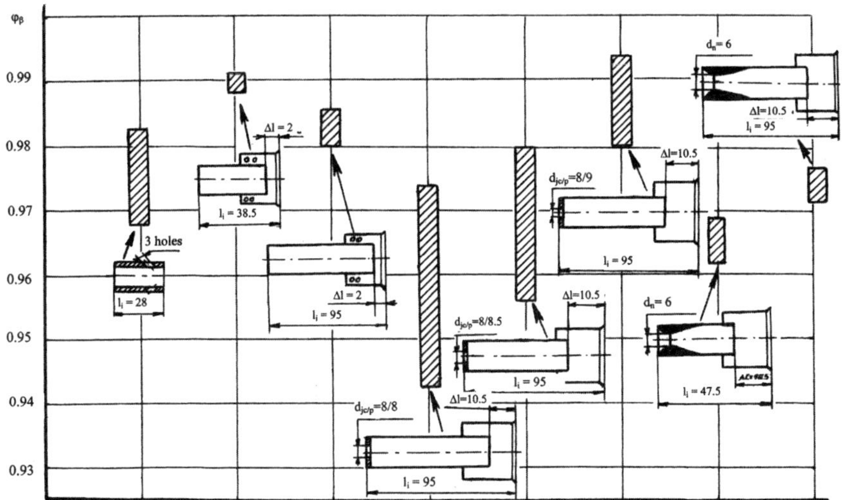

Fig. 15.10 Coefficients of flow-rate complex φ_β (K_m = 2.6 ± 0.4) for different mixing-head versions.

The mixing heads with emulsion injectors (version 1) and spray–centrifugal short (version 5) and long (version 7) injectors without jets, which have reduced values of oscillation decrements δT_n determined from noise, are characterized with large flow-rate coefficients: φ = 0.968–0.991.

The main head version 9 with jets d_j = 8 mm, which have the highest oscillation decrement δT_n determined from noise, has a wide scatter of φ with lower values of φ = 0.942–0.973. The increase in jet diameter for peripheral injectors with d_j = 8–8.5 mm slightly increased the combustion efficiency. The values of φ were 0.956–0.980. A larger jet diameter of 9 mm in the peripheral row of injectors significantly increased the efficiency indices to φ = 0.980–0.993.

Regretfully, the noted tendency for improvement of combustion efficiency with increase in the jet diameter of peripheral injectors under nominal conditions is accompanied, as mentioned earlier, with a decrease in indices of spontaneous stability. Apparently, a decrease in stability also takes place during startup because, for the head with d_j = 9 mm, large amplitudes of pressure pulsations were observed during startup.

It can be expected that increasing the jet diameter in the injectors of the peripheral row simultaneously results in two consequences worsening the stability: 1) the damping effect of jets is reduced; and 2) the mixing parameter π^* increases approximately by 30%, which, in accordance with the tendency to decreasing δT_n with increasing π^* noted earlier, should reduce the stability characteristics.

An interesting result was obtained for mixing heads with spray-centrifugal injectors of different lengths, which have sonic nozzles instead of jets at the inlet of injectors, that is, versions 14 and 15 (see Fig. 15.10). For this type of injector, the replacement of short injectors with $\bar{l}_{\text{ief}} \approx 0.25$ by long ones with $\bar{l}_{\text{ief}} \approx 0.5$ results in simultaneous increase in the value of φ (from $\varphi = 0.962$–0.969 to

0.971–0.977) and stability characteristics. The value of oscillation decrement determined from noise increases from $\delta T_n = 0.175$ to 0.195–0.220 at the frequency of the first tangential mode. This result, unusual at first glance, can be connected with transformation of an injector channel acoustically open at both ends into a channel acoustically closed at one end and the ensuing change of the standing-wave field in the combustion chamber. The pressure surge in a sonic nozzle at the inlets of injector channels also can introduce additional losses of acoustic energy.

Evaluation of stability under stable operating conditions of a combustion chamber has shown that there exists a large group of mixing head versions that provide sufficient stability margin. The highest indices were achieved by a head version accepted as the main one. This head is equipped with spray–centrifugal injectors adjusted by selection of an appropriate length to provide the maximum acoustic losses in the gas cavity of the mixing head (for oscillations at the frequency of the first tangential mode). The injectors also have jets at the inlets with $d_j =$ 8 mm for damping intra-injector oscillations. The outlet sections of injectors are made with undercutting (i.e., recess) of the separation bushings to $l = 10.5$ mm.

The influence of some design and operating parameters on stability characteristics has been revealed. At the primary frequency corresponding to the first tangential mode of oscillations, δT_n decreases, and the stability deteriorates if the following conditions occur: 1) reduction in the combustion chamber pressure $P_{\rm ch}$ and increase in the oxidizer-to-fuel ratio K, that is, shifting the operating mode to the right lower corner of operating conditions in the $P_{\rm ch}$-K coordinates; 2) increase in the jet diameter in peripheral injectors above $d_{\rm jp} = 8$ mm; 3) use of spray–centrifugal injectors with a jet diameter of $d_{\rm jp} = 8$ mm and adjustment of injector length so that they can provide maximum acoustic energy losses for the frequencies of the first and second modes of tangential oscillations simultaneously, that is, $l_i = \lambda_1 T/2 = 95$ mm and $\lambda_{2T}/2 = 57$ mm, as compared to injectors of the same type with a length of $l_i = \lambda_{1T}/2 = 95$ mm; and 4) decrease in the length of spray–centrifugal injectors having sonic nozzles instead of jets from $l_i = \lambda_{1T}/2$.

The oscillation decrements determined from noise increase and the stability are improved if the following conditions occur: 1) organizing coarse spray of fuel by means of spray–centrifugal injectors (e.g., eight tangential fuel holes instead of 12 holes with the same flow area); 2) equipping the mixing head with smaller emulsion injectors on the periphery rows and larger injectors in the center region (as compared to the baseline head version having injectors of the same size); and 3) increasing the length of spray-centrifugal injectors with a small undercutting of separation bushings to $l_i = \lambda_{1T}/2 = 95$ mm.

VII. Improvement of Stability in Unit 2UKS

Separate improvement of the RD-170 engine combustion chamber continued in unit 2UKS. Efforts were made to select the mixing head to improve the operating process stability under steady and transient conditions, specific impulse, and resistance of structural elements to thermal and mechanical loadings.

Evaluation of operating process stability in the startup mode (as well as under steady conditions) was made separately for soft and hard excitations. Analysis of operating process stability with respect to soft excitation was based on in-chamber natural noise spectra and the stability boundaries in the cases when oscillations occurred during the engine startup, even if the duration was short and the amplitude was small. The sequence of instantaneous spectrograms (for $\Delta\tau = 17$ ms) was examined in the startup phase with a duration of 0.5–1.0 s.

Evaluation of stability with respect to hard excitation during the start up transient was based on the same criteria as their counterpart under steady-state conditions. The following mixing heads were studied in unit 2UKS: mixing heads with a flat jet at the inlet of injectors, mixing heads with a tangential oxidizer inlet into the injectors, and mixing heads with vibration baffles made of protruded injectors.

In units 2UKS, the pressure range was brought closer to that set by technical specifications for the engine, and an actual oxidizing-gas temperature range was maintained. The main geometrical dimensions of the combustion chamber and gas passage of unit 2UKS are shown in Fig. 15.11. The dimensions of injectors tested on units 2UKS are shown in Fig. 15.12, and a general view of the mixing head with vibration baffles made of protruding injectors is shown in Fig. 15.13.

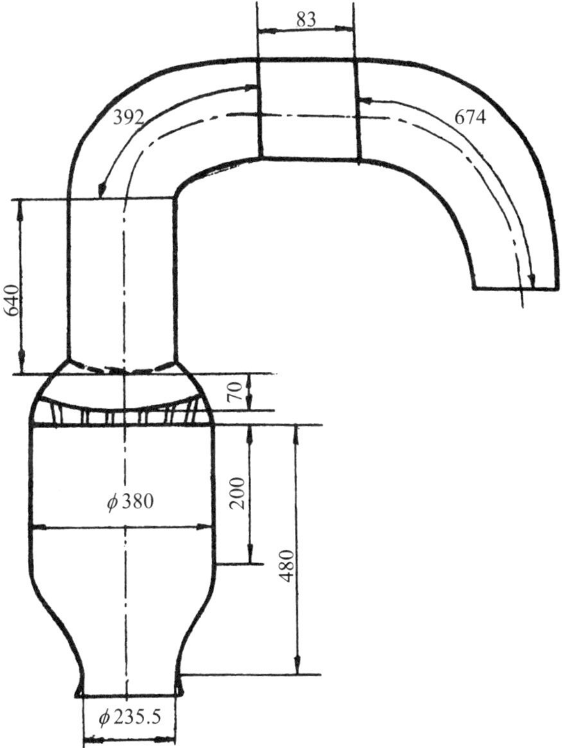

Fig. 15.11 Main geometric dimensions of unit 2UKS gas passage and combustion chambers.

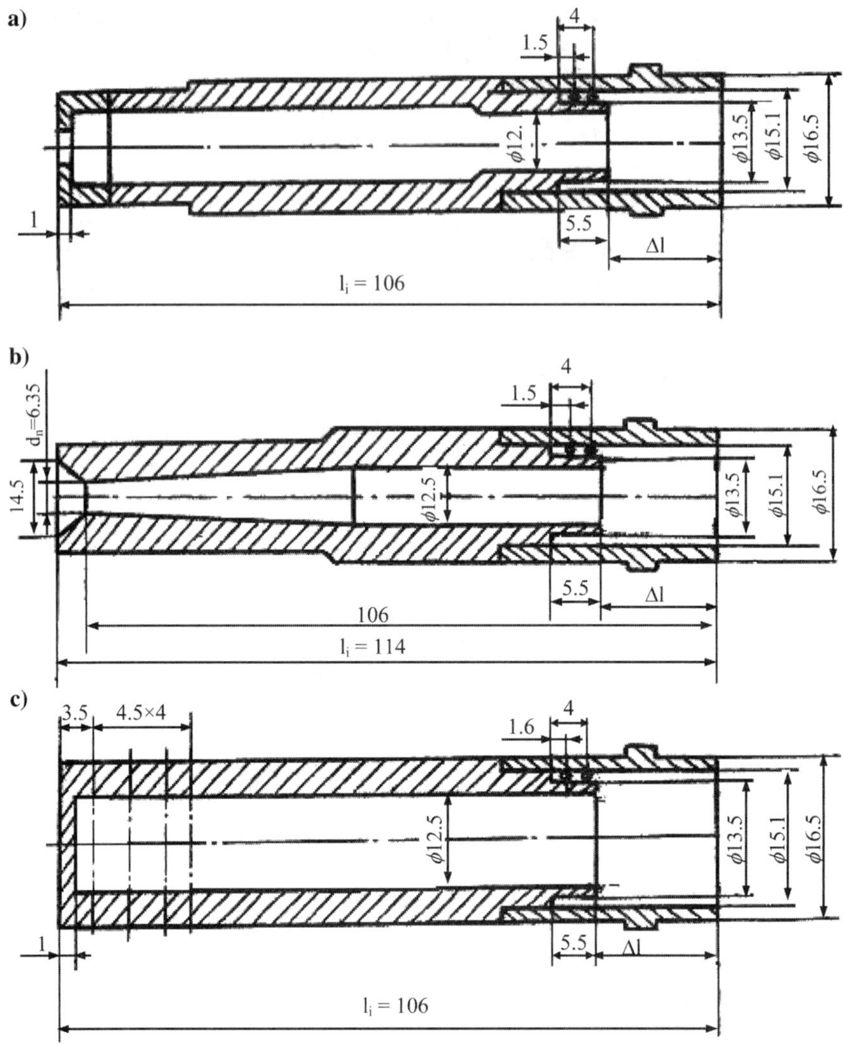

Fig. 15.12 Main geometric dimensions of unit 2UKS injectors.

A. Mixing Heads with Contoured Jets at Injector Inlet

Six bench tests on unit 2UKS were carried out with this version of mixing head. The magnetograms of four bench tests have shown that in the main operating modes the oscillation decrements at the frequency $f = 1.8$ kHz are quite high (see Fig. 15.14), and at the frequency $f = 2.5$–2.0 kHz the decrements δT_n vary in a quite wide range of 0.03–0.25. Instability indicated by a small oscillation decrement at the frequency of 2.15–2.30 kHz was observed in the area around the nominal mixture ratio $K_m = 2.6$. At the frequency $f = 1.1$ kHz under nominal

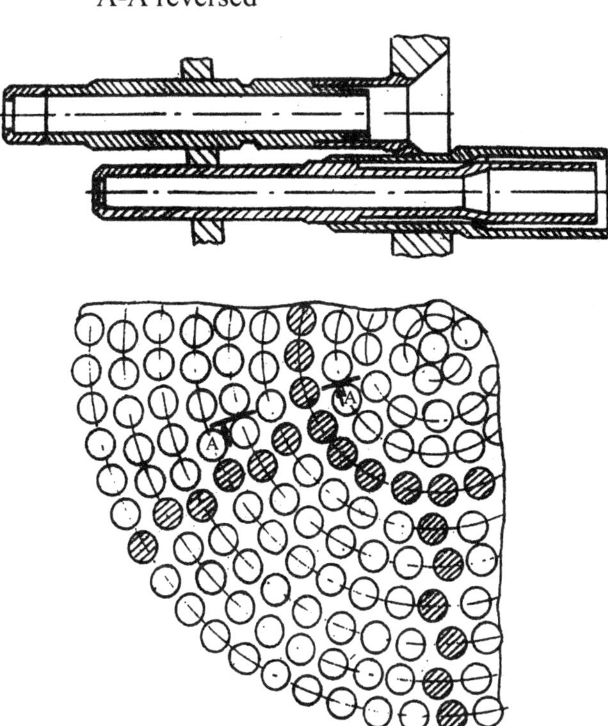

Fig. 15.13 General view of a mixing head with pulsation baffles made of manufactured injectors.

conditions, the oscillation decrement was quite small with $\delta T_n = 0$–0.02, that is, self-oscillations exist at a noise level. The oscillation decrements δT_n determined from noise indicate a large stability margin at the frequency of the first tangential mode f_{1T}, but an insufficient margin at frequencies of 2.15–2.30 and 1.1 kHz.

Tests with pulse disturbances with a 4-g explosive charge in the combustion chamber under steady operating conditions have shown that the operating process with this mixing head version has the stability margin with respect to hard excitation: $n = 5.45$, $\delta T = 0.087$, and $\tau_r = 6.9$ ms. At the same time, with the increase in the charge weight to 5 g with $n > 5.5$, the high-frequency oscillations of the first tangential mode got excited.

During the startup test with the given mixing head, practically from the very beginning of the operating process setup in the combustion chambers with $P_{\text{ch}} = 20$ bar and t_0 from −160 to −120°C, low-frequency oscillations at $f = 0.155$–0.280 kHz and high-frequency oscillations of low amplitudes ($P' = 1$–3%) were excited. The latter changed to large amplitudes with $P'_{\text{ch}} = 50$–70% at the frequency of the first tangential mode during the increases of the chamber pressure P_{ch} and the

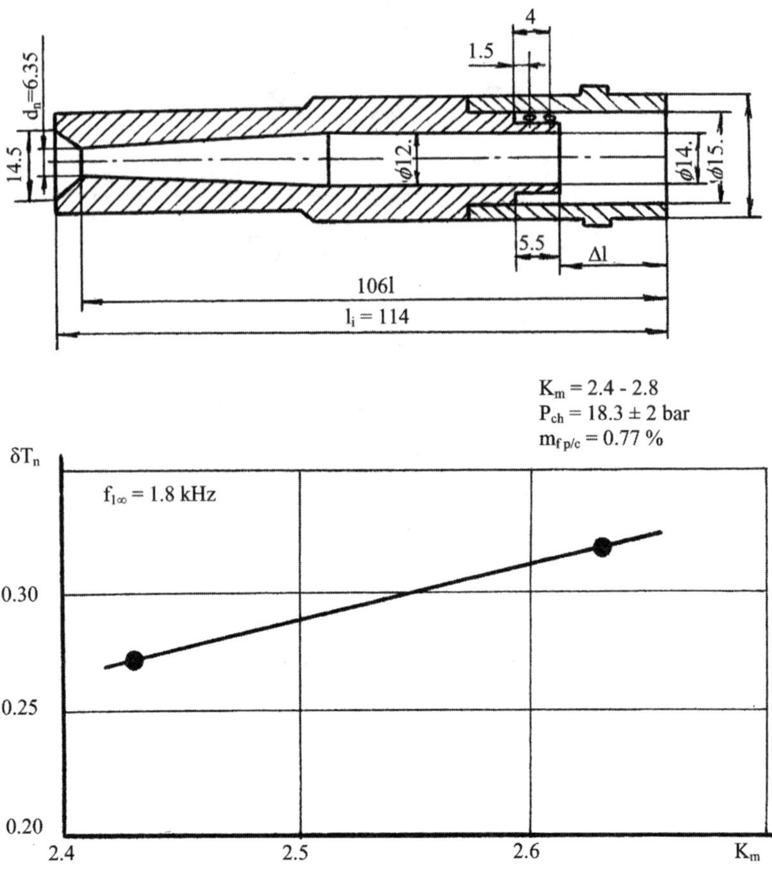

Fig. 15.14 Dependence of oscillation decrement δT_n on mixture ratio K_m.

oxidizing-gas temperature t_o. When $P_{ch} = 100$ bar and $t_o = 40$–60°C, oscillations of large amplitudes ceased. Further increase in P_{ch} and t_o to the nominal operating conditions renders the operating process stable. Such a pattern of excitation and cessation of oscillations indicates that the operating process passes the instability region during the engine startup. The instability region was quite extensive in the startup period with the given mixing-head version. High-frequency oscillations were also excited in the shutdown period of the combustion chamber and under throttling conditions.

During startup with low oxidizing-gas temperatures, the local velocity in the throat section of the contoured jet is much lower than the sound velocity. Under this condition, the acoustic-energy damping capability of a contoured jet is weaker than that of a planar jet. In light of this, as well as the complexity of the design and manufacturing of injectors with a contoured jet, mixing heads with a flat jet at the injector inlet were preferred.

B. Mixing Heads with Flat Jets at Injector Inlet

Various versions of mixing heads with flat jets at the injector inlet were employed. They differed in the jet diameter in the peripheral, preperipheral, and center-region injectors.

Also varied are the recess length of the separation bushing in the injector outlet section ($\Delta l = 10.5$, 14.5, 17.5, and 20 mm), the angle of inclination of the peripheral row of injectors ($\alpha = 0°$, 3°30′, and 4°30′), the ratio of the fuel flow rate through the peripheral injectors to that through the central injectors (G_f^p / G_f^c), and the fuel flow rate for the sheet ($G_s = 1.9$–3.5%).

Figure 15.15 shows the oscillation decrements determined from noise under nominal operating conditions in unit 2UKS, with $P_{ch} = 183 \pm 2$ bar, $K = 2.62$, and $G_{fm} = 3$–3.5%. The oscillation decrements as a function of the jet diameter in the peripheral row of injectors d_{jp} for three different oscillation frequencies are presented. For heads equipped with injectors having a separation bushing recess $\Delta l = 14.5$ mm, an increase in the jet diameter in the peripheral injectors from

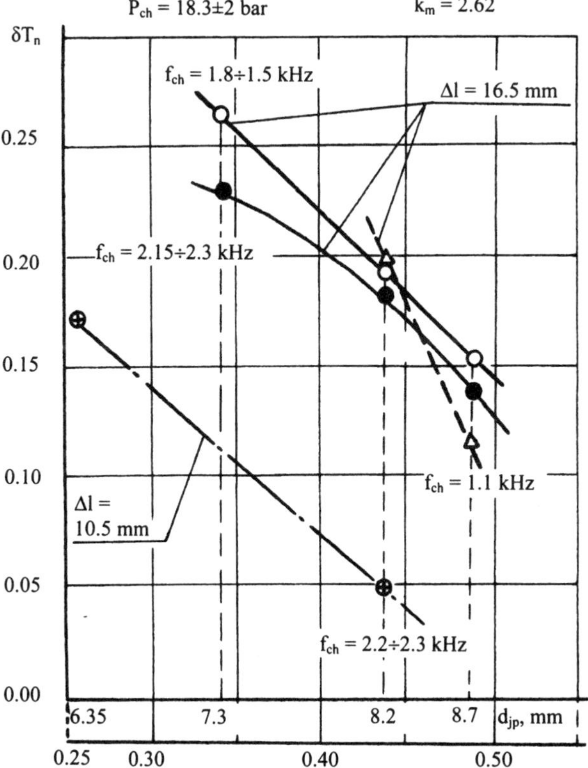

Fig. 15.15 Dependence of oscillation decrement determined from δT_n on jet diameter d_j in the peripheral injector row and on a value of separation bushing recess Δl.

$d_{jp} = 7.3$ to 8.7 mm results in a decrease in oscillation decrements at the three characteristic frequencies. In this case, a noise level increases accordingly. The maximum amplitude of noise for the lowest oscillation decrement δT_n does not exceed $A_n \approx 0.5$ bar, and in most cases is $A_n = 0.1$–0.3 bar.

The increase in oscillation decrement determined from noise, with decrease in the jet diameter in the peripheral rows of injectors, can be attributed to the stronger damping effect of the jet and a lower total fuel flow rate through the peripheral rows of injectors, which are most responsible for excitation of the first tangential mode of oscillation.

In addition, it can be seen from the plot in Fig. 15.15 that the oscillation decrement rises sharply with an increase of the separation bushing recess from $\Delta l = 10.5$ to 14.5 mm. It was impossible to trace the effect of further undercutting (i.e., more recess) on the oscillation decrement because of the lack of comparative tests differing in the parameter Δl only.

For mixing heads with a flat or contoured jet, the excitation of high-frequency oscillations at the frequency $f = 1.5$–1.6 kHz occurred during engine startup. In most cases, these oscillations were preceded by low-frequency oscillations. A significant improvement of stability with respect to excitation of high-frequency oscillations during engine startup was achieved by emulsification of fuel using a certain rate of gaseous nitrogen fed through the jets to the mixing head inlet ($d_{j.N2} = 2$ mm and $P_{N2} = 150$ bar).

Stability with respect to hard excitation for mixing heads with a flat jet was estimated for two different head versions: version 1 with $d_j = 9.1$ mm, $\Delta l = 14.5$ mm, and $\alpha = 4°30'$; and version 2 with $d_{j,p} = 7.3$ mm, $d_{j,\,p/p} = 10$ mm, $d_c = 9.3$ mm, and $\Delta l = 20$ mm.

For version 1, two pulse disturbances were applied from an explosive charge of 2 g and two pulse disturbances from an explosive charge of 3 g under nominal conditions of unit 2UKS. The diameter of the combustion-chamber nipple connected with the IDD outlet channel was 6 mm. All four pulses decayed quickly. The combustion-chamber relaxation time was $\tau_r = 2.92$–4.12 ms. The average oscillation decrement was $\delta T = 0.19$–0.22. The maximum relative disturbance value was $n = 1.62$. For mixing-head version 2, stability with respect to hard excitation was evaluated in the startup mode. In these and subsequent tests, pulse disturbances were applied through the 10-mm-diam nipple of the combustion chamber. During the engine startup, two pulse disturbances were applied from an explosive charge of 1 g and two disturbances from a charge of 2 g. All four pulses damped quickly, with relaxation time of $\tau_r = 1$–3 ms. The average oscillation decrement lies in the range of $\delta T = 0.20$–0.47. The maximum relative disturbance value is $n = 4.02$.

Stability characteristics obtained under startup conditions (with fuel emulsification: $d_{j,N2} = 3$ mm and $P_{N2} = 150$ bar) indicate a high stability margin for head no. 2 with respect to excitation of high-frequency oscillations within the oxidizing-gas temperature range of T_o from −5 to 180°C.

Studies have shown that a large group of mixing-head versions with a flat jet at the injector inlet provide high stability for both soft and hard excitations of high-frequency oscillations under nominal conditions. Those head versions, however, did not provide stability during the engine startup. Additional measures aimed at stability during the engine startup must be worked out.

C. Mixing Head with Tangential Feed of Oxidizer into Injector

It was shown experimentally that fuel emulsification by means of a metered flow of gaseous nitrogen during the engine startup, which intensifies the operating process, has a favorable effect on stability. The mixing process can also be intensified by swirling the oxidizer. Thus, various injector versions with swirling oxidizing gas flows were studied on a single-injector model unit, and one version was also tested in actual unit 2UKS. The main geometrical characteristics of this version are $\Delta l = 14.5$ mm, $l_i = 106$ mm, $n_o = 30$, $d_o = 2$ mm, and $\delta_w = 1.75$ mm as shown by the bottom plot in Fig. 15.12. The center region of the mixing head was equipped with 271 injectors of this version, and 60 injectors were installed on the peripheral row. They differed from the central injectors having smaller diameter fuel holes. As a result, the fuel flow through those injectors was 30% lower than that through the central injectors.

Model tests of injectors under simulated startup conditions have shown that, as compared to an earlier injector version with spray feed of oxidizer ($l_i = 106$ mm, $d_j = 9.1$ mm, and $\Delta l = 10.5$ mm), the injector version with swirling oxidizer (in the center region of an actual mixing head) provides a stable operating process in the operating regime with chamber pressures P_{ch} and oxidizing gas T_0 at which high-frequency pressure oscillations occur and persist in the case of spray feed of oxidizer.

One actual test of the just-mentioned mixing head with tangential feed of oxidizer into injectors also has shown the absence of high-frequency oscillations during engine startup in the operating regime of oxidizing gas pressures and temperatures under which high-frequency oscillations occur and persist during the test of mixing heads with spray feed of oxidizer without emulsification and without vibration baffles in unit 2UKS. More specifically, self-oscillations were excited, as a rule, at $P_{ch} = 20$–30 bar and $T_0 = -160$–120°C (the lower boundary of the instability region). They existed as the pressure rose and the oxidizing gas temperature increased. When $P_{ch} = 100$ bar and $T_0 = -20$–40°C were attained, oscillations ceased (the hysteresis boundary). During this test, high-frequency oscillations occurred during the engine startup at much higher pressure P_{ch} and temperature T_0. Thus, the first burst of high-frequency oscillations occurred at $P_{ch} = 140$ bar and $T_0 = 120$°C, and after 0.05 s, these oscillations decayed. Excitation of oscillations took place at $P_{ch} = 169$ bar and $T_0 = 220$°C, that is, at the parameters close to their values under nominal operating conditions of unit 2UKS ($P_{ch} = 180$ bar, $T_0 = 250$°C). Hence, swirling of oxidizing gas, while intensifying the mixing process during the engine startup stage, shifted the instability region to higher pressures and temperatures of oxidizing gas. The switch from a spray injector on the oxidizer line to a centrifugal one changed the acoustic properties of the injector channel and subsequently reduced the damping action of gas channels, which were previously adjusted to the frequency of the first tangential mode of oscillations.

D. Mixing Head with Vibration Baffles Made of Protruded Injectors

Sixteen tests of combustion chambers with mixing heads having vibration baffles made of injectors protruding by 40 mm into the combustion chamber were carried out on unit 2UKS, as shown in Fig. 15.16. The protruded injectors forming baffles were made of steel or bronze. The tests of mixing heads with baffles have

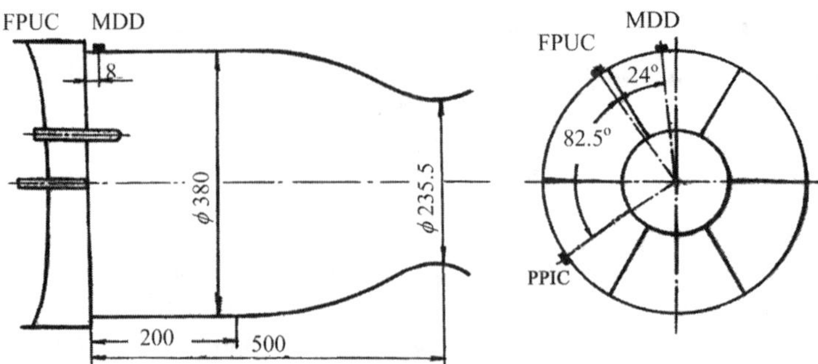

Fig. 15.16 Geometric characteristics of combustion chamber and layout of measuring devices and nipple for disturbance introduction.

shown that the oscillation decrements δT_n determined from noise at the frequency of the first tangential mode, $f_{1T} = 1.6 \pm 0.1$ kHz, have quite high values of $\delta T_n = 0.14$–0.40. This indicates a high stability margin with respect to soft excitation of oscillations at this frequency under nominal operating conditions. Distinct harmonic in the noise spectrum at the frequency of the first longitudinal mode, $f_{1L} = 1.1$ kHz, was practically absent for mixing heads with baffles.

Stability with respect to hard excitation in combustion chambers with vibration baffles was evaluated in four tests by applying pulse disturbances with charges of 2 and 3 g under nominal and startup conditions.

The following stability characteristics with respect to hard excitation were obtained from tests for mixing heads with baffles made of steel injectors. Disturbances were generated from explosive charges of 2 and 3 g.

Under startup conditions:

$$n = 3.23\text{–}7.7; \quad \tau_r = 1.6\text{–}7.8 \text{ ms}; \quad \delta T = 0.10\text{–}0.35$$

Under nominal conditions:

$$n > 2.84; \quad \tau_r = 1.99\text{–}3.94 \text{ ms}; \quad \delta T = 0.14\text{–}0.27$$

where $n = A_0/2A_{nm}$. The stability characteristics obtained provide evidence of a sufficient margin of stability with respect to hard excitation under nominal and startup conditions for mixing heads with baffles made of steel injectors.

The following stability characteristics with respect to hard excitation were obtained for mixing heads with baffles made of bronze injectors.

Under startup conditions:

$$n = 2.5\text{–}20; \quad \tau_r = 1.32 \text{ ms}; \quad \delta T = 0.4$$

Under nominal conditions:

$$n = 8\text{–}9; \quad \tau_r = 2.26\text{–}2.39 \text{ ms}; \quad \delta T = 0.26 \pm 0.01$$

The preceding results indicate that application of cooled vibration baffles from protruding injectors eliminate high-frequency oscillations during the startup period and provide a sufficient stability margin with respect to hard excitation under nominal and startup conditions.

Thus, the use of accepted quantitative stability characteristics made it possible to select the best version of mixing head in terms of operating process stability with respect to high-frequency oscillations and to evaluate the influences of design and operating parameters on stability based on tests of combustion chamber on units 1UKS, 2UK, and 2UKS. Among all tested mixing heads, the one with acoustically adjusted spray–centrifugal injectors having flat jets at the inlet of the gas cavity has the best characteristics with respect to soft and hard excitations.

The following stability characteristics with respect to soft excitation (i.e., oscillation decrements δT_n determined from noise) and to hard excitation (i.e., relative disturbance value, $n = A_0/2A_{nm}$; relaxation time τ_r; average decrement of damping of artificially initiated oscillations δT) were obtained for most combustion chambers with afterburning of the oxidizing products from gas generators.

Under nominal conditions—stability to soft excitation:

$$\delta T_n = 0.14\text{–}0.40 \quad \text{at} \quad f_{1T} = 1.6 \pm 0.1 \text{ kHz}$$

$$\delta T_n = 0.12\text{–}0.22 \quad \text{at} \quad f_{1T1L} = 2.2 \pm 0.1 \text{ kHz}$$

$$\delta T_n > 0.4 \quad \text{at} \quad f_{1L} = 1.1 \text{ kHz}$$

Under nominal conditions—stability to hard excitation:

$$n = 8\text{–}9$$

$$\tau_r = 2.3\text{–}2.4 \text{ ms}$$

$$\delta T = 0.26 \pm 0.01$$

Under startup conditions:

$$n = 2.5\text{–}20$$

$$\tau_r = 1.35 \text{ ms}$$

$$\delta T > 0.4$$

This mixing head with baffles provides a sufficient stability margin, results in a prespecified value of the specific impulse, and provides design reliability in strength characteristics and cooling. As a consequence, this type of mixing head was selected as the baseline version for tests in the RD-170 engine and for serial production.

VIII. Evaluation of Stability in Combustion Chamber as Part of Engine

The results of selecting the optimal injector head for the RD-170 engine are presented in Secs. I–VII of this chapter. Test results for this type of mixing head in an engine, including stability with respect to hard excitation, are briefly described in this section.

As mentioned earlier, the mixing head with spray–centrifugal injectors and vibration baffles formed by injectors protruding into the combustion bottom was selected as the baseline version. A schematic diagram of the mixing head, chamber flow scheme, and layouts of pressure-pulsation and vibration transducers is shown in Fig. 15.16. A total of 331 bipropellant injectors are installed in the mixing head, one at the center and the others on 10 concentric circles with an increasing number of injectors from row to row by six injectors. Fuel is fed through the centrifugal stage of an injector, and oxidizing gas is fed over the 12.5-mm-diam central channel with a flat jet having diameter of 9.1 mm and a thickness of 1 mm at the inlet. For injectors not used to form baffles, fuel swirling is provided by tangential holes with a diameter of 1.2 mm arranged in two rows with six holes in each row. The fuel swirling cavity is separated from the oxidizer by a separation bushing, with its end-face being recessed by 10.5 mm with respect to the injector outlet section. The swirling cavity ends with countersink with a vertex angle of 80 deg made on the combustion bottom body. The total length of injectors is 106 mm. The injectors forming vibration baffles protrude into the combustion bottom surface by 40 mm. They differ from other injectors only by the organization of fuel swirling with the help of spiral channels between the casing and the separation bushing and by the absence of separation bushing recess. At the same time, the spiral channels are the cooling lines of the injectors. The necessity for accurate checking of stability in the RD-170 engine was based on the following reasons: 1) tests of an operational combustion chamber in the unit 2UKS were carried out at a pressure of 85% of the nominal condition, and 2) in the RD-170 engine the oxidizing gas from the turbine exhaust manifold enters into four combustion chambers.

Concerns were thus raised that such an arrangement would result in acoustic coupling between the chambers and thereby would worsen stability, especially in the region of low frequencies of 80–320 Hz. As already noted, in this frequency range, short time-pressure pulsations occurred for the following reasons: 1) differences in stability characteristics under startup and other transient conditions; and 2) during engine tests, the combustion chambers are subject to pulsations and vibrations from the turbopump unit with amplitude and frequency ranges different from those in unit 2UKS, which could also affect the overall stability characteristics.

Stability was evaluated for typical conditions: engine startup, steady nominal conditions, smooth staged throttling conditions with a total of 3–13 stages and duration of 24 s, final-stage conditions, and engine shutdown.

In addition, stability was evaluated under high-power conditions at a pressure of up to 105%, under nominal conditions with variation of the oxidizer-to-fuel ratio within ±7% of the nominal value of 2.6, at different fuel temperatures at the engine inlet, at different fuel pressure drops on injectors beyond the tolerance limit (±20%), and under the influence of other factors.

Let us consider some estimation data on stability with respect to soft excitation of oscillations. Stability was estimated from oscillation decrements determined from noise in the combustion chamber. The oscillations and pressure-pulsation decrements were measured by a transducer mounted in the fuel cavity of the injector head. The measurements in the combustion chamber correlate well with the measurements upstream of injectors on the fuel line. During engine startup, similar to the tests of combustion chambers on unit 2UKS, there occurred low-frequency oscillations, which ceased when the engine reached its steady operation mode. The oscillation frequency was 100–170 Hz, and the maximum amplitude is 4–8 bar at the turbopump unit rotational speed of 5500–6500 rev/min.

This condition corresponds to the combustion-chamber pressure of 30–60 bar at the oxidizing gas temperature downstream of the turbine of –60–60°C. The patterns of development of low-frequency oscillations in different combustion chambers of the same engine differed. This is an indicator of weak acoustic coupling between the chambers over the same oxidizing gas line. Both soft and hard excitations of low-frequency pulsations were realized. As the engine reached nominal conditions, the minimum values of oscillation decrements at acoustic frequencies approached the average value of ~0.1 for all chambers, and the frequencies with minimum decrements approached the cutoff frequency range of the 33-blade turbine rotor. Oscillation at the natural acoustic frequency of the first tangential mode of ~1.8 kHz, which was regarded as the most dangerous in the startup mode, was not observed in the combustion chamber. As a result of engine tests under steady operating conditions for stability evaluation, depending on the preceding factors, it was shown that stability characteristics within the measurement accuracy practically do not change. The spectral distribution of pressure pulsations had a pattern of broadband noise with the maximum amplitude components of 0.5–0.6 bar at the frequency of 7–8 kHz, which is in the range of the blade frequency of the turbine wheel. Among a list of parameters to be checked for their influence on stability, only the pressure drop over the fuel line showed a statistically significant effect.

Stability with respect to hard excitation of pressure oscillations in the combustion chambers as part of the RD-170 engine was evaluated on two engine specimens. The combustion chambers of these two engines were additionally equipped with a nipple with an internal diameter of 10 mm for disturbance introduction and with an assembly for attaching an external disturbance device. Five-charge MDD with charges operating in low-speed detonation regime were used for disturbance introduction (see Chapter 10). The MDD used in this study was designed for an operating pressure of 600 bar. The mass of explosive charges in all tests was 0.6 g. During the tests, the channels for introduction of the disturbance were purged with inert gas with a pressure drop of 2–3 bar. The relative positions of the disturbance-device nipple, pressure-pulsation and vibration transducers, and vibration baffles are shown in Fig. 15.16. The location of the disturbance introduction nipple at 8 mm from the combustion bottom is determined by the combustor design features and the necessity to install the nipple at a distance of 40–45 mm, as required for optimal disturbance introduction. In the first test, all chambers were equipped with disturbance devices. In the second test, only one chamber was equipped with the disturbance device. A total of 17 disturbance pulses were applied to combustion chambers, of which 11 pulses were applied at different stages during the startup period and six pulses under different steady conditions.

The data characterizing the stability margin with respect to hard excitation were determined from pressure measurements upstream of fuel injectors. The use of such measurement instead of its in-chamber counterpart is justified by the fact that, based on the experience of unit 2UKS tests, the stability characteristics with respect to hard excitation of pressure oscillations determined from the pre-injector and in-chamber measurements were similar. Particular emphasis was placed on the estimation of stability with respect to hard excitation during the engine startup. To improve the reliability of the stability estimation for different conditions, the

Table 15.7 Results of stability characteristics with respect to hard excitation of oscillations

Date	No. of chamber	Time of disturbance introduction, s	G_3, g	P_{ch}, bar	T_o, °C	K'_m	A_0, bar	$2A_n$, bar	$N = A_0/2A_n$	τ_r, ms	A_{max}, bar	$N_1 = A_{max}/2A_n$
13 Jan 1984	II	2.419	0.6	20	−50	6.11	16.8	4.20	4.00	3.37	16.79	4.00
	III	2.468	0.6	30	−59	6.32	19.8	6.67	3.50	1.54	22.68	4.00
	II	2.568	0.6	52	0	4.44	47.2	5.20	9.00	2.62	47.21	9.00
	III	2.668	0.6	62	61	3.93	28.4	5.67	5.00	4.71	33.08	5.83
	II	2.717	0.6	63	61	4.36	39.9	5.20	7.60	3.27	41.96	8.00
	III	2.767	0.6	64	61	4.22	29.3	5.67	5.17	3.28	32.14	5.67
	II	2.867	0.6	76	69	4.76	40.9	4.20	9.70	4.58	40.91	9.75
	III	2.917	0.6	85	83	4.70	34.0	4.73	7.20	4.71	40.65	8.60
	II	2.970	0.6	93	104	4.41	31.5	5.20	6.00	2.95	43.01	8.20
	III	3016	0.6	98.5	121	3.99	32.7	4.73	6.80	2.99	41.60	8.80
	IV	4.118	0.6	216	386	2.10	8.8	2.93	3.00	2.14	28.50	9.75
3 Aug 1984	II	25.027	0.6	251.3	394	2.59	8.2	4.4	1.86	9.51	9.42	2.14
	II	31.027	0.6	250.1	433	2.40	6.7	3.77	1.78	3.53	8.37	2.22
	II	39.027	0.6	253.6	385	2.86	6.5	4.10	1.60	10.7	10.0	2.44

disturbance introduction to different chambers was duplicated. The interval between disturbances was 0.05 s. To avoid imposition of a new disturbance onto an undamped pressure oscillation arising from the previous disturbance and to maintain an interval of 0.05 s, the disturbances were introduced in turn to different chambers. Table 15.7 summarizes the characteristics of oscillations after the introduction of disturbances, along with the operating parameters, including the chamber pressure, the oxidizing gas temperature after the turbine, and the propellant mixture ratio at the moment of disturbance introduction. It is seen from Table 15.7 that all disturbances decayed. The relaxation time was 1.5–4.7 ms. The relative disturbance value is 3–9.7 for the first peak $n = A_0/2A_n$ and 4–17 for the maximum disturbance amplitude $n_1 = A_{max}/2A_n$. In the tests of the RD-170 engine, the relative value of the pressure disturbance was determined from the relation to the maximum noise amplitude in the test section before the disturbance introduction. If the disturbance values were recalculated not in terms of A_{rms}, but with $A_{n\,max}$, we would get $n = A_0/A_{rms} \approx 12$–38, and $n_1 = 16$–64, which generally meets the requirements of $15\,A_{rms} < A_0 < 25\,A_{rms}$ for the disturbance value (see Chapter 10). Therefore the combustion process in the chambers is stable with respect to hard excitation of high-frequency pressure oscillations during the entire startup period. Low-frequency oscillations excited by disturbances (~160 Hz) decayed with the pressure rise in the combustion chamber, as also observed under conditions without externally impressed disturbances.

The disturbances introduced into one combustion chamber were not seen in the background noise in pressure pulsations or in vibrations measured in other combustion chambers. Thus, the gas-path acoustic coupling, as well as structural mechanical coupling between separate chambers, is weak and practically does not affect the stability characteristics. The basic harmonic of ~1.6 kHz in the frequency spectrum of damped pressure pulsations corresponds to the frequency of the first tangential mode. The characteristic frequencies of 3.5 and 7 kHz for vibrations do not coincide with characteristic frequency of 1.6 kHz of pressure pulsations. The results confirmed the conclusion obtained in the tests with a standard mixing head as part of unit 2UKS about a good stability margin in the startup mode. The estimation of stability with respect to hard excitation under steady conditions was made from the testing of two engines. The parameters of stable operating conditions (chamber pressure, oxidizing-gas temperature, oxidizer-to-fuel ratio) and impressed disturbances are given in Table 15.7. The relaxation time is $\tau_r = 3$–10 ms under conditions of $P_{ch} = 126$–254 bar, $K_m = 2.10$–2.85, and the oxidizing gas temperature of 385–433°C. In this case, the relative disturbance n is 2–3.6 and ~10 in the test with $P_{ch} = 126$ bar. The increase in relative disturbance with the pressure rise in combustion chambers is connected with the use of charges of the same weight for all operating conditions of engines. Thus, the quantitative characteristics of stability in the combustion chambers of the RD-170 engine with standard mixing heads provide a sufficient stability margin in serial production. All statistical data of subsequent engine tests have confirmed the conclusions obtained.

Chapter 16

Stability Characteristics of Engines with Adjustable Injectors

HIGH stability has been achieved for the RD-120 and RD-170 engines using oxygen and kerosene. The mixing heads have adjustable injectors and vibration baffles made of injectors protruding into the chamber (see Chapter 15). Those features allowed application of the same type of mixing-head circuit for engines running on NT UDMH components.

LRE RD-0208 of the Proton launcher was used for the experiments. Engines RD-0208/RD-0209 are designed for the second stage of the three-stage launcher Proton [12]. Three RD-0208 engines (Fig. 16.1) and one RD-0209 are installed.

RD-0208 and RD-0209 are single-chamber liquid rocket engines. The propellant components are nitrogen tetroxide and UDNM. The engines have the following parameters: $P_v = 575$ kN (58.6 tf); $I_v = 3200$ m/s; $P_{ch} = 14.7$ MPa; $t = 150$ s; $M_{en} = 540$ kg (RD-0208), 560 kg (RD-0209); propellant mixture ratio is 2.6; thrust range is ±6%; and mixture ratio range is ±10%.

The engine is made on a closed circuit (i.e., staged-combustion cycle). Gas produced in the gas generator with a high excess of oxidizer is used as a turbine working fluid. A standard mixing-head design for engine RD-208 is shown in Chapter 17.

A second-stage engine RD-120 of launcher Zenit, which has a chamber with a mixing system similar to that of engine RD-170 and closer in dimensions to engine RD-0208, was chosen as a prototype of a mixing-head design. RD-120 was the first Russian LRE to pass stand firing tests in the United States, in October 1995. The propellants of engine RD-120 [12] (see Fig. 16.2) are liquid oxygen and kerosene, and the following: $P_v = 833$ kN (85 tf); I_v =350 s; $t =$ up to 360 s; and $P_{ch} = 162$ MPa.

The mixing-head design (Fig. 16.3) was similar for engines RD-170 and RD-120. It comprised spray–centrifugal injectors with a length of the gas channel equal to half the wavelength of the tangential mode. The gas channel had a jet at the inlet. Some injectors protrude into the combustion cavity and serve as baffles. The calculations and design of the head were made by KBHA and NIIHIMMASh.

However, the requirement for maintaining the injector pressure drop for oxidizing gas and fuel precluded full implementation of the proposed design. To provide optimal damping of oscillations, the oxidizing gas injectors should have

Fig. 16.1 Liquid rocket engine RD-0208 [12].

an effective length l_{ef} equal to half the wavelength of the acoustic wave in the generator gas and a jet at the inlet: $l_{ie} = \lambda/2$.

An injector is adjusted to damp the first tangential mode of oscillations arising in the combustion chamber of engine RD-0208. For high-frequency pressure oscillations in the chamber, the frequency and wavelength are $f_{1T} = 2480$ Hz (see Chapter 17) and $\lambda = C_o/f_{1T} = 444/2480 = 0.179$ m, where C_o is the sound velocity in the oxidizing gas in the injector (see Table 16.1). Hence $l_{ef} = \lambda/2 = 179/2 = 89.5$ mm.

The attached gas volume at the injector ends also participates in oscillations in the injector channel and should be taken into account. The geometrical length l_i of the injector channel required for damping a respective oscillation mode will be shorter than the effective length l_{ef} by a value of ~0.8 for the injector diameter:

$$l_i = l_{ef} - 0.8\, d_o = 89.5 - 0.8 \times 10 = 81.5 \text{ mm}$$

According to studies [45], to provide the damping, an optimal jet area ratio at the injector inlet should be

$$\overline{F}_n = \frac{F_n}{F_0} = 0.4 \cdots 0.55$$

Fig. 16.2 Liquid rocket engine RD-120 [12].

For the given injector

$$D_n = \sqrt{d_0^2 F_n} = \sqrt{10^2 \times 0.5} \approx 7\,\text{mm}$$

The injector gas channel diameter $d_0 = 10$ mm was selected on the basis of the chamber diameter ratio of engines RD-120 and RD-0208.

$$d_0 = d_{0\text{RD-120}} \times \frac{D_{\text{chRD-0208}}}{D_{\text{chRD-120}}} = 11.5 \times \frac{276}{320} = 9.93 \approx 10\,\text{mm}$$

Engineering development of the injector head made in KBHA, with consideration for installation in engine RD-0208, is shown in Fig. 16.3 and in Table 16.1. Two-hundred-seventeen injectors are mounted on the injector head; of them, 48 injectors protrude 30 mm into the combustion chamber and form baffles.

A gap between the external surfaces of the protruding injectors is 0.3–10 mm. Only these 48 injectors are adjusted to damp high-frequency oscillations in the combustion chamber. To provide uniform distribution of propellant components throughout the injector head, the fuel flow through the 48 protruding injectors is reduced by half, proportional to the reduction in the oxidizer flow caused by installation of jets. To improve power characteristics of the combustion chamber, the fuel flow through the wall row injectors is reduced by approximately 25%.

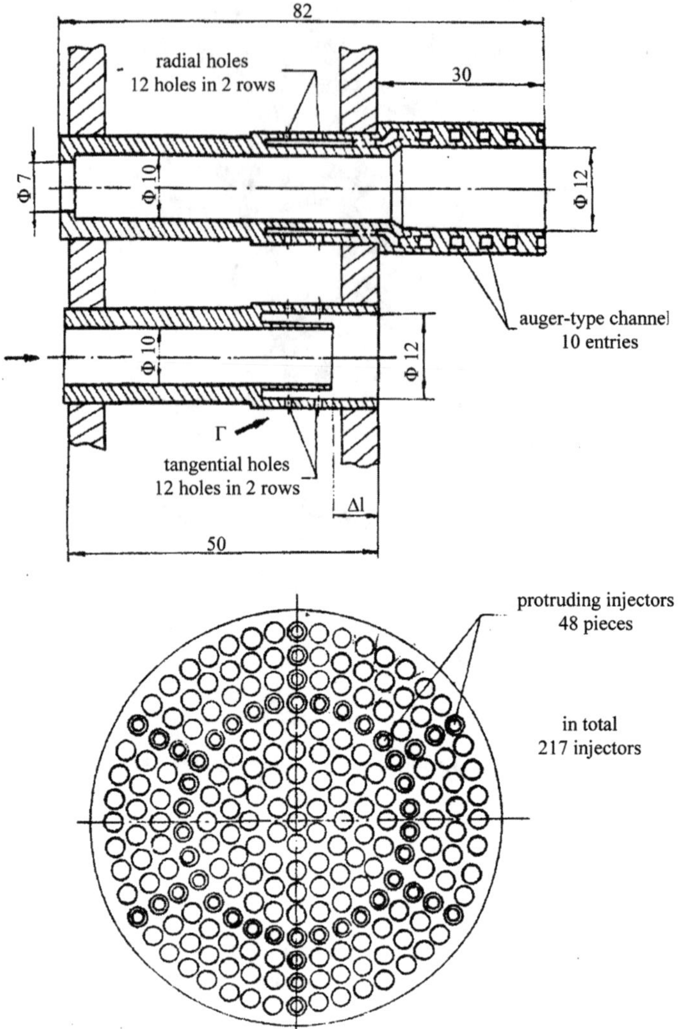

Fig. 16.3 Diagram of injectors and their arrangement in the mixing head.

Fuel injectors are made as follows. The main injectors are fitted with tangential inlet channels and with an annular nozzle; the elongated injectors are made with auger-type outlet channels up to the edge of the injector. To ensure effective protection of the protruding injectors, the spray angle of the main injectors is $2\alpha = 90$ deg, corresponding to the geometrical characteristic $A = 2.8$ and flow coefficient $\mu = 0.23$. A length of the area of contact between fuel and oxidizer inside the injector has been selected by recalculation of the $\Delta L/d_0$ relation for engine RD-120 and made up of about 9 mm. The bottom of the injector head was made from BRH-08-type bronze alloy; the injectors were made from the suitable

Table 16.1 Main operating and design parameters of combustion chambers of engines RD-0208 and RD-120 and experimental mixing head for chamber RD-0208

Description	Designation	Parameter units	Parameter value	
			RD-0208	RD-120
1	2	3	4	5
Diameter of combustion chamber	d_{ch}	mm	276	320
Length of chamber: cylindrical part	$L_{ch,\ cyl}$	mm	179	187.2
Up to critical section (throat)	$L_{cr,\ \Sigma}$	mm	356.7	490
Pressure in chamber	P_{ch}	MPa	14.7	16.2
Specific pressure pulse (relative velocity)	β(s)	s	170.7	181
Parameters of combustion products in chamber: Temperature	T_{ch}	K	3485	3860
Gas constant	R_{ch}	kgm/kg·deg	34.5	34.9
Polytropic index	χ		1.166	1.14
Sound velocity	C_{ch}	m/s	1170	1190
Acoustic frequency of 1st tangential mode	F	kHz	2.48	2.18
Total flow of propellant components	M_{Σ}	kg/s	181.8	242.9
Relative flow ratio	q	g/s: cm^2×bar	2.03	183
Parameters of oxidizing gas:				
Oxidizer temperature upstream of chamber injectors	T_o(OTUCI)	K	721	670
Gas constant	R_0	kgm/kg·deg	23.2	27.6
Sound velocity	C_O	m/s	444	455
Total number of injectors	n_i	pieces	217	271
Number of injectors in vibration baffles	N_{baffle}	pieces	48	54
Length of injectors-vibration baffles	L_{i1}	mm	81.5	95
Length of main injectors	L_{i2}	mm	50 (169 pcs)	95
Diameter of oxidizing gas injectors	d_0	mm	10	11.5
Amount of separator sleeve cutting	Δl	mm	9	10.5

steel. The external surfaces of the injectors protruding into the combustion cavity were coated with zirconium dioxide. In other respects, the combustion chamber was of standard design.

Two tests were carried out on this engine: the first test had a duration of 421 s; the second test lasted 261 s. The values of the relative velocity calculated from the measured pressure in the gas passage were obtained during the first test. It was 3.2 s lower than the values specified in the engine design specification. Before the second test the chamber was modified. The pressure measurement directly near the mixing-head hot side was introduced, so that the axis of the 4-mm measuring channel was made perpendicular to the hot cylindrical wall of the chamber. With the aim of increasing the engine specific thrust, the flow of fuel through the belt of the second sheet was reduced by closing 24 holes of 49. The value of relative velocity obtained in the second test, averaged for three nominal operating conditions and repeated during startup, was equal to 172 s. This value meets the engine design specification requirements, 170.7 ± 2.7 s. The data obtained have shown that the value of relative velocity calculated from the pressure in the gas passage is 3.5 s lower than the value directly measured in the chamber. It has also been found from measurements that the reduction in the flow of fuel in the second sheet belt resulted in an increase of 1.0 s in the relative velocity. Also, the reduction in flow of fuel per sheet increased its heating in the combustion chamber cooling line under nominal conditions by 6°C.

A five-pulse disturbance device was installed in the combustion chamber for evaluation of stability to hard excitation. The weight of the explosive charge ranged from 0.6 to 1.4 g. The quantitative characteristics of the operating process stability obtained during the first and the second tests are shown in Tables 16.2–16.4.

In the first test (Table 16.2), 16 engine stable operating conditions (2–17) were realized for various combinations of P_{ch} and K. As mentioned earlier, the oscillations in engine RD-0208 occurred at a frequency of 2400 Hz. It can be seen from Table 16.2 that there is practically no rise in the spectrum at this frequency. Decrements of oscillations in the chamber, according to measurements of HFPC (high-frequency pulsations in the chamber) by transducers, were 0.36 to >0.4; according to measurements of FPUCI (fuel pulsations upstream of chamber injectors) by transducers upstream of fuel injectors, the decrements were 0.22 to >0.35.

It can be seen from the table that as a result of extension of the injector channels and installation of baffles, the frequency of the first tangential mode decreased to 2100 Hz. The data shown are also in agreement with experimental data obtained on other engines.

The values of decrements at a frequency of 2100 Hz obtained from pressure pulsation measurements in the combustion chamber varied in the range of 0.16 to 0.26. Measurements of pressure pulsations upstream of fuel injectors gave decrements in the range of 0.07 to 0.15. Eleven stable operating conditions were realized. In this case the values of decrements at a frequency of 2100 Hz obtained from measurements by the transducer on the combustion chamber varied in the range 0.15 to 0.21 (Table 16.3). The values obtained from measurements by the pressure transducer upstream of the fuel injectors were in the range 0. 12 to 0.19. For the standard design of the combustion-chamber bead (see Chapter 17), the decrements obtained from measurements by the pressure transducer upstream of the fuel injectors were approximately 0.05 to 0.10 at a frequency of 2400 Hz. Test conditions were varied within the limits of the guaranteed square. It follows from Tables 16.2 and 16.3 that the presence of 30-mm vibration baffles made from the adjusted injectors caused deformation of the acoustic field. In addition to the

Table 16.2 Stability parameters ($\delta T, A, f$) under different operating conditions of engine RD-0208 with mixing heads with adjustable injectors (test no. 1)

						$f = 2.1$ kHz						$f = 2.4$ kHz					
						HFPC[a]			PFUCI[b]			HFPCI			PFUCI		
Ref. no.	$\tau_I – \tau_f$	P_{ch}	K_m	P_{ch}	K_m	f	δT	A	f	δT	A	f	δT	A	f	δT	A
2	23–40	148.5	2.60	N	N	2.1	0.19	0.252	2.1	0.1	0.218	2.42	0.38	0.402	2.2	0.35	0.122
3	42–60	138.8	2.60	–	N	2.0	0.16	0.217	2.0	0.1	0.223	2.42	0.39	0.268	2.5	0.35	0.106
4	64–80	160.0	2.59	+	N	2.1	0.21	0.301	2.1	0.2	0.213	2.42	0.38	0.463	2.3	0.35	0.128
5	85–100	159.8	2.79	+	+	2.1	0.19	0.267	2.1	0.2	0.203	2.40	0.37	0.474	2.4	0.35	0.141
6	111–120	160.6	2.34	+	–	2.1	0.24	0.405	2.1	0.1	0.25	2.35	0.39	0.494	2.2	0.34	0.183
7	123–145	139.4	2.33	–	–	2.1	0.18	0.311	2.1	0.1	0.298	2.38	0.40	0.268	2.1	0.22	0.162
8	155–165	138.8	2.80	–	+	2.0	0.22	0.207	2.0	0.1	0.198	2.42	0.38	0.216	2.6	0.35	0.104
9	167.5–180	148.8	2.80	N	+	2.1	0.21	0.227	2.1	0.1	0.213	2.42	0.36	0.34	2.4	0.35	0.107
10	191–200	149.7	2.33	N	-	2.1	0.22	0.37	2.1	0.1	0.297	2.30	0.38	0.391	2.2	0.22	0.187
11	206–230	149.3	2.58	N	N	2.1	0.2	0.252	2.1	0.1	0.238	2.40	0.39	0.299	2.2	0.35	0.125
12	311–325	149.0	2.58	N	N	2.1	0.23	0.267	2.1	0.1	0.238	2.40	0.38	0.288	2.2	0.35	0.125
13	328–345	155.2	2.45	+	–	2.1	0.25	0.434	2.1	0.1	0.26	2.38	0.39	0.36	2.8	0.30	0.15
14	352–375	154.5	2.79	+	+	2.1	0.37	0.35	2.1	0.2	0.205	2.40	0.37	0.35	2.4	0.35	0.11
15	383–385	148.9	2.44	N	–	2.1	0.26	0.335	2.0	0.1	0.26	2.42	0.40	0.23	2.5	0.35	0.107
16	393–405	148.6	2.80	N	+	2.0	0.3	0.232	2.0	0.1	0.2	2.42	0.37	0.24	2.7	0.35	0.107
17	411.4–421	149.1	2.58	N	N	2.1	0.26	0.286	2.1	0.1	0.238	2.42	0.40	0.27	2.2	0.35	0.122

[a]High-frequency pulsations in the chamber. [b]Fuel pulsations upstream of chamber injectors.

Table 16.3 Stability parameters ($\delta T, A, f$) under different operating conditions of engine RD-0208 with mixing heads with adjustable injectors (test no. 2)

Ref. no.	$\tau_H - \tau_K$, s	P_{ch}, MPa	K	P_{ch}	K	HFPC[a] f	HFPC[a] δT	HFPC[a] A	FPUCI[b] f	FPUCI[b] δT	FPUCI[b] A	OPUCI[c] f	OPUCI[c] δT	OPUCI[c] A
2	20–40	15	2.6	N	N	2.1	0.2	0.2	2.1	0.14	0.19	2.1	0.3	0.1
3	40–60	16	2.8	+	+	2.1	0.2	0.2	2.1	0.16	0.18	2.1	0.3	0.2
4	60–85	16	2.3	+	–	2.1	0.2	0.2	2.1	0.16	0.19	2.1	0.3	0.1
5	85–105	14	2.3	–	–	2.1	0.2	0.3	2.0	0.10	0.21	2.0	0.3	0.1
6	105–130	14	2.8	–	+	2.0	0.2	0.2	2.0	0.17	0.16	2.0	0.3	0.1
7	130–155	15	2.6	N	N	2.1	0.2	0.2	2.1	0.13	0.19	2.1	0.3	0.1
8	155–175	16	2.8	+	+	2.1	0.2	0.2	2.1	0.15	0.18	2.1	0.3	0.2
9	175–195	16	2.3	+	–	2.1	0.2	0.2	2.1	0.13	0.20	2.1	0.3	0.1
10	195–215	15	2.3	–	–	2.1	0.2	0.2	2.1	0.10	0.22	2.1	0.3	0.1
11	215–235	15	2.8	–	+	2.0	0.2	0.2	2.0	0.16	0.17	2.0	0.3	0.1
12	235–260	13	2.6	N	N	2.1	0.2	0.2	2.1	0.13	0.18	2.1	0.3	0.1

[a]High-frequency pulsations in the chamber. [b]Fuel pulsations upstream of chamber injectors. [c]Oxidizer pulsations upstream of chamber injectors.

Table 16.4 Characteristics of operating process stability to hard excitation of engine RD-0208 with adjustable injectors (test no. 2)

Parameter	G_{expl}, g	τ, s	P_{ch}, MPa	K	A_{mr}, MPa	A_M, MPa	A_M/A_{mr}	$\tau \times 10^{-3}$, s	F_l, kHz	A_1, MPa	F_2, kHz	A_2, MPa	F_3, kHz	A_3, MPa
HFPC	0.6	150	15.3	2.57	0.033	3.5	105	3.68	0.8	0.675	2.15	0.587	2.5	0.54
	1.0	170	15.9	2.77	0.029	2.2	74	1.55	1.7	0.245	2.3	0.214	—	—
	1.0	210	14.7	2.32	0.021	4.0	190	2.28	0.55	1.67	2.3	1.31	5.3	0.35
	1.0	250	15.3	2.57	0.034	3.0	88	0.736	1.25	0.77	2.15	0.6	2.3	0.6
	1.4	255	15.3	2.57	0.034	4.0	118	2.28	0.55	1.69	2.2	1.31	5.0	0.29

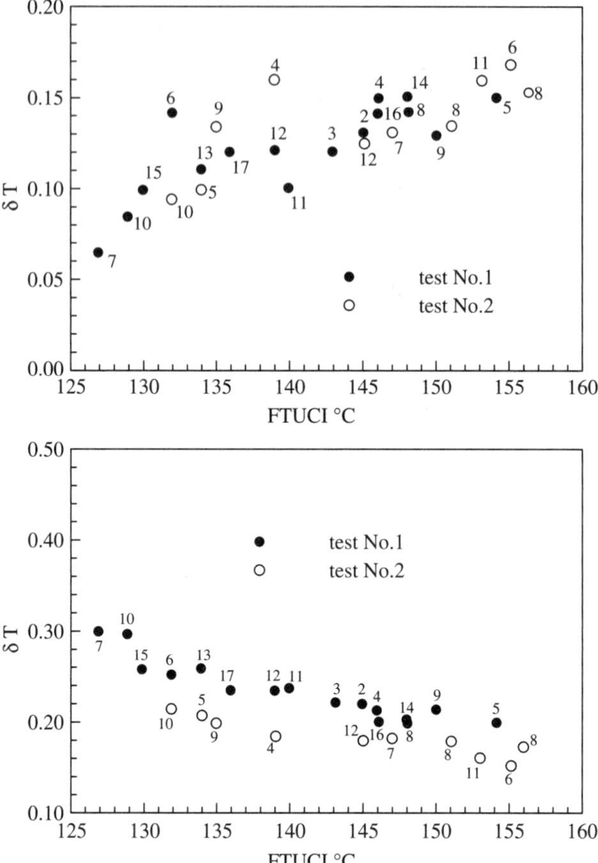

Fig. 16.4 Oscillation decrements and amplitudes as functions of fuel temperature, $f = 2.1$ KHz. The numbers correspond to the operating conditions in Table 16.2.

tangential mode at a frequency of 2100 Hz, oscillations persisted at a frequency of 2400 Hz with a high decrement. The increase of pressure oscillation decrements at 2100 Hz during the second startup results from elevated temperature of the fuel fed to the mixing head; 6°C was typical for this engine.

The pressure oscillation decrements and oscillation amplitude as a function of a fuel temperature at the chamber mixing bead inlet is shown in Fig. 16.4. A similar relation has been obtained for the standard mixing head. The characteristics of stability to hard excitation in engine RD-0208 with adjusted injectors are shown in Table 16.4. The stability estimation was made with increasing weight of explosive: 0.6, 1.0, and 1.4 g under conditions with nominal values of parameters; and 1.0 g under conditions of P_{chmin}, K_{min}, and P_{chmax}, K_{max}. In this case it was recorded that the relaxation time was lower than 3.7 ms, which is less than that of standard injector head with emulsion injectors. The test results obtained proved that the mixing system used in engines RD-170, RD-172, RD-180, and RD-120 for oxygen–kerosene components can be applied successfully to gas–liquid engines running on NT-UDMH components. The quantitative characteristics obtained in the tests described in this chapter can be significantly improved by adjusting all injectors to the appropriate frequencies in the combustion chamber.

Chapter 17

Control of Stability in Production of the Proton Engine

LET us consider the methods for estimating stability during production of LRE. According to experience, changes in the stability margin are observed in production of LRE, consisting of both combustion chambers and gas generators. These changes, as a rule, are associated with the changes of manufacturing techniques, design improvement of combustion chambers and gas generators, and, in some cases, modifications of other LRE components during production. Therefore, in production of engines, the control of parameters characterizing the stability margin during engine development is required. As an example, we consider the second-stage rocket for insertion of the spacecraft Proton into an Earth satellite orbit. The engine works in a closed loop with a staged-combustion cycle. The combustion products from the gas generator with excess oxidizer are used as the working medium for the TPU turbine.

The propellant components are nitrogen tetroxide (NTO) as oxidizer and unsymmetrical dimethylhydrazine (UDMH) as fuel. Chapter 16 gives the RD-0208 engine parameters. The RD-0212 engine [12] of the third-stage rocket Proton consists of the main RD-0213 engine and control engine RD-214. The RD-0213 engine is similar to the RD-0211 engine developed on the basis of the RD-0209 engine. The engines for the second and third stages of the Proton rocket have the same design of the injection head. The combustion-chamber injection head accepted for production comprises 61 injectors of an umbrella type, with the main fuel flow introduced into the oxidizing gas flow.

The engines have operated since 1965. In the period of manufacturing and operation of the engines from 1965 to 1987, over 1100 qualification tests were carried out, out of which 536 engines have passed flight tests. There was no high-frequency instability during the development tests of engines of standard design. Special control of the stability margin by existing procedures was not performed during engineering development. For the first time, high-frequency instability of the engine appeared during a flight test in 1972 after 332 qualification tests had been carried out. Over the period of manufacturing and tests of engines of the baseline design, eight cases of high-frequency pressure oscillations in the chambers were noted, of which four cases occurred during flight tests, one case during an investigation bench test of the engine and three cases during inspection sampling test (IST). In the period from the end of 1985 to November 1986, instabilities appeared more frequently.

There were four cases, one during a flight test and three during inspection sampling tests. Excitation of high-frequency oscillations occurred in the main-stage operation at a frequency of 2400 Hz, corresponding to the first tangential mode of pressure oscillations in the chamber. High-frequency instability has been recorded in the last six out of eight cases after the engine attained its main-stage operation ($\tau = 1.62–7.24$ s). As studies have shown, these cases have a common cause. Efforts were applied to investigate the causes of reduced stability margin, take measures to increase the stability margin, and make relevant modifications to ensure consistency in production of combustion chambers. The excitation of high-frequency oscillations in two cases at $\tau = 191.77$ s in 1972 and $\tau = 202.92$ s in 1975 was caused by leakage from soldered connections in the chamber mixing head. After some design and technical iterations on improvement of the mixing head, the soldering quality had been improved. There were no remarks on the condition of soldered connections during the production check and troubleshooting of engine equipment after tests. Engine serviceability was checked during ISTs and special tests (SVT) from an engine lot. Analysis of pressure pulsations and vibrations based on amplitude characteristics was carried out according to the operation specification and did not reveal any changes of the operating process. To study the factors responsible for the occurrence of high-frequency pressure oscillations in combustion chambers, control of the stability margin by the following parameters was implemented: control of oscillation decrements by pressure pulsations upstream of fuel injectors at a frequency of ~2.4 kHz under stable operating conditions of the engine. The pressure pulsations upstream of fuel injectors for this combustion chamber design correspond to pulsations in the gas cavity of the combustion chamber. To provide the required stability margin, the oscillation decrement δT should not be less than 0.06 during the inspection sampling test over the range of operating parameters and 0.05 with the change of operating conditions over a wider range during special tests. The minimum values of oscillation decrements determined from noise comply with the design requirements. In addition, control of oscillation amplitudes by the same parameter was introduced during the initial 10 s of engine operation after attaining the main stage with limitation of the maximum value of A less than 1.5 kgf/cm^2. The decrease of oscillation decrements is followed by the amplitude rise at this frequency.

The investigation of variations of pressure oscillation decrements made it possible to reveal the factors responsible for the increased occurrence of high-frequency oscillations during flight and bench tests, as well as to select a variant of the mixing head, which has a sufficient stability margin. Determination of pressure oscillation decrements from measurements in the fuel cavity upstream of injectors gave the following results:

1) Analysis of measured data during inspection sampling tests and special tests of engines has shown a reduction in the average minimum value of pressure oscillation decrement at a frequency of 2.4 kHz from $\delta T = 0.046$ in the period from 1983 through November 1985 to 0.033 in the period from December 1985 through 1986. The measured results characterizing the reduction in the stability margin are in full agreement with the increasing number of cases in which high-frequency oscillations occurred during flight and bench tests. The lowest values of oscillation decrements were obtained under conditions with minimum oxidizer-to-fuel ratios at the nominal and elevated pressures in the combustion chamber.

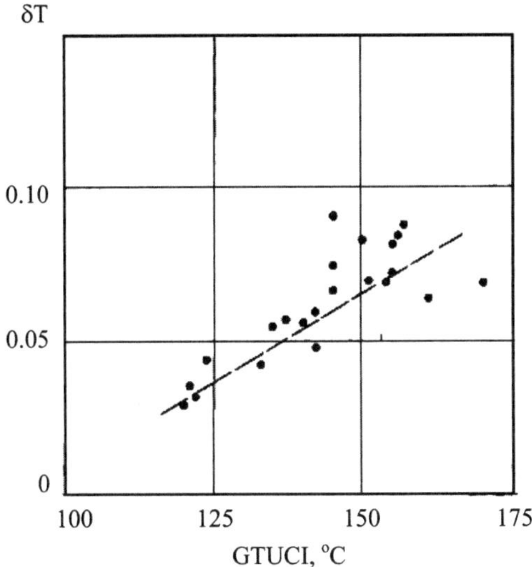

Fig. 17.1 Dependence of oscillation decrement on fuel temperature at mixing head inlet.

2) Analysis of oscillation decrements as a function of the fuel temperature at the mixing head inlet was carried out. On engine operation with a minimum oxidizer-to-fuel ratio K_m and chamber pressure (i.e., operation with the lowest high-frequency stability margin), the dependence of δT on the fuel temperature at the mixing head inlet is observed, as shown in Fig. 17.1.

With reduction in fuel temperature, the oscillation decrement decreases essentially. This dependence is less pronounced when the engine operates under nominal conditions, and the influence of fuel temperature on oscillation decrement does not manifest itself under most stable operating conditions. The measured data for the reduction in stability margin in the course of production offered an explanation of more frequent occurrence of high-frequency oscillations in combustion chambers during flight and bench tests. Estimations have shown a high sensitivity to disturbances introduced into the combustion chamber, and the entire range of engine operating conditions lies in the hysteresis region. Hence the observed cases of instability resulted from an insufficient stability margin because the excitation of oscillations has probabilistic mechanisms (see Sec. III of Chapter 3 and Sec. II of Chapter 12). The phenomenon manifests itself under unfavorable combination of external and internal factors.

Excitation of oscillations by probabilistic mechanisms is confirmed by the following factors:

1) The operating process has low values of oscillation decrements, that is, excitation of oscillations occurs near the instability boundary.

2) The conditions under which the transition from noise to spontaneous oscillations occurs are located in the hysteresis zone.

3) The reduction in oscillation decrement is not observed until spontaneous oscillations start developing.

4) The amplitude rise has a "tulip" shape.

Let us discuss the factors responsible for reduction in the stability margin during production. The following modifications in the engine design were made from 1977 to 1985, which could affect the process in the combustion chamber and reduce the stability margin. To reduce the difference of stability margin between combustion-chamber specimens, the following alterations were made in the design and production specifications for combustion chambers:

1) 1977—Orientation of injectors in the mixing head was introduced; the tolerance for the angle of fuel-feed spray holes in injectors was made more stringent from +4°C to +2°C.

2) 1978—A water temperature control system was introduced during water-flow tests of injectors.

3) 1979—Refinement of propellant holes in injectors was performed on a holder stabilizing the position of injectors.

4) 1984—The water-flow test of injectors was transferred to a semi-automatic stand.

5) 1984—Checkout water-flow test was introduced of blanket rings after applying thermal-protective coating and moisture-protective varnish.

The introduction of design and technical modifications concerning the injectors and mixing heads generally reduced the statistical value of pressure drop over the fuel line in the mixing head by 0.26 bar. The introduction of the checkout water-flow test of the blanket rings after applying the coating stabilized their operation and avoided the reduction in the coolant flow rate due to clogging of the holes and grooves of blanket rings.

In addition, modifications in the TPU design were made as follows:

1) 1981—An automatic device for the TPU axial unloading was introduced.

2) 1985—Thickening of the blade leading edge of pump impeller 0 from $0.9_{-0,2}$ mm to $1.5_{-0,3}$ mm and reduction of the pump impeller G from 183 to 182 mm were made.

The TPU modifications resulted in a drop of the pump head by 2%, which caused an increase in the temperature of the oxidizing gas upstream of the injectors by 25°C and the rotary speed by 300 rev/min. In addition, after the TPU modifications had been made, the error of the oxidizer-to-fuel ratio setting of the engine increased by 1.5% toward the side of decreasing K_m.

A detailed study of the combustor mixing system with modifications confirmed the improved consistency in production. These modifications, however, resulted in higher excess fuel in the near-wall region. That happens because of flow of fuel into this layer from the centrifugal stage of injectors, from the pair of spray holes oriented toward the chamber wall in peripheral injectors, and through the blanket ring located near the propellant combustion zone.

With the presence of those additional random factors, such as the flow of additional fuel into the near-wall layer, stability was lost irrespective of the engine operating conditions. Occurrences of high-frequency oscillations are confirmed by engine tests. Design parameters responsible for this phenomenon are the holes in the peripheral part of the mixing head, for feeding additional fuel and distributing it within the combustion zone in the peripheral part of the chamber. During tests of engines, high-frequency oscillations occurred in the chamber. Hence, one way to increase the stability margin was reduction of the excess fuel in the near-wall region. The change of the oxidizer-to-fuel ratio in the near-wall region by plugging one of the four fuel spray holes oriented toward the chamber wall in the 12 injectors (next but one) of the peripheral row of the mixing head resulted in a significant increase in the stability margin as compared to the baseline design. Decrements of oscillations determined from the noise in the combustion chamber increased to 0.062–0.145 within the engine operation range with required performance (see Fig. 17.13). Although the modifications made in TPU during serial production resulted in changes of the operating process (e.g., increases in the oxidizing gas temperature by 25°C, the TPU rotational speed by 300 rev/min, and hence the flicker frequency of blades), these studies gave no way of finding a clear relation to stability characteristics. The stability characteristics selected for this engine made it possible to choose the head designs: 1) for change of a serial head (from the stock of manufactured combustion chambers and engines); and 2) for introduction of changes into the design specifications for further serial production.

Through practical examples, descriptions will be given of the investigation of various modified combustion chambers with the baseline mixing head for the purpose of increasing the stability margin. All design modifications were based on

Table 17.1 Methods for increasing stability in chambers having the baseline mixing head[a]

Ref. no.	Measure	Change of stability margin[b]	Change of economical efficiency
	Distribution of fuel within the combustion zone		
1	Holes for fuel spray feed into the combustion zone are made in the central part of the head in the combustion bottom between standard two-stage injectors (Fig. 17.2)	Stability margin did not increase (Fig. 17.3)	Reduction in economical efficiency
2	Spray injectors in the peripheral zone of the mixing head between the peripheral and the second row of injectors (Fig. 17.4)	Stability margin decreased. Out of five tests in two cases the high-frequency oscillations occurred.	Reduction in economical efficiency
	Distribution of oxidizing gas within the combustion zone		
3	Increasing the oxidizer concentration in the flame of some injectors from 2.94 to 3.53 by closing one of four spray holes in injectors for fuel feed (Fig. 17.5)	Stability characteristics did not improve (Fig. 17.6)	Significant reduction in economical efficiency

(Continued)

Table 17.1 (Continued) Methods for increasing stability in chambers having the baseline mixing head[a]

Ref. no.	Measure	Change of stability margin[b]	Change of economical efficiency
	Distribution of fuel and oxidizing gas within the combustion zone over the mixing head zone		
4	Extension of the combustion zone by organization of flames enriched with oxidizer from injectors of the peripheral or the second row (plugging of one spray hole in injectors) and drilling of holes in the combustion bottom between peripheral and second rows (Fig. 17.7)	During the first test, low values of oscillation decrements were obtained (Fig. 17.8). During the second test, HFO in transient conditions occurred.	Low economical efficiency
	Transfer of the fuel mixing and burning zone beyond the injectors of the peripheral row		
5	Sleeves forming with the internal wall an annular fuel cavity were installed in the injectors of peripheral row. Oxidizing gas and fuel entered into the chamber from injectors in the form of concurrent sprays that resulted in the extension of the combustion zone (Fig. 17.9)	Stability margin increased as compared to the baseline design (Fig. 17.10)	Significant reduction in a level of economical efficiency
	Change of oxidizer-to-fuel ratio in the near-wall region of the chamber		
6	One of four fuel feed spray holes oriented on the chamber wal l was closed in 12 injectors (next but one) of the peripheral row of the mixing head (Fig. 17.11).	Essential increase in the margin of operating process stability (Fig. 17.12 and 17.13)	Economical efficiency is somewhat higher than that of a baseline mixing head.

[a]Only the main designs of the tested mixing heads, which defined the decision making, are shown in the table.

[b]Stability margin was estimated from the change of oscillation decrements at a frequency of ~2400 Hz (on the first tangential mode) in the control range during IST and SVT.

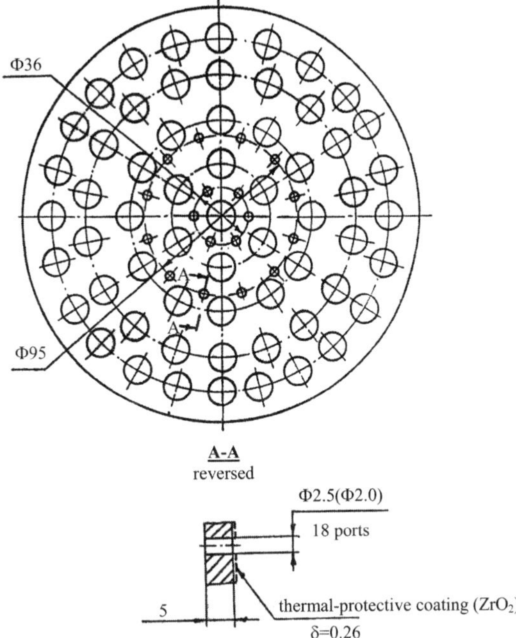

Fig. 17.2 Distribution of fuel in depth at the head center.

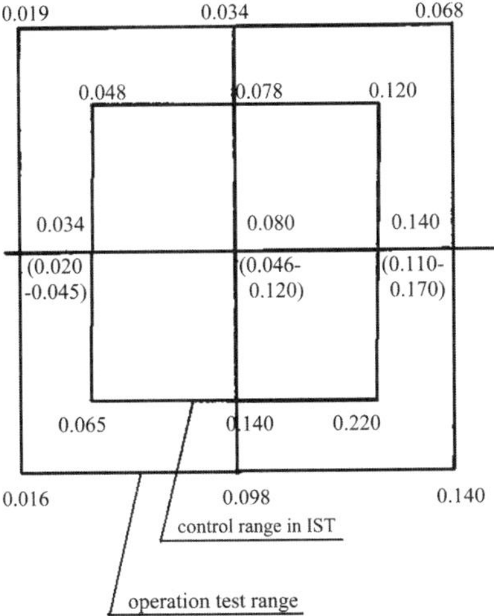

Fig. 17.3 Values of oscillation decrements at f = 2400 Hz with distribution of fuel in depth at the head center.

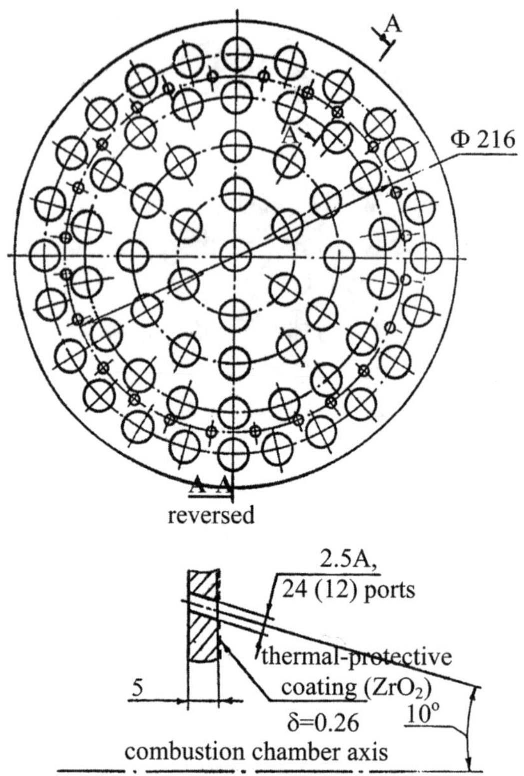

Fig. 17.4 Distribution of fuel in depth in the peripheral zone of the mixing head.

the following two methods of increasing the stability margin: 1) extension of the combustion zone and 2) change of the oxidizer-to-fuel ratio in the near-wall region of the chamber.

Table 17.1 summarizes the methods for increasing the stability margin in combustion chambers with the baseline mixing head. Thus, according to the results of engine tests with different variants of measures aimed at increasing the stability margin, a modified mixing head with a plugged fuel feed hole oriented toward the chamber wall in the 12 injectors of the peripheral row (next but one) was considered optimal. Engines with this mixing-head design were tested over a wide range of external and internal factors: $P_{\text{ch}} = 141.3–161.5$ kgf/cm^2; $K_m = 2.2–2.7$; oxidizer temperature at an inlet 1.1–31.7°C; and fuel temperature 2–31.3°C. This mixing-head design was tested for stability by injecting pulse pressure disturbances. The damping time τ_r did not exceed 4 ms. In addition, a number of combustion-chamber designs with measures aimed at increasing the stability margin have been developed and tested for subsequent production. From the analysis of stability characteristics, the best results were obtained for engines with the following chamber designs: 1) a chamber with a mixing head having injectors

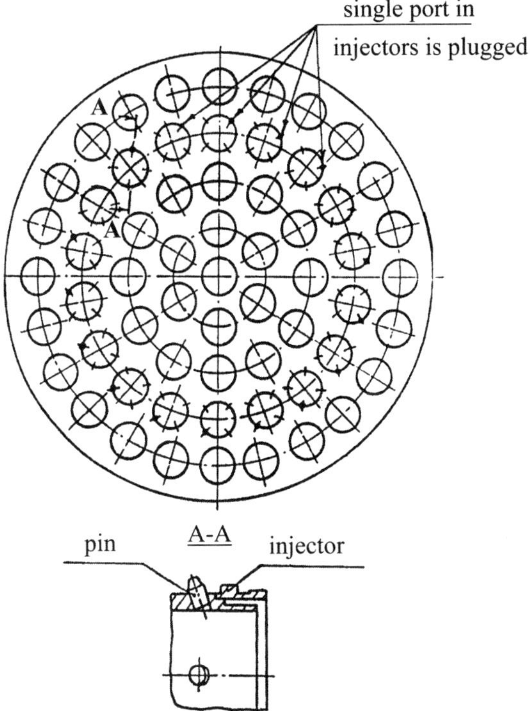

Fig. 17.5 Distribution of oxidizing gas in depth.

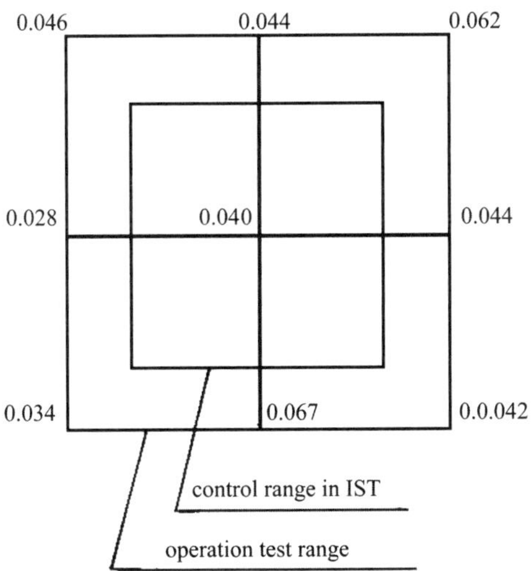

Fig. 17.6 Values of oscillation decrements at $f = 2400$ Hz with distribution of oxidizing gas in depth.

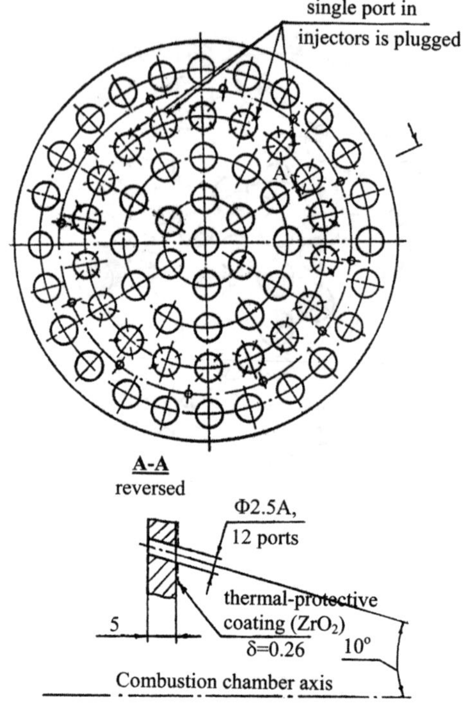

Fig. 17.7 Distributions of fuel and oxidizing gas in depth over zones.

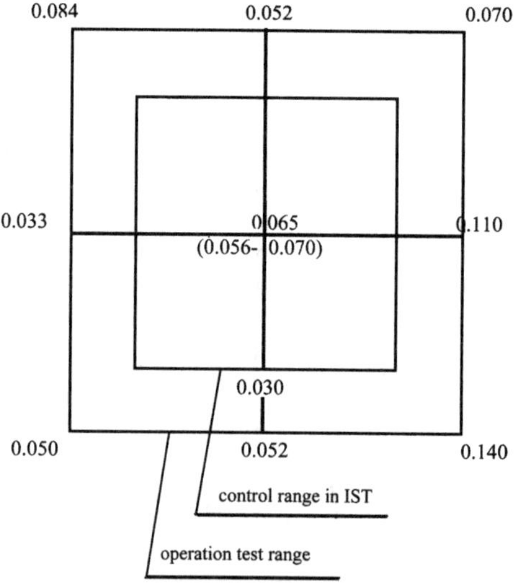

Fig. 17.8 Oscillation decrements at $f = 2400$ Hz with distributions of fuel and oxidizing gas in depth over zones.

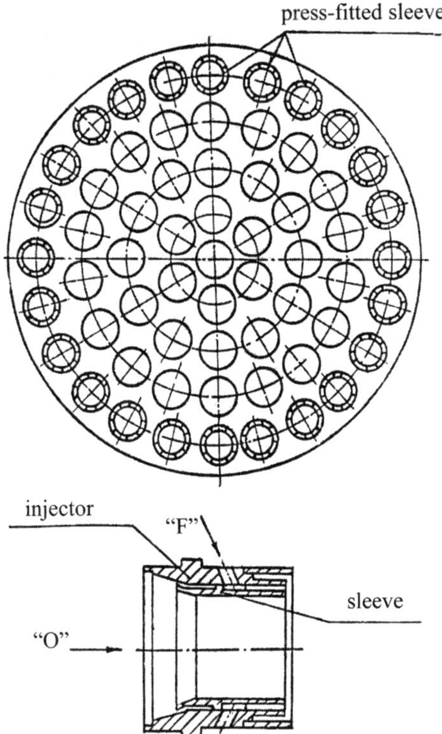

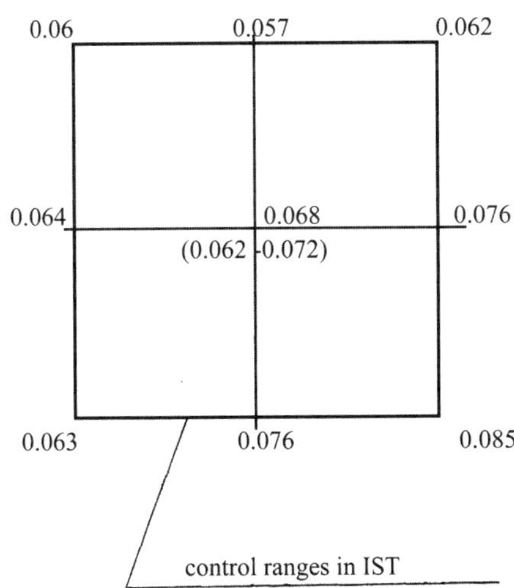

Fig. 17.9 Mixing of propellants beyond injectors of the peripheral row.

Fig. 17.10 Values of oscillation decrement at $f = 2400\,\mathrm{Hz}$ in tests with mixing of propellants beyond injectors of the peripheral row.

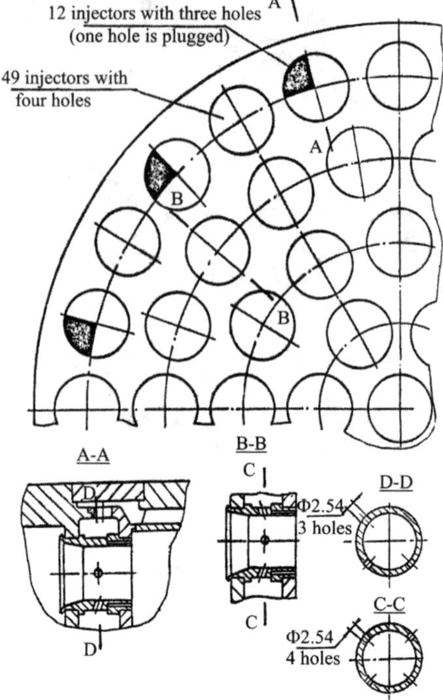

Fig. 17.11 Mixing head.

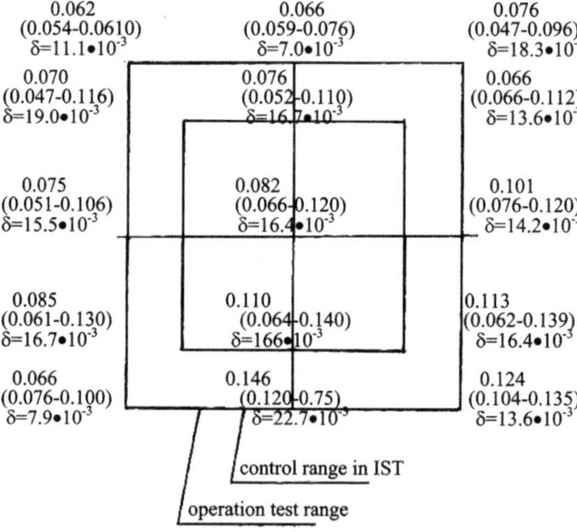

Fig. 17.12 Oscillation decrements at $f = 2400$ Hz with one of four fuel feed spray holes in 12 injectors of the peripheral row being plugged.

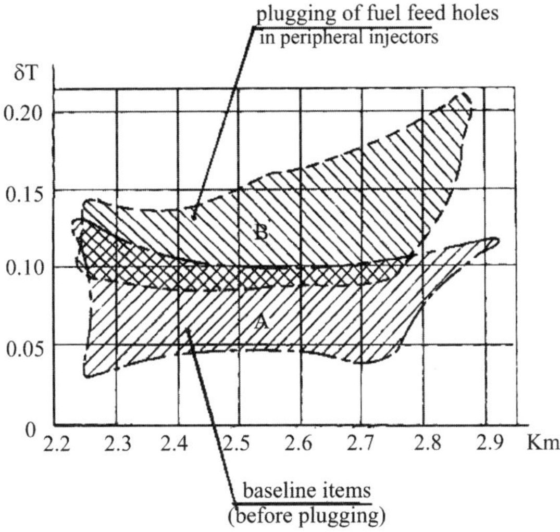

Fig. 17.13 Variation of oscillation decrement during IST (SVT) before and after plugging one of the fuel feed holes in 12 injectors of the peripheral row.

adjusted as quarter-wave absorbers to the frequency of the first tangential mode of acoustic oscillations in the chamber and with decreased fuel concentration in the near-wall region of the chamber; and 2) a chamber with a mixing head consisting of 217 low-flow bipropellant spray-centrifugal injectors, some of which protruded into the combustion zone and form vibration baffles (see Chapter 16).

This chapter confirms the effectiveness of using the quantitative stability characteristics proposed in this book in the production of engines.

References

[1] Isayev A. M., *First Steps to Space Engines*, Mashinostroyeniye, Moscow, 1979.

[2] "Cosmonautics," *Sovietskaya, Encyclopedia*, Sovietskaya Encyclopedia Publishing House, Moscow, 1985.

[3] Rahmanin, V. F. and Sternin, L. E. (eds.), *Once and Forever*, Mashinostroyeniye, Moscow, 1998.

[4] Ilchenko, M. A., Kryuchenko, V. V., Mnatsakayan, Yu. S., Pinke, I. M., Rudakov, A. S., Rudenko, A. N., Folomeyev, E. A., and Epshtein V. A., *The Stability of Operation Process in Flying Vehicle Engines*, Mashinostroyeniye, Moscow, 1995.

[5] *50 Years Ahead of Its Age*, International Program of Education, Russian Space Agency, Moscow, 1998.

[6] *Liquid Rocket Engines, Terms, and Definitions*. GOST 17.655-89. State USSR Committee of Product Quality and Standards Control, 1989.

[7] *Vibration. Terms and Definitions*. GOST 24.346-80. State USSE Committee of Product Quality and Standards Control, 1980.

[8] Gelfand, B. E., Dranovsky, M. L., and Shcherbakov, E. N., "About the Possibility of Detonation Combustion Mode in LRE Combustion Chamber," *NIICHIMMASH Proceedings*, 1967.

[9] *Reliability in Engineering. Basic Notions, Terms, and Definitions*, GOST 27.002-89. State USSR Committee of Product Quality and Standards Control, 1989.

[10] Harrje, D. T., and Readon, F. H. (eds.), *Liquid Propellant Rocket Combustion Instability,* NASA SP-194, 1972.

[11] Natanzon, M. S., *Combustion Instability*, Mashinostroyeniye, Moscow, 1986.

[12] *Aircraft, Rocket, Marine, Industrial Engines 1944–2000*, AKS-Conversalt, Moscow, 2000.

[13] Belyakov, V. P., Dranovsky, M. L. et al., "Stabilization of Hydraulic Characteristics of LRE Mixing Elements," *NIICHIMMASH Proceedings*, 1967.

[14] Malyavin, I. I., and Sulzhenko, S. I., Control of an Atomization Cone Angle of the Centrifugal Injector with Retaining Its Geometrical Characteristics, *NIICHIMMASH Proceedings*, 1967.

[15] Mikhaylov, V. V., and Bazarov, V. G., *Throttled Liquid Rocket Engines*, Mashinostroyeniye, Moscow, 1985.

[16] Dityatkin, U. A., Klyachko, V. V., and Novikov, V. V., *Atomization of Liquids*, Mashinostroyeniye, Moscow, 1967.

[17] Khavkin, Yu. I., *Centrifugal Injectors*, Mashinostroyeniye, Moscow, 1976.

[18] Ishchenko, V. I., and Selivanov, A. N., "Study of Thermal Processes in Rocket Propellant RG-1 on a Heated Surface Under Supercritical Pressure," *NIICHIMMASH Proceedings*, 1967.

[19] Belyakov, V. P., and Dranovsky, M. L., "About the Interaction of Liquid Drops with Shock Waves," *NIICHIMMASH Proceedings*, 1967.
[20] Rachuk, V., "New Millennium Rocket Engines from Chemical Industry Automatics Design Bureau," *Aerospace Journal,* May–June 2000.
[21] Murray, I. F., "Modeling Acoustically Induced Oscillations of Droplets," AIAA Paper 1997-0014, 1997.
[22] Dranovsky, M. L., et al., "On The Behavior of Liquid Droplets in a Gaseous Medium with Acoustic Oscillations," *NIICHIMMASH Proceedings*, 1966.
[23] Belyakov, V. P., Dranovsky, M. L., et al., "Occurrence of High-Frequency Instability with Hard Natural and Artificial Excitation," *NIICHIMMASH Proceedings*, 1967.
[24] Gakhun, G. G., *Design and Development of Liquid Rocket Engines.*
[25] Illarionov, V., "Energomash Engines Serve Russia and America," *Air Fleet Messenger – Aerospace Review*, 2002.
[26] Dubovkin, V. G., Malanicheva, N. F., Massur, Yu. P., and Fedorov, E. P., "Physical, Chemical and Operating Properties of Aviation Turbine Fuels," *Reference Book, Chemistry*, Moscow, 1985.
[27] Bendat, D., *Principles and Applications of Random Noise Theory*, Nauka, Moscow, 1965.
[28] Dranovsky, M. L., Kochanov, V. Ya., and Puhov, V. A., "Study of LRE Process Stability by a Disturbance Method," *NIICHIMMASH Proceedings*, 1965.
[29] Isakovich, M. A., *General Acoustics*, Nauka, Moscow, 1973.
[30] Rzhevkin, S. N., *A Course of Lectures on Sound Theory*, Moscow State Univ. Publishing House, Moscow, 1960.
[31] Vasiliev, A. P., Kudryavtsev, V. M., Kuznetsov, V. A., Kurpatenkov, V. D., et al., *Principles of Theory and Design of Liquid Rocket Engines*, Vysshaya Shkola, Moscow, 1967.
[32] Bolshakov, G. F., "Chemistry and Technology of Liquid Rocket Fuel Components," *Chemistry*, L., 1983.
[33] Norton, M. P., *Fundamentals of Noise and Vibration Analysis for Engineers*, Cambridge Univ. Press, London, 1994.
[34] Shevelyuk, M. I., "Theoretical Principles of Designing the Liquid Rocket Engines," *Oborongiz*, M, 1960.
[35] Zarembo, L. K., and Krasilnikov, V. A., *Introduction to Nonlinear Acoustics*, Nauka, Moscow, 1966.
[36] Kuznetsov, D. S., *Special Functions,* 2nd ed., Vysshaya Shkola, Moscow, 1965.
[37] Blokhintsev, D. I., *Acoustics of Heterogeneous Flowing Medium*, 2nd ed., Nauka, Moscow, 1981.
[38] Yang, V., and Anderson, W., *Liquid Rocket Engine Combustion Instability*, Progress in Astronautics and Aeronautics, Vol. 169, AIAA, Washington, D. C., 1995.
[39] Dranovsky, M. L., et al., "Study of Gas Medium Disturbance Induced by a Shock Tube," *NIICHIMMASH Proceedings*, 1967.
[40] Sulimov, A. A., and Yermolayev, B. S., "Low-Velocity Detonation in Solid Explosives," *Proceedings of Symposium on Combustion and Explosion*, Chernogolovka, 1977.

[41] Hasainov, B. A., Yermolayev, B. S., Borisov, A. A., and Korotkov, A. I., "Low-Velocity Detonation of High-Density Explosives," *Proceedings of Symposium on Combustion and Explosion*, Chernogolovka, 1977.

[42] Belyakov, V. P., Dranovsky, M. L., et al., "Study of the Possibility of Quantitative Evaluation of High-Frequency Pressure Oscillation Self-Excitation Modes in Closed-Circuit LRE Combustion Chamber on the Basis of Model Tests Under a Low Pressure," *NIICHIMMASH Proceedings*, 1967.

[43] Dexter, C. E., Fisher, M. F., Denisov, K. P., Shibanov, A. A., and Agarkov, A. F., "Scaling Techniques in Liquid Rocket Engine Combustion Devices Testing," *Proceedings of the Second International Symposium on Liquid Rocket Propulsion*, 1995.

[44] Belyakov, V. P., Dranovsky, M. L., and Nedashkovsky, A. K., "The Effect of Height of Baffles on High-Frequency Process Stability in LRE Combustion Chambers," *NIICHIMMASH Proceedings*, 1967.

[45] Zrelov, V. A., and Kartashov, G. G., *NK Engines*, Samarsky Publishing House, Samara, 1999.

[46] Raushenbah, B. V., *Vibrational Combustion*, Fizmatgiz, Moscow, 1961.

[47] Katorgin, B. I., *A Road in Rocket Engineering*, Mashinostroyeniye and Mashinostroyeniye-Polyot, Moscow, 2004.

Bibliography

Alemasov, V. E., et al., "Thermodynamic and Thermophysical Properties of Combustion Products," *Reference Book of VINITI,* edited by V. P. Glushko, Moscow, 1980.

Awad, E., and Culick, F. E. C., "On the Existence and Stability of Limit Cycles for Longitudinal Acoustic Modes in a Combustion Chamber," *Combustion Science and Technology*, Vol. 46, 1986, pp. 195–222.

Belyakov, V. P., et al., "Study of the Characteristics of a Device for Artificial Excitation of High-Frequency Pressure Oscillations in LRE Combustion Chamber," *NIICHIMMASH Proceedings*, 1966.

Belyakov, V. P., et al., "Study of the Influence of Combustion Chamber Physical and Design Parameters on Some Quantitative Characteristics of Operation Process Stability in Closed-Circuit LRE with Afterburning of Oxidizing Generator Gas," *NIICHIMMASH Proceedings*, 1967.

Belyakov, V. P., Gelfand, B. E., and Dranovsky, M. L., "Study of LRE Process Dynamic Stability by a Disturbance Method," *NIICHIMMASH Proceedings*, 1966.

Borisenko, A. I., *Gas Dynamics of Engines*, Oborongiz, Moscow, 1962.

Dranovsky, M. L., "Amplitude Characteristic of a Disturbing Device with Small-Size Explosive Charges," *Collection of RKT Papers*, Series IV, Issue 9, 1985.

Dranovsky, M. L., and Nedashkovsky, A. K., "Experimental Study of Pressure Oscillation Damping from a Single Pulse in LRE Chambers During Cold Purging," *NIICHIMMASH Proceedings*, 1967.

Dranovsky, M. L., and Nedashkovsky, A. K., "On a Mode of Oscillatory Process in LRE Combustion Chamber When Applying the External Disturbance," *NIICHIMMASH Proceedings*, 1967.

Dranovsky, M. L., et al., "About Dynamic Transients of Combustion Process in a Through-Flow Chamber," *Proceedings of Tenth Symposium on Combustion and Explosion*, Chernogolovka, 1994.

Furletov, V. I., "About Some Peculiarities of Gas Diffusion Flame Transition to Vibrational Combustion," *Proceedings of Tenth Symposium on Combustion and Explosion*, Chernogolovska, 1994.

Furletov, V. I., "Variations of Heat Release Rate During Vibrational Combustion," *Proceedings of Eighth National Symposium on Combustion and Explosion*, Chernogolovska, 1986.

Gelfand, B. E., "Droplet Breakup Phenomena in Flows with Velocity Lag," *Energy Combustion*, 1966.

Gelfand, B. E., and Dranovsky, M. L., "Intensification of Liquid Jet Atomization with Gas Injection into an Atomizer Outlet," *Reports of Academy of Sciences of the USSR*, 1991.

Goldstein, M. E., *Aeroacoustics*, McGraw–Hill, New York, 1976.

Konstantinov, B. L., *Hydrodynamic Sound Formation and Propagation in a Confined Medium*, Nauka, Moscow, 1974.

Landau, L. D., and Lifshitz, E. M., *Hydrodynamics*, 3rd ed., Nauka, Moscow, 1986.

Lependin, L. F., *Acoustics*, Vysshaya Shkola, Moscow, 1978.

Melchior, A., "A New Bipropellant Rocket Engine for Orbit Maneuvering," Joint Conference on Engine Units, July 1990.

Oefelein, J. C., and Yang, V., "Comprehensive Review of Liquid Propellant Combustion Instabilities in F-1 Engines," *Journal of Propulsion and Power*, Vol. 9, No. 5, 1993, pp. 657–677.

Zalmanzon, L. A., *Theory of Pneumonics Elements*, Nauka, Moscow, 1965.

Index

Supporting Materials

A complete listing of titles in Progress in Astronautics and Aeronautics and other AIAA publications is available at http://www.aiaa.org.